Based on comparative case studies, this important addition to the literature on environmental justice exposes the conflicts arising from plans for ecological improvements. The authors address how the seemingly positive effect of producing green amenities in low-income neighborhoods often leads to displacement of residents. They show how neoliberal competition for creative industries contributes to injustice, which its progenitors rationalize as producing generally desirable benefits to the urban landscape. They also describe ways by which mobilized, working-class citizens can confront the forces of gentrification while also bringing about improvements in their neighborhoods. Bringing together careful analyses of greening efforts in twenty-one European and North American cities, this book points to the variety of greening policies and their consequences for spatial equity.

Susan S. Fainstein, Harvard Graduate School of Design; Author, The Just City

As sustainable urbanism gathers pace, a troubling question arises – is greening the city doing more harm than good? This book cuts to the heart of the issue, drawing on stories of urban sustainability and its potential injustices from around the world. Together, these cases demonstrate the vital importance of facing the uncomfortable truth that urban sustainability can lead to widening social inequalities and gives us the means to start ensuring that we are building just transitions.

Harriet Bulkeley, Professor of Geography, Durham University

This impressive, international comparison of twenty-one cases of urban greening systematically illustrates the social costs of glitzy green urbanism, the compounded impacts of urban greening in post-industrial environments, the re-creation of unjust racialized landscapes, the complexities of intertwining gentrification processes, and green justice victories. This is essential reading for activists and scholars interested in creating a new tale for urban greening that incorporates all people's rights to a healthy environment.

Professor Kenneth A. Gould and Professor Tammy L. Lewis, Graduate Center and Brooklyn College, City University of New York

THE GREEN CITY AND SOCIAL INJUSTICE

The Green City and Social Injustice examines the recent urban environmental trajectory of 21 cities in Europe and North America over a 20-year period. It analyses the circumstances under which greening interventions can create a new set of inequalities for socially vulnerable residents while also failing to eliminate other environmental risks and impacts.

Based on fieldwork in ten countries and on the analysis of core planning, policy and activist documents and data, the book offers a critical view of the growing green planning orthodoxy in the Global North. It highlights the entanglements of this tenet with neoliberal municipal policies including budget cuts for community initiatives, long-term green spaces and housing for the most fragile residents; and the focus on large-scale urban redevelopment and high-end real estate investment. It also discusses hopeful experiences from cities where urban greening has long been accompanied by social equity policies or managed by community groups organizing around environmental justice goals and strategies.

The book examines how displacement and gentrification in the context of greening are not only physical but also socio-cultural, creating new forms of social erasure and trauma for vulnerable residents. Its breadth and diversity allow students, scholars and researchers to debunk the often-depoliticized branding and selling of green cities and reinsert core equity and justice issues into green city planning—a much-needed perspective. Building from this critical view, the book also shows how cities that prioritize equity in green access, in secure housing and in bold social policies can achieve both environmental and social gains for all.

Isabelle Anguelovski is the director of Barcelona Laboratory for Urban Environmental Justice and Sustainability (B.C.N.U.E.J.) and an I.C.R.E.A. Research Professor at the Universitat Autònoma de Barcelona. Her research examines the processes and dynamics behind urban environmental (in)justices in the Global North and South.

James J. T. Connolly is codirector and affiliated researcher of Barcelona Laboratory for Urban Environmental Justice and Sustainability (B.C.N.U.E.J.) and Assistant Professor of Community and Regional Planning at the University of British Columbia. His research examines how cities are made greener and more socially just.

Routledge Equity, Justice and the Sustainable City
Series editors: **Julian Agyeman** and **Stephen Zavestoski**

This series positions equity and justice as central elements of the transition toward sustainable cities. The series introduces critical perspectives and new approaches to the practice and theory of urban planning and policy that ask how the world's cities can become 'greener' while becoming more fair, equitable and just.

The *Routledge Equity Justice and the Sustainable City* series addresses sustainable city trends in the global North and South and investigates them for their potential to ensure a transition to urban sustainability that is equitable and just for all. These trends include municipal climate action plans; resource scarcity as tipping points into a vortex of urban dysfunction; inclusive urbanization; "complete streets" as a tool for realizing more "livable cities"; the use of information and analytics toward the creation of "smart cities".

The series welcomes submissions for high-level cutting-edge research books that push thinking about sustainability, cities, justice and equity in new directions by challenging current conceptualizations and developing new ones. The series offers theoretical, methodological, and empirical advances that can be used by professionals and as supplementary reading in courses in urban geography, urban sociology, urban policy, environment and sustainability, development studies, planning, and a wide range of academic disciplines.

Urban Ecosystem Justice
Strategies for Equitable Sustainability and Ecological Literacy in the City
Scott Kellogg

Living Detroit
Environmental Activism in an Age of Urban Crisis
Brandon M. Ward

The Green City and Social Injustice
21 Tales from North America and Europe
Edited by Isabelle Anguelovski and James J. T. Connolly

For more information about this series, please visit www.routledge.com/Routledge-Equity-Justice-and-the-Sustainable-City-series/book-series/EJSC

THE GREEN CITY AND SOCIAL INJUSTICE

21 Tales from North America and Europe

Edited by Isabelle Anguelovski and James J. T. Connolly

Routledge
Taylor & Francis Group
LONDON AND NEW YORK

earthscan
from Routledge

First published 2022
by Routledge
2 Park Square, Milton Park, Abingdon, Oxon OX14 4RN

and by Routledge
605 Third Avenue, New York, NY 10158

Routledge is an imprint of the Taylor & Francis Group, an informa business

British Library Cataloguing-in-Publication Data
A catalogue record for this book is available from the British Library

Library of Congress Cataloging-in-Publication Data
Names: Anguelovski, Isabelle, editor. | Connolly, James J. T., editor.
Title: The green city and social injustice: 21 tales from North America and Europe/edited by Isabelle Anguelovski and James Connolly.
Description: Abingdon, Oxon; New York, NY: Routledge, 2022. | Series: Routledge equity, justice and the sustainable city | Includes bibliographical references and index. |
Identifiers: LCCN 2021023708 (print) | LCCN 2021023709 (ebook)
Subjects: LCSH: Social justice–North America. | Social justice–Europe. | Sustainability–North America. | Sustainability–Europe. | Equality–North America. | Equality–Europe.
Classification: LCC HM671 .G73 2022 (print) | LCC HM671 (ebook) | DDC 303.3/72097–dc23
LC record available at https://lccn.loc.gov/2021023708
LC ebook record available at https://lccn.loc.gov/2021023709

ISBN: 978-1-032-02413-4 (hbk)
ISBN: 978-1-032-02411-0 (pbk)
ISBN: 978-1-003-18327-3 (ebk)

DOI: 10.4324/9781003183273

Typeset in Bembo
by Deanta Global Publishing Services, Chennai, India

CONTENTS

ILLUSTRATIONS

Figures

Tables

CONTRIBUTORS

Troy D. Abel is a Professor of Environmental Policy in the Urban and Environmental Planning and Policy department at Western Washington University. His research and teaching focuses on environmental justice, urban riskscapes and industrial environmental performance.

Isabelle Anguelovski is the director of Barcelona Laboratory for Urban Environmental Justice and Sustainability (B.C.N.U.E.J.) and an I.C.R.E.A. Research Professor at the Universitat Autònoma de Barcelona. Her research examines the processes and dynamics behind urban environmental (in)justices in the Global North and South.

Lucía Argüelles is a Juan de la Cierva postdoctoral fellow at the Urban Transformations and Global Change Laboratory (T.U.R.B.A.) at the Universitat Oberta de Catalunya. She is also an affiliated researcher at B.C.N.U.E.J. Her research lies at the intersection of social theory and environmental politics and explores issues such as urban greening and agro-food systems.

Francesc Baró is Assistant Professor of Urban Ecology at the Geography and Sociology Departments at the Vrije Universiteit Brussel (V.U.B.), Belgium, and affiliated researcher at B.C.N.U.E.J. His interdisciplinary research approach combines geospatial modeling, advanced quantitative and qualitative data analysis and participatory methods to study the complexity of urban social-ecological systems.

Helen V.S. Cole is the co-coordinator for research on the urban environment, health and equity and a postdoctoral researcher at B.C.N.U.E.J. Her research challenges traditional public health perspectives by questioning and evaluating the long-term social justice impacts of structural urban interventions.

James J.T. Connolly is co-director and affiliated researcher of Barcelona Laboratory for Urban Environmental Justice and Sustainability (B.C.N.U.E.J.) and Assistant Professor of Community and Regional Planning at the University of British Columbia. His research examines how cities are made greener and more socially just.

Lucia Di Paola is Just Transition Officer at I.C.L.E.I. Europe – Local Governments for Sustainability. With her team, she advocates for cities and European institutions to integrate social equity and justice in urban sustainable development.

Taliah Dommerholt holds a Master's in Sociology of Development from Wageningen University and Research, where she is completing a second Master's in Urban Environmental Management. Her current research explores environmental justice as it relates to housing affordability and access to sustainable amenities.

Mario Fontán Vela is a public health and preventive medicine doctor and a doctoral researcher at the University of Alcalá. His research explores the relationship between the urban environment and the population's health, focused on how cities can promote physical activity through green spaces, based on the social determinants of health frameworks.

Melissa García-Lamarca is a postdoctoral researcher and the coordinator of urban housing and justice research at B.C.N.U.E.J. Her work unpacks structural and lived urban (green/housing) injustices and how collective urban struggle can generate more egalitarian alternatives. She is a core editorial collective member of the *Radical Housing Journal*.

Neil Gray is an urban researcher and housing activist based at the University of Glasgow. His work primarily focuses on urban regeneration, devaluation, social housing and class composition. He is an active member of Living Rent Tenants' Union and the author of *Rent and Its Discontents: A Century of Housing Struggle* (Rowman & Littlefield, 2018).

Sarka Hajtmarova graduated from the Master's Degree in Interdisciplinary Studies in Environmental, Economic and Social Sustainability at the Institute for Environmental Science and Technology UAB, specializing in Global Change, and was an intern researcher at the BCNUEJ lab until Fall 2021.

Jordi Honey-Rosés is an Associate Professor at the School of Community and Regional Planning at the University of British Columbia in Vancouver, Canada. He is an environmental planner with research interests in impact evaluation and public space.

Derek Hyra is Professor in the Department of Public Administration and Policy and Founding Director of the Metropolitan Policy Center, American University.

Dan Immergluck is Professor of Urban Studies at Georgia State University, whose research and policy work covers issues of housing affordability and stability, race, gentrification, neighborhood change, housing and development finance and related urban issues. His latest book, *Red-Hot City: Housing, Race, and Exclusion in Twenty-First Century Atlanta,* is expected to be published in 2022.

Wendy A. Kellogg (Ph.D. City and Regional Planning, Cornell University) is a Professor of Urban Planning and Environmental Studies at the Levin College of Urban Affairs, Cleveland State University. Her research focuses on sustainability planning and governance.

Panagiota Kotsila is a Juan de la Cierva postdoctoral researcher at the Institute for Environmental Science and Technology (I.C.T.A.-U.A.B.) and coordinator of research on urban political ecology at B.C.N.U.E.J. Her research examines the unequal distribution of health risks and intersectional vulnerabilities and how the politics of urban re-naturing informs (in)justice.

Mick Lennon researches and teaches at the U.C.D. School of Architecture, Planning and Environmental Policy, Dublin, Ireland. His work focuses on green-blue infrastructures, sustainable communities and cultivating resilience in the context of socio-ecological transitions.

Mateus Lira is an urban planner at the United Nations Human Settlements Programme (U.N.-Habitat), where he addresses social inclusion and human rights issues in urban planning and design. He has been a researcher at B.C.N.U.E.J. (I.C.T.A.-U.A.B.) and T.U.R.B.A. Lab (U.O.C.), and his research focuses on housing justice and grassroots urbanism.

Austin Matheney is a doctoral researcher at I.C.T.A.-U.A.B. and B.C.N.U.E.J. From a transdisciplinary lens, his research explores the impacts of urban greening projects on the (re)construction of identities, place and community.

Dave Moore is a community and food activist who returned to academia as a (very) mature learner at University College Dublin. He pursues action research in the areas of environmental and social justice.

Emilia Oscilowicz holds a Master's in Community and Regional Planning from the University of British Columbia in Vancouver, Canada. Her research interests focus on the intersection of policy interventions relevant to green gentrification and equitable housing.

Carmen Pérez del Pulgar is a doctoral researcher at I.C.T.A.-U.A.B. and B.C.N.U.E.J. Her research explores the political and social production of green and playful cities. She holds an M.Sc. in Human Geography from the Universiteit van Amsterdam and a Bachelor's in Political Science from the Universidad Complutense de Madrid.

Malini Ranganathan is Associate Professor in the School of International Service, American University.

Rebecca Rutt is Assistant Professor in the Department of Food and Resource Economics (IFRO) at Copenhagen University.

Filka Sekulova is a postdoctoral researcher at the Institute of Environmental Science and Technology (I.C.T.A.) at the Universitat Autònoma de Barcelona. Her field of research centers on the thematic of degrowth, wellbeing and social inclusion. She is working on the application of nature-based solutions in cities and explores the role and impacts of community-based initiatives in building an ecologically sustainable and socially just society.

Galia Shokry is a doctoral researcher at I.C.T.A.-U.A.B. and B.C.N.U.E.J. Her research examines intersections between climate adaptation planning, green gentrification and urban inequities, focusing on how these processes shape new social vulnerabilities, sense of belonging and struggles for social and racial justice in the city.

Margarita Triguero-Mas is co-coordinator for research on urban environment, health and equity and associated researcher at B.C.N.U.E.J., and an affiliated research scientist at Mariana Arcaya's research lab of the Department of Urban Studies from the Massachusetts Institute of Technology (D.U.S.P.-M.I.T.). Her research examines how green cities can be healthier for all.

Aaron Vansintjan is a Ph.D. candidate at the Department of Film, Media, and Cultural Studies in Birkbeck, University of London. His research focuses on how gentrification shapes the everyday food strategies of marginalized long-term residents. He is also a co-editor of Uneven Earth, a website focusing on ecological politics, and a member of Anti-Eviction Montreal.

ACKNOWLEDGMENTS

All books take a team to produce, but this one took an (international, mostly peaceful) army. Dedicated researchers, with the support of practitioners, activists and guides on the ground, carried out intensive and coordinated fieldwork in 29 cities across 10 countries and 2 continents. It was no easy feat, but one that our authors executed with a high standard of professionalism, ethics and quality that extended through their analysis and writing of this book. It is this collective effort—rather than any individual contribution—that generated the distinctive value of this book. We sincerely appreciate the work of all of the authors and collaborators, especially their continued willingness to revisit the chapters throughout the editing process in order to ensure that the project as a whole was coherent and that we included important new developments in their city.

We especially want to highlight the central role played by those who were part of the European Research Council (E.R.C.)-funded grant, "GreenLULUs." In addition to Isabelle Anguelovski and James Connolly, they include (in alphabetical order): Lucia Argüelles Ramos, Francesc Baró, Helen Cole, Melissa Garcia-Lamarca, Emilia Oscilowicz, Carmen Pérez del Pulgar, Galia Shokry and Margarita Triguero-Mas. Each of these authors aided in the conceptualization and development of this book overall, above and beyond their role as chapter authors. Much of the fieldwork preparation and analysis was also made possible by the numerous dedicated and enthusiastic students, interns and collaborators hosted by the Barcelona Lab for Urban Environmental Justice and Sustainability: Ana Terra Amorim-Maia, Amalia Calderón Argelich, David Camacho, Helene Christensen, Kaitlin Dietz, Taliah Dommerholt, Mario Fontán-Vela, Elsa Gallez, Sarka Hajtmarova, Jordi Honey-Rosés, Panagiota Kotsila, Katrina Lehman-Grube, Emilia Lewartowska, Ariella Levitch, Mateus Lira, Jonathan Luger, Julia Mangione, Austin Matheney, Erin Monroe, Nick Martin, Julia Neidig, Ella O'Neill, Aina Planas Carbonell, Alicia Ramos, Carla Rivera Blanco, Filka

Sekulova, Marina Skarpeti, Pratika Taleagonkar, Sara Ullstrom and Madeleine Wahlund. At different points, each of these people made some part of this book possible.

Beyond the researchers who wrote the chapters, at the heart of this book are the practitioners, activists and guides who showed us the reality of greening and social dynamics in their neighborhoods. From the start, the main goal of this book was to elevate their perspectives—especially those often overlooked—and make their voices heard in decision-making processes by amplifying their civic mobilizations, progressive practices and daily struggles. Without their support, critical thinking and willingness to share their experiences, this book could not exist. We thank everyone who spoke to us for an interview, took us on a tour and welcomed us into their homes and workplaces. We hope that our efforts have honored their time and contribution.

This book could also not have been possible without the financial support and collaboration of various research groups and centers. We thank our supporters from the European Research Council (GreenLULUs, GA678034), the European Union Research and Innovation Programme Horizon 2020 (Naturvation, 730243; Urbana, 822357) and the María de Maeztu Programme for Units of Excellence of the Spanish Ministry of Science and Innovation (CEX2019-000940-M). We also thank our colleagues at our academic home, the Institute for Environmental Sciences and Technology (I.C.T.A.) within the Universitat Autònoma de Barcelona (U.A.B.). Specifically, we owe much of our productivity and sanity to our administrative colleagues at I.C.T.A., including Laura Jimenez, Rosa Laura Sans, Isabel Lopera, Cristina Durán, Maica Nogales and so many others. As well, during the five years of the GreenLULUs project, we were lucky to be hosted by the Medical Research Institute of the Hospital del Mar (I.M.I.M.) in Barcelona, where so many meetings were held, analysis conducted, chapters drafted and endless chocolate and coffee consumed.

Finally, we thank the steady hand of our editor and communications officer Ana Cañizares for weaving everything into a clean, finished piece. To the extent that we were able to produce tight and similarly structured chapters, we have Ana to thank. Her incisive edits were invaluable as we made the frenzied push toward a complete manuscript.

INTRODUCTION

Positioning urban green injustices

Vancouver Greenest City Action Plan. Boston Municipal Harbor Plan. GreenPlan Philadelphia. Imagine Austin. Copenhagen Urban Nature. Barcelona Green and Biodiversity Plan. Bristol One City Plan. The list of ambitious policies and plans that municipalities around the Global North have in place to achieve their climate and other environmental goals is long and reflects the growing global orthodoxy of urban greening. This prevailing attitude contributes to city branding and establishes a competitive framework for green and sustainable urbanism (Connolly 2019, Anguelovski, Connolly, and Brand 2018). Yet, within an ever-expanding list of interventions, not all "green cities" are alike. There are multiple paths toward bringing nature back into the urban landscape and different incarnations of equity within the green city. While some neighborhoods benefit from greening projects and other nature-centered projects (Jennings, Browning, and Rigolon 2019), others remain neglected—the still-marginalized receiving points for displaced residents from other recently upgraded (and greened) neighborhoods. In this context, the resulting landscape for green planning can seem contradictory. While some historically green-deprived neighborhoods are mobilizing for new environmental amenities, others are more wary of the possible social impacts and push back against green locally unwanted land uses (GreenL.U.L.U.s) (Anguelovski 2016).

Reflecting on these trends, *The Green City and Social Injustice* is about the compounded injustices and inequalities embedded in or following from urban green interventions across and within 21 cities in 10 countries of North America and Europe. We focus on the opportunities offered by radical or progressive planning and civic urbanism. We ask: to what extent does green planning exacerbate historic social and racial inequalities in access to the benefits of urban green amenities, or in contrast, how does it address them? What are the overlapping

DOI: 10.4324/9781003183273-101

environmental risks and impacts that residents living in greener neighborhoods might be facing, and what environmental interventions are left behind? What are promising activist organizing frames, strategies and tactics as these actors navigate through an environmental justice conundrum of greening neighborhoods whose new livability features might be captured by a few living in a privileged green utopia? What are the main policy and community interventions that can mitigate the negative impacts of greening in historically marginalized neighborhoods, and what are their limitations? As we attempt to answer these questions, we recognize that, most obviously, both local and national urban politics shape the extent and characteristics of green tensions—and alternative green and just futures.

Four shades of green in one city

In *To Green or Not to Green: Four Stories of Urban (In)Justice* (2020),[1] filmmaker Alberto Bougleux illustrates the tensions that ambitious greening agendas can create in a city, using Barcelona, Spain, as an example. The documentary portrays four historically working-class and migrant neighborhoods whose recent environmental fates could not be more different in the new self-branded green Barcelona. In the neighborhoods of Tres Turons and Poblenou, new parks and redevelopment projects are exacerbating gentrification, driving prices up and pushing out vulnerable residents. Some of those interventions are underpinned by the 2012–2020 *Pla del Vert i de la Biodiversitat* (2020 Green and Biodiversity Plan), a plan meant to create green spaces within 300 meters of every resident's home, incorporate ecologically sound agriculture in urban and peri-urban spaces, and build a network of green corridors, among others. In contrast, other areas such as El Raval or Sants lack green space altogether, depriving certain communities of essential physical and mental health benefits while their residents have to endure the negative social impacts of an intensive tourism- and visitor-centered economy.

In Barcelona's Sants-Montjuïc district, *To Green or Not to Green* offers a testimony about the city's decision in February 2019 to renew the contract of the Fira de Barcelona, the city's international trade fair, without public consultation. This choice effectively allocated 380 million euros to bolster tourism and attract large events—excluding the possibility of an alternative use of existing buildings and spaces in a central area encompassing 27 hectares or 21 city blocks. In response, as David Bravo explains in the film, citizens have mobilized to create a platform called "La Fira o la Vida," which proposes the alternative development of a new neighborhood with green space and social housing centered on the needs of residents—a radical and progressive shift away from megaproject urbanism and toward a more balanced approach to the provision of amenities and housing in Barcelona. As of early 2021, debates about the future of the Fira continue and future planning decisions about the fairground remain unknown.

A 15-minute walk away, El Raval in the Old Town has historically been the home of North African, Pakistani and Latinx residents together with working-class Spanish residents. However, in the 1990s and 2000s, the city-driven revitalization tore apart the existing built and social environment and sparked heavy gentrification in the neighborhood through the construction of new hotels, short-term rentals and cultural venues. In 2013, after years of community organizing pushing the construction of a new and much-needed healthcare center, local resident Elisenda Puchades explains how the city instead granted the intended space called the Capella de la Misericòrdia to the M.A.C.B.A. museum. Outraged, neighbors and healthcare workers occupied the space in 2019. Thanks to the protests, the contract with M.A.C.B.A. was terminated in early 2020, but the construction of the healthcare center still has not advanced. Organized through the C.A.P. Raval Nord platform, residents are also asking for new green space and the protection of informal gardens in the city's densest and least green neighborhood—a push toward access to health for those residents who have not only been deprived of green infrastructure, but also made more vulnerable through state-sanctioned gentrification.

In contrast, Barcelona's northern neighborhood of El Carmel is bordered by the Parc del Guinardó, Parc de la Creueta del Coll and the famous Gaudí-designed Parc Güell. It also hosts some of the last row houses in Barcelona, self-constructed by immigrants in the early mid-20th century. Despite their architectural heritage and cultural memory, the municipality aims to develop a central park that connects these three green vantage points as means to decentralize tourism and provide new green space to the city as a whole, demolishing the historic row houses in the process. In response, residents have mobilized and drawn media attention to their struggle in an effort to preserve the homes and plan an alternative development of the park. Much of their distress, as resident Aisha Pallerés relates, concerns the eviction and displacement of row house residents as well as the impending property cost spikes as the new flagship park project materializes.

Last, *To Green or Not to Green* features the neighborhood of Poblenou, next to Barcelona's northern waterfront and home to some of the city's most substantial land use transformation since the 2000s. Underpinned by the 22@ Plan, the area is being remade in order to attract tech and startup economies, design firms, offices and hotels, favoring these uses over new local businesses. In 2016, it was in Poblenou that the city launched its pilot Superblock[2] project, rerouting traffic around 3×3 blocks to create pockets of urban green within the so-called "pacified" street. In 2019, it also inaugurated the Cristobal de Moura "green axis" in an effort to create linear green space, integrate sustainable urban drainage systems (S.U.D.S.) and remediate climate-driven flooding risks in the city. As resident and organizer Albert Valencià relates, these transformations have caused speculation and displacement, with new skyscraper hotels built around those projects, the closure of traditional shops and the ejection of long-term neighbors. The promise of high-paying jobs has instead largely materialized as low-wage call center or service jobs. In response, neighbors have created the Ens Plantem and Observatori dels Barris de Poblenou organizing platforms to call for more protected housing and protest a tourism- and office-centered economy, now both under threat due to work-from-home practices and the loss of tourism as a result of the COVID-19 pandemic in 2020–2021.

These four cases demonstrate the web of social tensions in cities that follows when greening is deployed without being accompanied by bold, progressive social policies. Those tensions also emerge when the primary development model of a city—tourism in the case of Barcelona—remains mostly untouched, even if often questioned by social movements and progressive mayors, such as the current Mayor Ada Colau. What remains clear from the Barcelona case—but also Vancouver, Boston, Austin or Copenhagen—is that greening is the easy part. Often underpinned by a list of ecological, health, social and economic benefits, when residents listen to public officials, planners, investors and even health and ecology experts, there are told that greening is a win-win strategy. No one is left behind by the trickle-down of benefits from green infrastructure. Green increases value, helps you connect with your neighbors, encourages physical activity and recreation and fulfills climate goals. It's a 100% fail-proof package, a drumbeat that seems to have cities across the globe marching in unison.

Indeed, this drumbeat makes transforming a former industrial waterfront into a new resilient shoreline, an unused railway into a linear park or vacant land into new community gardens a straightforward political proposition—why would we not? The environmental, technical and financial equations add up especially so because the green city sells in the global market and the common metrics used to describe outcomes appeal broadly. Indeed, the promotional materials tell us that green sells property, it sells new investments and markets and it sells a new urban brand (Anguelovski, Connolly, and Brand 2018, Liebelt, Bartke, and Schwarz 2018, Garcia Lamarca et al. 2021). Given all the benefits, the headaches—namely the social, equity and justice questions—are often put aside.

Put differently, though urban greening and green spaces are vital to climate, ecological and human health, achieving equity requires questioning the social exclusion and white privilege embedded in new greening landscapes and how those social dynamics reshape security and vulnerability. Green planning can indeed obscure the latent racialized and working-class geographies within which it proposes solutions and spatial tools, and thereby exacerbates minority and low-income residents' claims to space and new amenities, and increases their vulnerability to health and climate impacts. In contrast, planning for the just green city requires considering housing and land rights alongside greening to ensure the ability of lower-income, minority and immigrant residents to remain in place.

The social impact of green interventions

In *The Green City and Social Injustice*, we define green interventions as small- or large-scale nature-based infrastructure and amenities such as parks, municipal or community gardens, greenbelts and greenways, raingardens and bioswales, green roofs and walls, green streets and alleys or restored waterways. We explore projects that embody the clean-up and/or restoration of degraded, abandoned or underused—post-industrial or not—urban landscapes and/or gray infrastructure (i.e., highways, bridges, railways), and their transformation into green, nature-centered sites (sometimes paired with high-tech environmental strategies) that redefine their purpose and are often part of a new overall vision for an area (Wachsmuth and Angelo 2018, Anguelovski, Connolly, and Brand 2018). Our perspective is both historic and multi-faceted as we integrate a grounded narrative of the legacy of environmental risks and impacts into our analysis and examine how socially vulnerable residents are exposed to physical and socio-cultural displacement from greening while often living next to untreated toxic sites or climate-vulnerable areas.

We start from the argument—now well-established in the literature—that the marriage of urban redevelopment with greening interventions creates a paradox wherein some elements provide widescale benefits even while others produce acute injustices. When what we have previously called "green gaps" (Anguelovski, Connolly, and Brand 2018) are exploited to generate a new potential "green rent" for municipalities, private investors and privileged residents, urban greening interventions targeting working-class neighborhoods often become a GreenL.U.L.U. (Anguelovski, 2016a) for low-income residents and minorities. This process involves municipalities, investors and privileged residents couching green projects in discourses of win-win benefits and public good for all while ignoring a reality wherein the social and health benefits of new environmental interventions do not seem to accrue over time for some long-term or socially vulnerable residents. Faced with this new green space paradox (Faber and Kimelberg 2014), residents and allied groups organize to contest the social effects of greening projects as a central part of efforts to create a just green city (Pearsall and Anguelovski, 2016; Connolly, 2019).

In addition, we examine the compounded and overlapping environmental impacts that socially vulnerable residents have to confront when they are faced with greening projects that might displace them while being forced to remain next to toxic facilities, contaminating land uses or climate-exposed areas, or when faced with having to move to areas with lower environmental quality as the only places where they can afford to live. Importantly, greening, especially of the consumption-oriented variety, does not eliminate other environmentally unsustainable practices or growing climate risks; those practices often co-inhabit and thrive side-by-side via glamorous greening projects and their accompanying real estate developments, thereby challenging the mobilization of environmental justice activists and their allies and complicating bold policy responses. In this book, we study how processes of long-term environmental injustices can come together with new green injustices to create cumulative and overlapping inequalities and impacts for minority and low-income residents in a variety of interventions, urban development pathways and political contexts. This analysis builds on a body of environmental justice (E.J.) studies demonstrating how low-income and minority residents suffer from greater environmental contamination and burdens and have less access to green space and livable, healthy neighborhoods (Boone et al. 2009, Grove et al. 2018) than whiter and wealthier groups, and today many of them also witness the greening and rebranding of their neighborhood for socially and racially privileged residents. The territorial scale and ecological characteristics of some urban environmental improvements in the book reveal a rather targeted and superficial "green treatment" rather than a more in-depth and wide-reaching environmental remediation in a neighborhood or district (Checker 2011).

Our book is thus about the now prominent strand of consumption-oriented urban greening initiatives, viewed as a political project grounded in technocratic principles and the apolitical assumption that greening will, unassisted, result in both more just and prosperous cities. Here, we argue that even if green practitioners and managers have the best social intentions, the broader urban greening orthodoxy leaves aside urban tensions, inequities, contradictions and trade-offs between different social groups and between environmental choices. As municipalities replicate urban greening in neighborhoods throughout a city, few of them begin with an equity lens and include concrete measures to ensure that greening solutions benefit all residents, and in particular residents and communities who are historically vulnerable to environmental racism and/or displacement. Furthermore, few of them hold contaminating factories or developers accelerating climate risks or climate citadels accountable for their social and environmental impacts.

Last, *The Green City and Social Injustice* highlights that the historic association between social groups and greening is not always linear. For example, postwar segregation patterns, characterized by whites leaving city centers in the U.S., have meant that Black residents who moved to formerly white neighborhoods inherited many green spaces. Also, the organizing of communities of color in the U.S. beginning in the 1980s resulted in the creation of many new green spaces in historically non-white neighborhoods. In Europe, the spatial pattern of cities can also mean that green inequities take on different forms around different types of greening initiatives and immigration patterns. Given this complex, non-linear history, we connect empirical studies on green space distributional inequities with long-term exclusionary processes embedded both in the political economy of development and in the (re)creation of urban nature. We thus expose the role of socio-environmental and cultural history in creating oppressive experiences, anxiety, chronic stress, socio-spatial segregation and overall poor access to protective green spaces for minority and low-income residents. We also expose how these practices get reproduced into the future.

Theorizing green environmental (in)justices

The cases we analyze demonstrate a variety of pathways that help us theorize green (in)justice. In this book, we explore three essential patterns of inequality at stake as the main analytical lenses: green locally unwanted land uses and the formation of green privilege; dispossession, displacement and the racialization of nature; and banking on nature. We then build on those lenses to propose a new understanding of justice in urban greening.

- Analytical lenses of green (in)justice

From the late 1970s onwards, environmental injustices have been traditionally studied and defined from the point of view of heavy toxics, industrial and waste facilities and other "brown" sites of contamination, neglect and abandonment (Bryant and Mohai 1992, Mohai and Saha 2015, Sze 2007, Lerner 2010, Chakraborty 2012). They were what activists, planners and academics defined as L.U.L.U.s, unequally exposing the bodies, lives and health of working-class, minority and immigrant residents to contaminants (Schively 2007). Much of this work also examined how structural racism embedded in redlining, racial covenants, neighborhood segregation, zoning, unequal applications of environmental regulations and racial capitalism contributed to the production and reproduction of those injustices (Pellow 2001, Checker 2008, Morello-Frosch and Lopez 2006, Taylor 2014, Pulido 2017, Maantay 2002, Campbell, Kim, and Eckerd 2014).

In *The Green City and Social Injustice*, we look at the other side of the environmental justice coin: the formation of green injustice and privilege through GreenL.U.L.U.s. In 2011, Park and Pellow defined environmental privilege as the disproportionate access and benefit harnessed by white, upper-class residents in regard to green and open spaces, waterfronts and coastlines, fresh food, healthy housing and recreational facilities (Park and Pellow 2011). While previous studies had already identified injustices in the provision of environmental amenities (Landry and Chakraborty 2009, Heynen, Perkins, and Roy 2006, Hastings 2007), Park and Pellow put emphasis on the racially exclusive manner in which environmental privilege is produced. Our goal here is to propose a further analysis of GreenL.U.L.U.s and the formation of green privilege. We argue here that green privilege exists because of continuing processes of land devaluation and the creation of a green rent gap through new value creation derived from consumption-oriented greening strategies (Dillon 2014, Anguelovski, Connolly, and Brand 2018), green dispossession and accumulation (Safransky 2014), green exclusion (Anguelovski et al. 2019) and green gentrification and displacement (Pearsall 2012, Gould and Lewis 2017). This green rent gap develops primarily through the marriage of sustainability policy and urban planning (Bunce 2017).

Second, our analysis of urban greening expands traditional understandings of dispossession and displacement, and ties these processes with a racialization of what is considered desirable nature. The tales of urban life described here point at displacement through loss of housing and socio-cultural erasure. We expose how the latter occurs through the invisibilization and silencing of environmental and social practices, voices and identities of long-term and/or vulnerable residents as neighborhoods become rebranded into green, smart and resilient places. The chapters expand green gentrification research about physical displacement (Immergluck 2009, Immergluck and Balan 2018) to further distill social, cultural and mental loss (Brand 2015, Anguelovski 2016), dispossession (Anguelovski 2015, Valli 2015), and states of displaceability (Yiftachel 2020)—the constant state of insecurities and losses (financial, housing, food, environmental, public space and institutional)—in which marginalized groups live. We also further highlight race and racism as the primary and ultimate predictor and indicator of unequal green inequities (Mohai and Saha 2015, Pulido 2017). We tie in how racialized minorities might see new green amenities—particularly those designed to remake an area in a different image of what is "good"—as GreenL.U.L.U.s because of the legacy of urban greening in a particular urban history or even national history.

Finally, the cases included throughout examine how concurrent processes of green accumulation and dispossession would not exist without financial models enabling urban growth and specifically intended to produce economic value through greening (Bakker 2007, Castree 2008). As greening becomes a branding tool for cities (Garcia Lamarca et al. 2021), we further argue that it is monetized, financialized and ends up creating speculation and rent capture for a few (Anguelovski et al. 2019). As greenways, parks or redeveloped waterfronts often act as anchors for development-oriented greening, investment remains concentrated in the hands of developers and investors and high-end real estate development. Those amenities help market neighborhoods (Krueger and Gibbs 2007, Tretter 2013) and elicit expectations of financial rewards, economic growth, neighborhood revitalization and increased competitiveness through business investments (Dooling 2009, Quastel 2009). Unpacking financial actors, their intermediaries and economic beneficiaries is one goal of the cases presented here. Finally, we also disentangle the role that greening plays in the process of urban development vis-à-vis the financialization of nature (Brand and Wissen 2014, Sullivan 2013), the financialization of urban development (Weber 2002, Rutland 2010) and the deployment of new financial instruments and finance capital (García-Lamarca and Ullström 2020).

- *Toward a new understanding of (in)justice in urban greening*

Based on the three analytical lenses described above, we aim to provide a new understanding of (in)justice in urban greening. In our view, the circumstances generating green inequalities faced by cities are perhaps best understood by putting aside the traditional analytical lens of distributional, recognition and procedural justice (Schlosberg 2013) and allowing for a more situated mode of registering (in)justice in urban greening. In part, this situated register of (in)justice is meant to surface a view of injustice from those affected—without a strict requirement to fit these views into a traditional philosophical box (Anguelovski et al. 2020). Here, an expanded sense of justice is meant to work in service of

(1) uncovering material and immaterial power embedded in urban greening and (2) building effective principles of equitable change needed for a more just greening.

Our 21 tales that unfold in the pages ahead highlight how urban greening injustice expands materially (in green infrastructures, projects, physical resources and ecologies) and immaterially (via explicit plans, discourses and branding) as municipalities transform existing landscapes into green and privileged utopias (Safransky 2014, Anguelovski, Connolly, and Brand 2018, Knuth 2016). In return, they illustrate how the material and immaterial (re)production of urban greening physically and symbolically delimits space within the green, sustainable, smart, resilient city orthodoxy (Connolly 2019), eventually producing injustices for vulnerable residents ignored within the hegemony of greening. Here, imaginaries (Castoriadis 1997), or the tales people tell of urban greening—which may include positive or negative impressions, discourses or renderings for a greened neighborhood or nature-centered intervention—play a central, yet often overlooked role in advancing praxis.

The tales people tell about urban greening are the focus of this book and connect persistent or new imaginaries and discourses to the dilemmas and tensions that urban greening produces and to historical linkages related to processes of domination and geographical racism (Pulido 2017). Neighborhoods inhabited by minorities or people of color are sometimes "rediscovered" in the process of urban greening and these locations, landscapes and cultural practices accrue new value in the process (Checker 2011, Summers 2019) through the way they are interpreted within the discourse and materiality of new interests. At the same time, these new green narratives and materialities can invisibilize the historic production of racialized, neglected, contaminated landscapes, coopting ideas or advancing certain visions over others. The result can be a process of green rent capture, by which greening adds economic and financial value to previously "invisible"—but often community-valued—spaces, such as informal gardens or community-based adaptation practices (Goodling, Green, and McClintock 2015). Many green injustices are indeed reproduced and exercised in a racialized and violent manner furthering the hegemony of color-blind advanced green capitalism and neoliberalism (Hardy, Milligan, and Heynen 2017).

In our analysis, we contrast such material and immaterial power at work in urban greening with antisubordination, intersectional and relational modes of greening as new urban justice theoretical propositions. These modes of greening represent a shift, sometimes a radical shift, away from that which displaces, racializes and financializes the integration of nature with urban environments. These urban green modes are not so much dependent on one form of justice—be it distributional, recognition or procedural—but rather direct our thinking toward a holistic and situated view of greening that accounts for ways that historical urbanization shapes the present and future realities of city residents.

First, we bring to the fore an *antisubordination* mode of green practice by drawing on critiques of whiteness (and white supremacy) as a spatial practice of domination and subordination (Steil 2018, Steil and Delgado 2019, Williams 2020). The practice of racializing space "orients bodies in specific directions, affecting how they take up space," even invisibly or unconsciously (Ahmed 2007, 150) and manifests as a set of multi-scaled and global geographical practices of domination (Bonds and Inwood 2016). Green justice as an antisubordination practice cannot be realized in a context of pervasive social and racial stratification and oppression—both in the U.S. and Europe. This approach necessitates the questioning of those institutions and practices that reproduce the subordinate social status of oppressed groups (Balkin and Siegel 2003).

Analytically, antisubordination green practice entails examining sustained, socially constructed categories of inequality and highlighting how they are produced geographically, spatially and within institutions. It also means understanding the elements of emancipatory and liberating spaces and geographical formations at the intersection of land, resources and nature. Next, this mode of greening involves new institutional arrangements, practices and policies to secure their control and use by marginalized groups, moving away from the ongoing "afterlives" of slavery (Harris, 2016) and colonizing approaches, or what McKittrick (2013) calls "plantation futures."

The analysis, understanding and institutional basis of antisubordination greening integrates a *reparative (or restorative) justice* perspective with urban green planning and development. This perspective-building work ensures that green interventions consistently acknowledge and address histories and geographies of oppression and exclusion through healing and spaces of refuge (Anguelovski 2014). In this vein, some have recently called for green reparations in cities suffering from historic harm, such as legacy pollution, land dereliction and institutional neglect (Draus et al. 2019). In addition, an antisubordination approach to greening is closely connected to the notion of *preventative justice*, through supporting practices that prevent or reduce the risk of possible future harm (Ashworth and Zedner 2014), and recognize differentiated risks by various social groups. Preventative justice is related, in part, to preventing the criminalization of racialized and gendered behaviors that inform green design solutions and policies, including those of groups coerced, criminalized and surveilled in public space (Pellow 2016)—many of those claims being articulated by recent Black Lives Matter organizing.

Second, we highlight the principle of *intersectional justice* in greening, which refers to exploring the "vexed dynamics of difference and the solidarities of sameness in the context of antidiscrimination and social movement politics" (Cho, Crenshaw, and McCall 2013) as related to environmental justice (Pellow 2016). Urban greening practice has particular intersectional effects that overlap gender, race, ethnicity, poverty, age, religion and others (Parker 2016). In this regard, in *The Green City and Social Injustice*, we attempt to examine how people understand, sense and live in neighborhoods given multiple, concurring identities expressed through place while being exposed to environmental inequalities, injustices and insecurities that magnify, overlap and at times conflict with one another. Here, urban greening research and practice can make important contributions by uncovering the intersectional experiences at work in all public spaces and neighborhood spaces and highlighting which non-dominant urban spaces can serve as refuges for non-dominant groups.

Third, we draw on the feminist principle of *relationality*, highlighting connections between sites, scales and subjectivities in order to emphasize how "everyday life" in the green city should strengthen the connection with spaces of home life, public life and work life. Greening that respects the everyday life of all citizens would explore the possibilities for challenging silences and violence, racialization, classism and patriarchy in ways that politicize and re-humanize place–life relationships (Gilroy 2018, Ranganathan and Bratman 2021). Relationality in greening reinforces antisubordination and intersectional greening by considering past and ongoing sites and experiences of violence and the emotional, political and social charges they still convey as greening linked to displacement and redevelopment plows through urban neighborhoods in Washington D.C., Dallas, Seattle, Portland, Milan, Glasgow or Dublin.

Mobilizing relationality within greening means highlighting issues and experiences of care and connection and of cultural recognition for residents in contact with urban greening (Derickson 2018, Ahmed 2017, Ranganathan and Bratman 2021). This approach re-situates urban greening away from property or economic valuation, highlighting the differentiated relational values (Himes and Muraca, 2018) embedded in and expressed by people's interaction with nature, while also noting that individualizing solutions for certain groups raise tensions and risks (Woods 2017). Here the voice of others (other ways of being, feeling, living or knowing spaces) is allowed to surface. Many of the cases in the book highlight the claims of activists for restored green amenities that can hold positive connotations for residents, building on their perception, interactions and use of green spaces (Checker 2011, Kabisch and Haase 2014, Haase et al. 2017, Guitart, Pickering, and Byrne 2012).

- *Radical practices for urban green justice*

From there, our theoretical contributions lead to propositions for new radical practices for urban green justice. Our tales of urban greening show that a different reality centered on wellbeing for socially vulnerable groups and democratization of urban space is possible. The instances where we see moves in this direction, though, require that community activists create the alliances that raise the political power of working-class and minority residents and allow them to stay in newly greened neighborhoods. Most cases included in this book examine activists' agency and their responses to (green) gentrification and displacement, with a focus on how they process the conundrum they face, and on the civic strategies they develop. A number of questions remain, however, about the most effective and long-lasting strategies for addressing community concerns around green gentrification and displacement.

For example, activists are potentially caught in a position where they voice grievances that planners and developers might use to justify siting urban greening interventions in more privileged and white neighborhoods. Here we examine which frames, tactics and other mobilizing tools might be more successful than others to create environmentally just places (Agyeman 2013, Anguelovski 2014, Agyeman et al. 2016, Anguelovski et al. 2019). We also explore how residents see themselves in the process of building alliances with "creative class" gentrifiers and how their identities, priorities and environmental strategies get rearticulated. We highlight the effect that such processes have on the sometimes-competing politics of environmental and community preservation in cities (Connolly 2019). Those cases also raise important questions about the role of traditional environmental movements in furthering green equity (Rigolon and Németh 2018, Pellow and Brulle 2005) and revisiting their traditional growth (Layzer 2015) positions in order to support greater environmental justice advocacy.

Lastly, a different, more just urban green reality is possible when it is developed alongside a political will and commitment to foster social housing, job training or economic development that lifts working-class residents and prioritizes equity over green halos and other glitzy green projects. In *Just Green Enough*, Curran and Hamilton already highlighted the need to build equity-driven urban greening accompanied by structural housing and anti-displacement policies (Curran and Hamilton 2017). Their book demonstrates that social justice goals need to be placed at the forefront of environmental regeneration and change with purposeful alternatives to prevent mass displacement. In *The Green City and Social Injustice*, we examine the role of municipal decision-makers and agencies in addressing or preventing inequities in greening.

Our 21 cases provide ample room for further analyzing the responses (or lack thereof) offered by planners and elected officials to community concerns over risks of physical and social displacement, especially so when they are deployed in the name of ecological risk. Our large international comparative approach—in the same vein as Susannah Bunce (Bunce 2017)—allows us to present experiences of bold, radical equity planning (Vienna, Nantes or even Copenhagen and Amsterdam) that have allowed for a right-based approach to housing and to neighborhood stability and for land rights for socially vulnerable groups, although some of these experiences are not without contradictions and threats to their survival. The book shows how cities that at once prioritize equity in green access, investment in secure and de-commodified housing and progressive social policies can achieve both environmental and social gains without becoming green museums for the few.

Research design behind 21 tales from North America and Europe

Our book draws on field research conducted between 2018 and 2020 in 21 cities across 10 countries: the United States, Canada, France, Ireland, the United

Kingdom, Denmark, the Netherlands, Austria, Italy and Spain. While we recognize that there are multiple forms through which the urban green orthodoxy and its practice have expanded throughout the Global South (Comelli, Anguelovski, and Chu 2018, Anguelovski, Irazábal, and Connolly 2019), this project was conceived as a North American and Western European collaborative research endeavor. We selected cities and neighborhoods from a larger E.U.-funded (European Research Council) research project which aimed to compare the social impact of new environmental amenities in North America and Western Europe: GreenL.U.L.U.s. The project also draws on field research conducted for a partner project for which we also led and coordinated international collaborative research called Naturvation.

Building on an initial set of 99 possible cities and the *Green Trajectories*[3] report we published in 2018, we selected 21 cities that represent a variety of geographic areas (Coastal, Midwest and Southern U.S.; Canada; Northern Europe; and Southern Europe), city types (i.e., industrial, post-industrial, economically growing or shrinking) and greening priorities (e.g., industrial transformations, climate-resilient infrastructure, biodiversity protection and enhancement and neighborhood redevelopment). We selected only cities with populations ranging from 500,000 to 1.5 million residents in order to compare cities with relatively similar exposure to globalization forces and whose greening trajectory over the last two or three decades could be rigorously identified and traced. Last, based on our review of city documents and media reporting on greening, demographic and social change, civic activism around greening and consultation with collaborators and experts in each city, we identified a specific neighborhood in which to conduct field work in each city. All the selected neighborhoods have experienced a history of environmental degradation, burden, loss and injustice together with recent greening, urban revitalization, social change and neighborhood organizing for green justice.

For each city and neighborhood, the author(s) spent between four and six weeks on field work, conducting semi-structured interviews with several types of people: neighborhood and citywide residents, municipal staff and elected officials, developers and other real estate stakeholders, representatives of community-based or nonprofit organizations and activists working in the neighborhood. We interviewed at least two respondents from each type through snowball sampling and conducted interviews until reaching saturation. We identified initial respondents through previous research, expert contacts, internet searches and review of local media articles about each case. All participants provided written informed consent for participation. For the large majority of respondents, interviews were audio-recorded and transcribed.

We conducted interviews using a collectively designed semi-structured interview protocol. We used the same interview guide for all participants, with modifications or adjustments by interviewee type. Each interview included specific questions on the following themes: history of local urban development, baseline and changes in environmental exposures for residents, neighborhood revitalization projects, development of new environmental amenities, inclusion and equity issues in regard to those changes and developments, health impacts for residents, policy and community responses and challenges in addressing social and health justice concerns. In each city, interviewees spoke to experiences, knowledge and expertise regarding a specific case neighborhood, and broader overlapping perspectives such as city- or region-wide policy or planning processes. Our final dataset included 515 transcribed interviews.

In addition to primary data, each field researcher collected relevant secondary data to complement knowledge and analysis of local urban development changes and equity issues for each city and neighborhood: (1) documents and fact sheets, reports or policy documents produced by local organizations (nonprofit, for-profit or government agencies), (2) media articles regarding our cases or other relevant events and (3) city planning documents concerning environmental sustainability, urban greening, climate change, urban revitalization, and health, wellbeing and social equity in each city. We used this additional data to corroborate and enrich the accounts of respondents and also to identify information such as specific sources of contamination, greening programs, ecological data or relevant events which respondents had mentioned. Gray literature, along with published academic literature on a particular city, also provided an understanding of the history of each area and its transformation.

In order to analyze our data, we created a detailed coding scheme based on the main conceptual and analytical themes we sought to understand. The coding scheme was developed by two of the primary researchers, and slight modifications and clarifications were made during its creation when suggested by one of the coders and agreed upon by all researchers. When the coding scheme was complete, we completed training to discuss the structure, content and meaning of the coding framework and its codes in order to ensure consistent understanding and intercoder reliability. While researchers used NVivo software to code interviews for each city independently, team members were in continuous communication to clarify doubts or questions about the coding process.

To ensure analytical consistency and ease comparability between our 21 tales, each chapter is broadly organized as follow: (1) key concepts and keywords that summarize the urban development pattern of the city and neighborhood (e.g., fast recent growing, recovering, long-term stable growth, industrial, tech-driven, etc.), the urban greening trajectory of the city and/or neighborhood (e.g., focus on new parks and gardens, green resilient infrastructure, food production), the inequalities at stake (e.g., green gentrification and/or displacement, deep inequities in access to green, climate resilience/unequal climate protection from greening, continued exposure to contamination); (2) a summary paragraph on the main storyline and theoretical specificity of the urban green inequality "tale" for that city; and (3) the body of the chapters with a description of historical and baseline situations for urban development and environmental conditions before the greening; the "tale" of urban greening (with a focus on a particular neighborhood or two) and on the inequalities exacerbated, addressed or grappled with along the way; and a description of governmental action and civil society response to inequalities. Our goal with each tale is to combine a historical and narrative approach with a thematic analysis of injustices in urban greening, allowing for greater comparability and discussion within both academic settings and urban practice. We kept the chapters short in order to make them accessible to a wide audience and allow for easier comparisons.

Overall, our comparative urbanism approach (McFarlane 2010, Robinson 2011, Lees 2012) allows us to examine with fine historic detail, empirical breadth and analytical rigor how urban greening has been adopted and transformed both discursively and in practice by a variety of cities across political contexts and histories of urbanization. It also allows us to tell a story about which cities seem to be faring better in terms of guaranteeing improved urban environments for all, especially so for socially vulnerable groups. Despite the increasingly theoretically sophisticated and empirically rich research on the displacement and exclusion processes embedded in urban sustainability planning, we see a need to distill the socio-spatial dynamics and ramifications of urban greening and green inequalities in a wide and comprehensive manner. Overall, we hope that instructors and students interested in urban greening, activists fighting for a more just green city and policy-makers planning and building the green and just cities of the future will find challenging, puzzling and comparative inspiration in the different tales presented here.

Book outline

The book is divided into 5 thematic sections comprised of 21 tales of greening in cities throughout Europe and North America. Each section offers an analysis of a type of urban greening "experience" or tension. The term *tales* is used deliberately to highlight how greening has often become or created a new imaginary—a fictional tale of sorts—of green infrastructure and its possibilities, while the full story includes its invisible, unspoken or under-treated tensions and injustices.

Part 1, "The social costs of glitzy green urbanism," examines the social impacts (e.g., green gentrification and displacement, real estate speculation and luxury developments) of implementing greening projects in working-class or historically marginalized neighborhoods. It analyzes how tradeoffs are viewed by various actors (community organizers, city policy-makers, real estate developers and lobbyists) under the policy and planning frameworks of green growth, smart growth and eco-urbanism. Through the cases of Milan, Amsterdam, Bristol and Valencia, this section brings to the surface the key contradictions that contemporary urban greening efforts embody but often obscure.

Part 2, "Compounded risks and impacts of urban greening in post-industrial environments," examines what greening means when deployed in neighborhoods and cities with a long history and legacy of environmental burdens and contamination. Here, socially vulnerable residents are confronted with dynamics of greening and unequal real estate redevelopment while living with lasting environmental impacts from contaminated sites and industries. The continued presence of a heavily industrial landscape gives these cities a distinct relationship to greening and a distinct set of social effects which are explored through the tales of Cleveland, Dallas, Glasgow, San Francisco and Seattle.

Part 3, "(Re)creating unjust racialized landscapes in the green city?" introduces how greening agendas create, re-create and sustain unjust racialized places in segregated neighborhoods and cities. It articulates the cultural link between whiteness and greening vis-à-vis the historical legacy of segregation and discrimination that those neighborhoods often had to endure. While greening is not solely linked with white culture, the cases of Atlanta, Austin, Washington, D.C., and Boston demonstrate how in particular times and places that link takes shape and facilitates an expression of white privilege in and through space.

Part 4, "The complex entanglement of greening and multiple other gentrification pressures," presents how greening is often not a single dimension of redevelopment and transformation but is rather embedded in a web of dynamics that create additional pressures on the lives of marginalized residents. These might also include food gentrification, tourism and commercial gentrification or student gentrification. Especially in circumstances where there are visibly associated drivers as in Montreal, Dublin, Barcelona and Philadelphia, greening must be understood as part of the process of generating new inequities and gentrification in cities, but not as the sole driver.

In our final section, "(Fragile) green justice victories and grey zones in the just green city," we offer alternative scenarios and lenses on more equal and just green futures. This section of chapters shows what greening can mean when it goes hand in hand with social policies that guarantee and/or protect a right to place for fragile residents. Copenhagen, Nantes, Vienna and Portland show that such successes can be fleeting and are not exempt from contradictions but that there are examples to build from.

These 21 tales are followed by a chapter on the policy, planning and financial tools and pathways that have proved promising in securing greater justice in greening—many of which are responding to the demands of community activists and social movements—in North America and Europe. Last, we conclude the book with a discussion chapter of all our cases. Here, we examine the most salient green trends and pathways across cities, pointing to the different forms of injustices embedded in greening, their structural and individual drivers and alternative routes for more equitable, democratic and emancipatory green cities.

Notes

1 https://www.youtube.com/watch?v=Q8dUMucYN0w&t=28s
2 https://ajuntament.barcelona.cat/superilles/es/?utm_content=bufferd8070&utm_medium=social&utm_source=plus.google.com&utm_campaign=buffer
3 https://www.google.com/search?client=safari&rls=en&q=Green+Trajectories&ie=UTF-8&oe=UTF-8

References

Agyeman, Julian. 2013. *Introducing Just Sustainabilities*. London: Zed Books.

Agyeman, Julian, David Schlosberg, Luke Craven, and Caitlin Matthews. 2016. "Trends and Directions in Environmental Justice: From Inequity to Everyday Life, Community, and Just Sustainabilities." *Annual Review of Environment and Resources* 41:321–40.

Ahmed, Sara. 2007. "A Phenomenology of Whiteness." *Feminist Theory* 8 (2):149–168.

Ahmed, Sara. 2017. *Living a Feminist Life*. Duke University Press.

Anguelovski, Isabelle. 2014. *Neighborhood as Refuge: Environmental Justice, Community Reconstruction, and Place-Remaking in the City*. Cambridge: MIT Press.

Anguelovski, Isabelle. 2015. "Alternative Food Provision Conflicts in Cities: Contesting Food Privilege, Injustice, and Whiteness in Jamaica Plain, Boston." *Geoforum* 58:184–194.

Anguelovski, Isabelle. 2016. "From Toxic Sites to Parks as (Green) LULUs? New Challenges of Inequity, Privilege, Gentrification, and Exclusion for Urban Environmental Justice." *Journal of Planning Literature* 31 (1):1–14. doi: 10.1177/0885412215610491.

Anguelovski, Isabelle, Anna L. Brand, James J. T. Connolly, Esteve Corbera, Panagiota Kotsila, Justin Steil, Melissa Garcia Lamarca, Margarita Triguero-Mas, Helen Cole, Francesc Baró, Johannes Langemeyer, Carmen Perez del Pulgar, Galia Shokry, Filka Sekulova, and Lucia Arguelles. 2020. "Expanding the Boundaries of Justice in Urban Greening Scholarship: Towards an Emancipatory, Anti-Subordination, Intersectional, and Relational Approach." *Annals of the American Association of Geographers* 110:1743–1769.

Anguelovski, Isabelle, James Connolly, and Anna L. Brand. 2018. "From Landscapes of Utopia to the Margins of the Green Urban Life: For Whom Is the New Green City?" *City* 22 (3):417–436.

Anguelovski, Isabelle, James Connolly, Melissa Garcia Lamarca, Helen Cole, and Hamil Pearsall. 2019. "New Scholarly Pathways on Green Gentrification: What Does the Urban "Green Turn" Mean and Where Is It Going?" *Progress in Human Geography* 43 (6):1064–1086.

Anguelovski, Isabelle, Clara Irazábal, and James Connolly. 2019. "Grabbed Urban Landscapes: Sociospatial Tensions in Green Infrastructure Planning in Medellín." *International Journal of Urban and Regional Research* 43 (1):133–156.

Ashworth, Andrew, and Lucia Zedner. 2014. *Preventive Justice*. Oxford: OUP.

Bakker, Karen. 2007. "The "Commons" versus the "Commodity": Alter-Globalization, Anti-Privatization and the Human Right to Water in the Global South (2007)." *The Globalization and Environment Reader* 39 (3):430–455.

Balkin, Jack M., and Reva B Siegel. 2003. "The American Civil Rights Tradition: Anticlassification or Antisubordination." *Issues in Legal Scholarship* 2 (1).

Bonds, Anne, and Joshua Inwood. 2016. "Beyond White Privilege: Geographies of White Supremacy and Settler Colonialism." *Progress in Human Geography* 40 (6):715–733.

Boone, Christopher, Geoffrey Buckley, Morgan Grove, and Chona Sister. 2009. "Parks and People: An Environmental Justice Inquiry in Baltimore, Maryland." *Annals of the Association of American Geographers* 99 (4):767–787.

Brand, Anna Livia. 2015. "The Most Complete Street in the World: A Dream Deferred and Coopted." In *Incomplete Streets: Processes, Practices, and Possibilities*, edited by J. Agyeman and S. Zavetoski, 245–266. New York: Routledge.

Brand, Ulrich, and Markus Wissen. 2014. "Financialisation of Nature as Crisis Strategy." *Journal für Entwicklungspolitik* 30 (2):16–45. New York.

Bryant, Bunyan I., and Paul Mohai. 1992. *Race and the Incidence of Environmental Hazards: A Time for Discourse*. Boulder: Westview Press.

Bunce, Susannah. 2017. *Sustainability Policy, Planning and Gentrification in Cities*. New York: Routledge.

Campbell, Heather E., Yushim Kim, and Adam Eckerd. 2014. "Local Zoning and Environmental Justice: An Agent-Based Model Analysis." *Urban Affairs Review* 50 (4): 521–552. doi: 10.1177/1078087413505736.

Castoriadis, Cornelius. 1997. *The Imaginary Institution of Society*. Cambridge, MA: MIT Press.

Castree, Noel. 2008. "Neoliberalising Nature: The Logics of Deregulation and Reregulation." *Environment and Planning A* 40 (1):131–152.

Chakraborty, Jayajit. 2012. "Cancer Risk from Exposure to Hazardous Air Pollutants: Spatial and Social Inequities in Tampa Bay, Florida." *International Journal of Environmental Health Research* 22 (2):165–183.

Checker, Melissa. 2008. *Polluted Promises: Environmental Racism and the Search for Justice in a Southern Town*. New York City: New York University Press.

Checker, Melissa. 2011. "Wiped Out by the "Greenwave": Environmental Gentrification and the Paradoxical Politics of Urban Sustainability." *City & Society* 23 (2):210–229. doi: 10.1111/j.1548-744X.2011.01063.x.

Cho, Sumi, Kimberlé Williams Crenshaw, and Leslie McCall. 2013. "Toward a Field of Intersectionality Studies: Theory, Applications, and Praxis." *Signs: Journal of Women in Culture and Society* 38 (4):785–810.

Comelli, Thaisa, Isabelle Anguelovski, and Eric Chu. 2018. "Socio-Spatial Legibility, Discipline, and Gentrification through Favela Upgrading in Rio de Janeiro." *CITY* 22 (5–6):633–656.

Connolly, James. 2019. "From Jacobs to the Just City: A Foundation for Challenging the Green Planning Orthodoxy." *CITIES* 91:64–70.

Curran, Winifred, and Trina Hamilton. 2017. *Just Green Enough: Urban Development and Environmental Gentrification*. New York: Routledge.

Derickson, Kate. 2018. "Masters of the Universe." *Environment and Planning D: Society and Space* 36 (3):556–562.

Dillon, Lindsey. 2014. "Race, Waste, and Space: Brownfield Redevelopment and Environmental Justice at the Hunters Point Shipyard." *Antipode* 46 (5):1205–1221. doi: 10.1111/anti.12009.

Dooling, Sarah. 2009. "Ecological Gentrification: A Research Agenda Exploring Justice in the City." *International Journal of Urban and Regional Research* 33 (3):621–639.

Draus, Paul, Dagmar Haase, Jacob Napieralski, Juliette Roddy, and Salman Qureshi. 2019. "Wounds, Ghosts and Gardens: Historical Trauma and Green Reparations in Berlin and Detroit." *Cities* 93:153–163.

Faber, Daniel, and Shelley Kimelberg. 2014. "Sustainable Urban Development and Environmental Gentrification: The Paradox Confronting the U.S. Environmental Justice Movement." In *Uprooting Urban America: Multidisciplinary Perspectives on Race, Class & Gentrification*, edited by Horace R. Hall, Cynthia Robinson, and Amor Kohli, 77–92. New York: Peter Lang Publishers.

Garcia Lamarca, Melissa, Isabelle Anguelovski, Helen Cole, James J. T. Connolly, Lucía Arguelles, Francesc Baró, Carmen Perez del Pulgar, and Galia Shokry. 2021. "Urban Green Boosterism and City Affordability: For Whom is the 'Branded' Green City?" *Urban Studies* 58 (1).

García-Lamarca, Melissa, and Sara Ullström. 2020. ""Everyone Wants This Market to Grow": The Affective Post-Politics of Municipal Green Bonds." *Environment and Planning E: Nature and Space*. 2514848620973708.

Gilroy, Paul. 2018. ""Where Every Breeze Speaks of Courage and Liberty": Offshore Humanism and Marine Xenology, or, Racism and the Problem of Critique at Sea Level: The 2015 Antipode RGS-IBG Lecture." *Antipode* 50 (1):3–22.

Goodling, Erin, Jamaal Green, and Nathan McClintock. 2015. "Uneven Development of the Sustainable City: Shifting Capital in Portland, Oregon." *Urban Geography* 36 (4):504–527.

Gould, Kenneth A., and Tammy L. Lewis. 2017. *Green Gentrification: Urban Sustainability and the Struggle for Environmental Justice*. New York: Routledge.

Grove, Morgan, Laura Ogden, Steward Pickett, Chris Boone, Geoff Buckley, Dexter H Locke, Charlie Lord, and Billy Hall. 2018. "The Legacy Effect: Understanding How Segregation and Environmental Injustice Unfold Over Time in Baltimore." *Annals of the American Association of Geographers* 108 (2):524–537.

Guitart, Daniela, Catherine Pickering, and Jason Byrne. 2012. "Past Results and Future Directions in Urban Community Gardens Research." *Urban Forestry & Urban Greening* 11 (4):364–373.

Haase, Dagmar, Sigrun Kabisch, Annegret Haase, Erik Andersson, Ellen Banzhaf, Francesc Baró, Miriam Brenck, Leonie K Fischer, Niki Frantzeskaki, and Nadja Kabisch. 2017. "Greening Cities–To Be Socially Inclusive? About the Alleged Paradox of Society and Ecology in Cities." *Habitat International* 64:41–48.

Hardy, R. Dean, Richard A. Milligan, and Nik Heynen. 2017. "Racial Coastal Formation: The Environmental Injustice of Colorblind Adaptation Planning for Sea-Level Rise." *Geoforum* 87:62–72.

Harris, C. I. 2016. *The Afterlife of Slavery: Markets, Property and Race*. Lecture at Artist Space. New York: Books & Talks.

Hastings, Annette. 2007. "Territorial Justice and Neighbourhood Environmental Services: A Comparison of Provision to Deprived and Better-Off Neighbourhoods in the UK." *Environment and Planning C* 25:896–917. doi: 10.1068/c0657.

Heynen, Nik, Harold Perkins, and Parama Roy. 2006. "The Political Ecology of Uneven Urban Green Space." *Urban Affairs Review* 42 (1):3–25. doi: 10.1177/1078087406290729.

Himes, A., and B. Muraca. 2018. "Relational Values: The Key to Pluralistic Valuation of Ecosystem Services." *Current Opinion in Environmental Sustainability* 35:1–7.

Immergluck, Dan. 2009. "Large Redevelopment Initiatives, Housing Values and Gentrification: The Case of the Atlanta Beltline." *Urban Studies* 46 (8):1723–1745.

Immergluck, Dan, and Tharunya Balan. 2018. "Sustainable for Whom? Green Urban Development, Environmental Gentrification, and the Atlanta Beltline." *Urban Geography* 39 (4):546–562.

Jennings, Viniece, Matthew Browning, and Alessandro Rigolon. 2019. *Urban Green Spaces: Public Health and Sustainability in the United States*. Cham, Switzerland: Springer.

Kabisch, Nadja, and Dagmar Haase. 2014. "Green Justice or Just Green? Provision of Urban Green Spaces in Berlin, Germany." *Landscape and Urban Planning* 122:129–139.

Knuth, Sarah. 2016. "Seeing Green in San Francisco: City as Resource Frontier." *Antipode* 48 (3):626–644.

Krueger, Rob, and David C. Gibbs. 2007. *The Sustainable Development Paradox: Urban Political Economy in the United States and Europe*. New York: Guilford Press.

Landry, Shawn, and Jayajit Chakraborty. 2009. "Street Trees and Equity: Evaluating the Spatial Distribution of an Urban Amenity." *Environment and Planning A* 41 (11):2651–2670.

Layzer, Judith A. 2015. *The Environmental Case*. Thousand Oaks, CA: Sage.

Lees, Loretta. 2012. "The Geography of Gentrification Thinking through Comparative Urbanism." *Progress in Human Geography* 36 (2):155–171.

Lerner, Steve. 2010. *Sacrifice Zones: The Front Lines of Toxic Chemical Exposure in the United States*. Cambridge, MA: MIT Press.

Liebelt, Veronika, Stephan Bartke, and Nina Schwarz. 2018. "Hedonic Pricing Analysis of the Influence of Urban Green Spaces onto Residential Prices: The Case of Leipzig, Germany." *European Planning Studies* 26 (1):133–157.

Maantay, Juliana. 2002. "Zoning Law, Health, and Environmental Justice: What's the Connection?" *The Journal of Law, Medicine & Ethics* 30 (4):572–593. doi: 10.1111/j.1748-720X.2002.tb00427.x.

McFarlane, Colin. 2010. "The Comparative City: Knowledge, Learning, Urbanism." *International Journal of Urban and Regional Research* 34 (4):725–742. doi: 10.1111/j.1468-2427.2010.00917.x.

McKittrick, K. 2013. "Plantation Futures." *Small Axe: A Caribbean Journal of Criticism* 17 3 (42):1–15.

Mohai, Paul, and Robin Saha. 2015. "Which Came First, People or Pollution? Assessing the Disparate Siting and Post-Siting Demographic Change Hypotheses of Environmental Injustice." *Environmental Research Letters* 10 (11):115008.

Morello-Frosch, Rachel, and Russ Lopez. 2006. "The Riskscape and the Color Line: Examining the Role of Segregation in Environmental Health Disparities." *Environmental Research* 102 (2):181–196.

Park, Lisa Sun-Hee, and David Pellow. 2011. *The Slums of Aspen: Immigrants vs. The Environment in America's Eden*. New York: New York University Press.

Parker, Brenda. 2016. "Feminist Forays in the City: Imbalance and Intervention in Urban Research Methods." *Antipode* 48 (5):1337–1358.

Pearsall, Hamil. 2012. "Moving Out or Moving In? Resilience to Environmental Gentrification in New York City." *Local Environment* 17 (9):1013–1026.

Pellow, David. 2001. "Environmental Justice and the Political Process: Movements, Corporations, and the State." *Sociological Quarterly* 42 (1):47–67.

Pellow, David, and Robert J. Brulle. 2005. *Power, Justice, and the Environment: A Critical Appraisal of the Environmental Justice Movement, Urban and Industrial Environments*. Cambridge, MA: MIT Press.

Pellow, D. N. 2016. "Toward a Critical Environmental Justice Studies: Black Lives Matter as an Environmental Justice Challenge." *Du Bois Review: Social Science Research on Race* 13 (2):221–236.

Pulido, Laura. 2016. "Geographies of Race and Ethnicity II: Environmental Racism, Racial Capitalism and State-Sanctioned Violence." *Progress in Human Geography* 41 (4):0309132516646495.

Pulido, Laura. 2017. "Geographies of Race and Ethnicity II: Environmental Racism, Racial Capitalism and State-Sanctioned Violence." *Progress in Human Geography* 41 (4):524–533.

Quastel, Noah. 2009. "Political Ecologies of Gentrification." *Urban Geography* 30 (7):694–725.

Ranganathan, Malini, and Eve Bratman. 2021. "From Urban Resilience to Abolitionist Climate Justice in Washington, DC." *Antipode* 53 (1):115–137.

Rigolon, Alessandro, and Jeremy Németh. 2018. ""We're not in the Business of Housing:" Environmental Gentrification and the Nonprofitization of Green Infrastructure Projects." *Cities* 81:71–80.

Robinson, Jennifer. 2011. "Cities in a World of Cities: The Comparative Gesture." *International Journal of Urban and Regional Research* 35 (1):1–23. doi: 10.1111/j.1468-2427.2010.00982.x.

Rutland, Ted. 2010. "The Financialization of Urban Redevelopment." *Geography Compass* 4 (8):1167–1178.

Safransky, Sara. 2014. "Greening the Urban Frontier: Race, Property, and Resettlement in Detroit." *Geoforum* 56:237–248.

Schively, Carissa. 2007. "Understanding the NIMBY and LULU Phenomena: Reassessing Our Knowledge Base and Informing Future Research." *Journal of Planning Literature* 21 (3):255–266.

Schlosberg, David. 2013. "Theorising Environmental Justice: The Expanding Sphere of a Discourse." *Environmental Politics* 22 (1):37–55.

Steil, Justin. 2018. "Antisubordination Planning." *Journal of Planning Education and Research* 0739456X18815739.

Steil, Justin P., and Laura Humm Delgado. 2019. "Limits of Diversity: Jane Jacobs, the Just City, and Anti-Subordination." *Cities* 91:39–48.

Sullivan, Sian. 2013. "Banking Nature? The Spectacular Financialisation of Environmental Conservation." *Antipode* 45 (1):198–217.

Summers, Brandi Thompson. 2019. *Black in Place: The Spatial Aesthetics of Race in a Post-Chocolate City*. Chapel Hill, NC: UNC Press Books.

Sze, Julie. 2007. *Noxious New York: The Racial Politics of Urban Health and Environmental Justice*. Cambridge: MIT Press.

Taylor, Dorceta. 2014. *Toxic Communities: Environmental Racism, Industrial Pollution, and Residential Mobility*. New York: NYU Press.

Tretter, Eliot M. 2013. "Contesting Sustainability: 'SMART Growth' and the Redevelopment of Austin's Eastside." *International Journal of Urban and Regional Research* 37 (1):297–310.

Valli, Chiara. 2015. "A Sense of Displacement: Long-Time Residents' Feelings of Displacement in Gentrifying Bushwick, New York." *International Journal of Urban and Regional Research* 39 (6):1191–1208.

Wachsmuth, David, and Hillary Angelo. 2018. "Green and Gray: New Ideologies of Nature in Urban Sustainability Policy." *Annals of the American Association of Geographers* 108 (4):1038–1056.

Weber, Rachel. 2002. "Extracting Value from the City: Neoliberalism and Urban Redevelopment." *Antipode* 34 (3):519–540.

Williams, Rashad Akeem. 2020. "From Racial to Reparative Planning: Confronting the White Side of Planning." *Journal of Planning Education and Research* 0739456X20946416.

Woods, Clyde. 2017. *Development Drowned and Reborn: The Blues and Bourbon Restorations in Post-Katrina New Orleans*. Vol. 35. Athens, GA: University of Georgia Press.

Yiftachel, Oren. 2020. "From Displacement to Displaceability." *CITY* 24 (1–2):151–165.

PART 1

The social costs of glitzy green urbanism

The chapters in the following part analyze what we call "The social costs of glitzy green urbanism," that is, the social impacts—including green gentrification, real estate speculation and luxury developments and loss of connection with place or displacement—on vulnerable residents when their neighborhood becomes the host of greening projects underpinned by green design, smart growth or growth-oriented eco-urbanism logics. In Milan, Amsterdam, Bristol and Valencia, green urbanism embodies key socio-environmental contradictions that often remain obscured or overlooked.

Milan, the largest smart city in Europe, has embraced a glitzy form of eco-urbanism through the redevelopment project Porta Nuova and its vertical forest tower, Bosco Verticale. Activists from Milan's *centri sociali* (the well-known *social centers*) have denounced how the overall project has privatized a historic public park and how the rental costs of the recent skyscrapers are prohibitively expensive for most Milan residents. However, this vertical greening is counterpoised by new horizontal green spaces, especially community gardens driven by neighborhood activists.

Bristol, despite being named the European Green Capital in 2015, has yet to be able to provide all its citizens with access to quality green space and affordable housing. With a 66% budget cut in funding for the maintenance and creation of new greenspace and the government's inability to deliver on promises of new affordable housing, both the city and its South Side communities have been faced with finding alternatives to creating a just and green city for all, with much of the burden falling on resident groups to both maintain green space and oversee plans for housing developments.

Valencia on the southeastern coast of Spain is one of the many cities transforming rail infrastructure into parks, as the Valencia Parc Central project illustrates. Here, we see the impact of an announcement effect through which gentrification

DOI: 10.4324/9781003183273-1

accelerates due to the construction of luxury skyscrapers built along with Parc Central. This facilitates the financial feasibility of the luxury projects and enacts a more complete realization of Valencia's new "modernity" in a competitive race across Europe for the most attention-worthy new infrastructure projects.

The tale of Amsterdam Noord's green redevelopment illustrates the growing extent to which what has been long portrayed as an ideal urban model of social justice in Northern Europe is being dismantled. Although Amsterdam preserves a relatively high level of social equality thanks to the residuals of a strong institutional social architecture, the perils of green design, neoliberal policies, hyper-segregation and gentrification in the redeveloped Amsterdam Noord's waterfront seem to have come to stay.

1

MILAN'S PRIVATE VERTICAL FORESTS VS. HORIZONTAL URBAN GREENING

Lucia Di Paola

Milan brands itself as the largest smart city in Europe, openly embracing eco-urbanism as a development modus operandum. Between 2004 and 2015, Milan completed Porta Nuova, its largest re-urbanization project to date, prioritizing urban greening and sustainability as symbolized by the new Vertical Forest (*Bosco Verticale* in Italian) skyscrapers. But activists from Milan's *centri sociali* (social centers) and resident groups have strongly opposed the self-branded green project, criticizing the privatization of a public park and denouncing prohibitively high rents. According to them, Porta Nuova and its Vertical Forests illustrate a top-down privatization project that leads to exclusion and gentrification. In response, residents and activist groups have developed a public community garden in the same neighborhood that serves as a new model for democratic horizontal urban greening.

From an isolated working-class neighborhood to a "Gate of Green"

As the sixth most polluted city in Italy (Legambiente, 2019), Milan has designated urban greening as one of the top priorities in its Government Plan for 2030 (P.G.T., 2018). The municipality's endorsement of the re-urbanization project of Porta Nuova heeds its commitment to placing greening and environmental sustainability at the center of redevelopment (Anselmi, 2014; Porta Nuova, n.d.). Developed, financed and managed by the global real estate firm Hines and led by Hines' C.E.O. Manfredi Catella, Porta Nuova was the largest redevelopment project in Europe at the time, covering an area of 290,000 m^2 which the municipality had vowed to redevelop for more than 60 years. The new infrastructures of Porta Nuova are partially built in the existing neighborhood of Isola, historically labeled by attracted developers as a somewhat degraded area crossed by the

DOI: 10.4324/9781003183273-2

railways of the Garibaldi train station and disconnected from the rest of the city (Porta Nuova, n.d.). Hereby it is worth mentioning Isola's historical background to understand how Porta Nuova has changed it.

The name Isola (island in Italian) characterizes the neighborhood's spatial isolation due to its location behind the extensive walls of the rail yards of the Garibaldi station, which used to make it a relatively unattractive and thus more affordable neighborhood (Brizioli, 2015). After World War I, a wave of Southern Italian immigrants moved to Milan to access the fast-emerging industrial jobs. Many of the new engineering industries were located in Isola, such as Brown Bover, Isaria and Gondand (Signorelli, 2015). For this reason, the large majority of inhabitants living in Isola were industry workers. It was here that the first Italian worker rights movements developed and fomented the concentration of socialist political activity. During World War II, Isola represented one of the cores of Italian partisans' resistance[1] (Brizioli, 2015). At the time, residents could afford to buy Isola's relatively cheap housing. Many of the historical residents owned these properties until not long ago. Additionally, Isola also hosted several social housing buildings, making poorer classes an integral part of the neighborhood.

Already 50–60 years ago, Isola was a lively place. The diverse cultures of immigrant workers brought to Isola different Italian regional restaurants and the political activism brought frequent public events. If on the one hand, Isola's

FIGURE 1.1 New building and infrastructure of Porta Nuova. The Library of Trees is a central park connecting the Isola neighborhood to the rest of the new Porta Nuova district. Elevated maintenance costs are due to its 90,000 m^2 surface area, which contains 450 trees and 90,000 plants. The Vertical Forest towers are the buildings covered by trees. (Photo by: Goldmund100, via Wikimedia Commons.)

urban fabric and connectivity were poor; on the other, it was a vibrant neighborhood and thus became an attractive location. It is worth mentioning that it was since World War II that the municipality and private investors had made attempts to redevelop the area of Isola. Yet, these projects had not been successful; in the 1990s, locals and activists of Isola successfully appealed against two urban plans not in full compliance with city regulations on permitted building cubage (Anselmi, 2014). Nonetheless, along with the European trend, Milan's industries started to shut down and the economy started to shift from the industrial to the service sector, leaving various residents unemployed and unable to afford more expensive living standards.

It the 1990s Isola also started hosting new *centri sociali*—political and cultural community centers usually characterized by left-wing politics, as well as artistic initiatives and boutiques. Yet, the newcomers were always very explicit about the wish of keeping Isola a diverse and inclusive neighborhood. In 2001, for instance, residents, artists and associations squatted an old factory located in the neighborhood's only public park. La Stecca, as it was called, went on to become an artistic, cultural and political hub that also denounced the municipality's scarce attention towards bottom-up movements, opposing its general top-down attitude towards re-urbanization (Brizioli, 2015). It was the symbol of Isola's identity: a neighborhood with a strong and engaged civil society.

Yet, Isola's unique profile made it increasingly attractive to the public and real estate developers (Brizioli, 2015), and despite the successful resident-led struggles against large-scale re-urbanization projects back in the 1990s, the municipality approved and supported the new re-urbanization project Porta Nuova in 2004. In 2007, local resistance and proposals for an alternative development plan failed to prevent the demolition of La Stecca and nearby old buildings to make space for Porta Nuova's Vertical Forest towers (Figure 1.3). This setback epitomized the largest defeat of resistance groups (Brizioli, 2015), especially since the green area around La Stecca was the only public park in Isola. Resistance against Porta Nuova's social and environmental impact continues today.

Meanwhile, Hines, the site developer, defined Porta Nuova as "Milan's Gate of Green," a "heaven for pedestrians and cyclists" and a unique example of urban greening and sustainability for its use of innovative technology and materials (Porta Nuova, n.d., p. 8). The main eco-symbols in Porta Nuova are the significantly vast park, the Library of Trees designed by the Dutch studio Inside Outside, and the Vertical Forest residential towers, designed by Boeri Studio. These tree- and plant-covered buildings are conceived as a way to mitigate climate change, increase bird and insect biodiversity and reduce urban sprawl (Boeri, n.d.).

Overall, residents attribute Isola's attractiveness at the end of the 1990s to its underground and charming character born from its diverse and tightly knit community. As construction began in Porta Nuova in 2006, Isola became "the place to be" at night for visitors, leading to the closure of many traditional retail stores like longtime *panineria* (sandwich) places that were replaced by trendy bars and burger restaurants. In interviews, residents and activists complained about noise at night, unaffordable home and store prices, privatization of street space for restaurant tables and loss of the traditional neighborhood identity.

Gentrification was felt almost immediately in the neighborhood, causing housing prices to soar and transforming the social fabric of the neighborhood. Fifteen years later, although still socially diverse, high rents and lack of sufficient social housing are compromising the ability of long-term residents to resist displacement. In their discourse, planners and developers have promoted urban sustainability through environmentally friendly spaces and buildings, but have neglected the social inclusion and well-being of existing residents.

FIGURE 1.2 The Vertical Forest towers. A juxtaposition between old and new characterized by the neighborhood of Isola versus Porta Nuova. (Photo by: Angelo Stara via Unsplash.)

Procedural injustice in the "green" transformation of Isola

In addition to increasing social inequalities, Isola's activists denounced what they called a fake participatory process that lacked meaningful community planning. Between 2006 and 2009, Hines organized a participatory process consisting of more than 100 meetings with locals, but some activists saw this as a strategy to gain approval, consensus and legitimacy for the project rather than to meet residents' demands. One activist said,

> The participation process was a method to create consent and not a method to go back to the project and repostulate the project together … Their thinking was "If I don't convince you, I have to find a way to split the groups and recreate consensus by organizing fun parties and stuff like that."

In the end, the participatory process was perceived more as a negotiation over details—like the height of skyscrapers facing social housing, one resident recalls—rather than an inclusive decision-making process.

Furthermore, the developers did not fulfill their initial promise to preserve La Stecca, having demolished it to build the Vertical Forest towers. One activist recalls,

> La Stecca was testimony of an industrial past … And they did not only destroy La Stecca, but also other beautiful ex-factories around it which were dismissed but were beautiful buildings. All those buildings were destroyed. They only kept one which became Catella's foundation, all the rest is gone. This is like violence, it is an act of violence. It is the deleting of history.

The loss of cultural and social patrimony eroded the trust of residents in the participatory process. Many also regret the absence of promised public cultural spaces to compensate for the disproportionately high number of Porta Nuova's private spaces. Put differently, residents saw the participatory process as a political move of "social washing" that masked the true nature of the project.

Lastly, activists denounce the municipality for disregarding their needs and instead prioritizing the interests of the private developer in the construction of the buildings and the park, and in the management of the green space. When Hines entered the site in 2003, our interviews revealed that the developer gained the support of the municipality for its ability to mobilize capital in order to unify a highly fragmented area under one single land owner (Anselmi, 2014). Moreover, in 2019, the municipality entrusted the management of the Porta Nuova's park, the Library of Trees, to the private developer Manfredi Catella in order to shift maintenance costs away from the municipality (although the park ought to be public). In exchange for maintaining the park, Catella is allowed to use the space to host events and place temporary kiosks and stands for commercial purposes. The final design and management of the Library of Trees has raised a common concern among activists and residents about the orientation of the space privileging Porta Nuova rather than Isola and about the privatization of parks.

Cosmetic green and fake sustainability in a forest of luxury constructions

Porta Nuova holds the potential to fulfill Milan's dream of becoming an international green capital within the global green orthodoxy of greening, livability and sustainability (Connolly, 2018) while leaving social vulnerability unattended. As one activist argues,

> Milan is trying to articulate this discourse of "now we are ready to present ourselves to the world," no longer as an Italian financial capital, which is parochial, but as a metropolis that winks at Europe and the world. To do this, the municipality needs to say that this transformation [Porta Nuova] is made with extreme attention to the national and global debates. We can guarantee that in the middle of all these skyscrapers we made a big park we call the Library of Trees, and that two of these skyscrapers, the smallest ones, will be covered by 200 plants [Vertical Forest].

In other words, Porta Nuova functions as an enabler of Milan's green innovation agenda.

While residents and activists alike concur that the Library of Trees and the Vertical Forests provide beautiful and pleasing scenery, they are nonetheless "cosmetic surgeries" unable to offer meaningful benefits to locals. At the end of the day, the Vertical Forest buildings are luxury skyscrapers that most Milanese cannot afford. A 275 m^2 apartment in the building costs 3,850,000 euros. "I don't care if they put trees on it, it is still a skyscraper," as one respondent regretted. To a similar extent, the formal design and aesthetics of the Library of Trees park lend it a semi-private, manicured appearance, making the area more of a green museum for visitors and tourists than a public space (as La Stecca used to be) for local recreational and social activities.

Furthermore, the environmental sustainability of the entire Porta Nuova district is questionable. The skyscrapers required a significant amount of natural resources to be dug up and built and significant levels of energy are required for their daily use. In addition, the terraces of the Vertical Forests hold hundreds of kilos of trees and plants, and therefore "require an excessive amount of concrete and steel compared to normal civil constructions," as one activist explained. Some residents also point out that if the municipality was truly interested in supporting an environmentally sustainable urban project, it would not deforest abandoned naturalized areas in the city. Such claims mirror those formulated by activists in Nantes opposing the Jardin Extraordinaire construction and the redevelopment of the Carrière Misery site. Simply put, many residents in Milan see a paradox in the municipality endorsing the construction of a building that produces "an oxygen cylinder per day" (Vertical Forest) while at the same time supporting the deforestation of Milan's existing green lungs.

Timid government responses to growing civic green mobilization and inequalities

Since 2001, the mobilization against Porta Nuova has involved thousands of people as well as numerous organized groups either focused on Isola or active in other parts of the city. The civil society groups resisting the redevelopment of Porta Nuova homogenously shared a vision to save Isola's identity from a large-scale re-urbanization project. However, for some the contestation goes beyond the neighborhood to challenge broader city politics and the economic model behind the project.

During the 15 years when Porta Nuova was under planning and construction, some activists employed tactics such as squatting, protests, boycotting and acts of civil disobedience in the attempt to stop construction. They did not take part in the participation processes because, as they explain it, their interests were irreconcilable with those of Hines, the private developer. Their actions represent a form of resistance as well as alternative proposals for the use and development of space. For example, back in 2012 they illegally squatted a vacant shop to create Piano Terra, an initiative still active today that hosts numerous associations and bottom-up groups engaged in feminism, environmentalism, precarious workers' rights and other social justice issues. Piano Terra, as the organizers described, is not simply a place for people to meet and discuss resistance, but to fill the absence of public cultural spaces in Isola that promote inclusion and equality.

Other civil society groups also engaged with Isola residents in the creation of alternative greening models for Porta Nuova, such as Isola Pepe Verde (IPV), a community garden in a vacant lot in the neighborhood. The space was initially squatted until the municipality authorized temporary use of the space. As one of the garden leaders explained,

> The most important aspect is sociability. We value doing things together. And even if someone tries to say that the Library of Trees green space is like Isola Pepe Verde, it is not; they are completely different. Theirs is a type of green where you can go and consume, but you cannot use or touch. You cannot organize an event like we do here; here we host birthday parties for children, because the neighborhood school does not have a place to celebrate.

This testimony reveals the value of usability over aesthetics. The garden is a place where members actively participate, maintain and co-create the garden, use their hands, touch natural elements and interact with friends. IPV also plays an important role in educating children and building their environmental consciousness and stewardship. Participants not only teach children how to take care of plants but also share the importance of broader values like respect, inclusion and shared responsibility.

Community activists defend a vision for urban green spaces in which participants democratically decide on the development and design of a space. In their statute, IPV's community states a belief in active participation as a way to govern the neighborhood and the city. This principle also applies to environmental goals: "IPV achieves sustainable development through environmental protection, it believes that environmental objectives can be pursued through democratic local action" (Isola Pepe Verde, 2018, p. 16). Decisions are meant to be taken all together. The alternative green space also embodies values of responsibility and duty-sharing and the co-construction of urban imaginaries. Participants place affordability, inclusion, transparency and equality at the center of sustainability. For the activists, a sustainable neighborhood is a neighborhood where everyone has the right to housing.

While the municipality of Milan is fairly transparent about the considerable number of vacant buildings in the city (Comune di Milano, 2019), these are mostly abandoned and uninhabitable. Some activists argue that the city seems more invested in projects like Porta Nuova than in renovating vacant buildings or damaged social housing structures. In other words, they believe urban sustainability is not about new, so-called sustainable buildings, but about making sure that existing housing is habitable and accessible.

FIGURE 1.3 A view of Isola Pepe Verde (IPV), one of Milan's community gardens in the Isola neighborhood located about 200 meters from the Vertical Forest towers. (Photo by: Barbara Barberis.)

Moreover, despite the large number of vacant buildings in Milan, authorities have not prevented evictions in the previous years, including of social centers reclaiming the right to public spaces, and poor families squatting vacant social housing because they are unable to afford rent. While squatters are willing to pay to become legal tenants, restrictive social housing regulations (e.g., waiting lists) do not allow such arrangements.

An uncertain future for alternative greening

Porta Nuova's imaginary does not honor Isola's historically working-class identity nor does it address the social and environmental needs of its residents. Today, Porta Nuova reflects the vision of its private developers, who, supported by the municipality, have wielded their power to shape and design a new green district. As the Library of Trees park demonstrates, private developers also took control of a park that was initially conceived as public. When green space is developed to fulfill private interests and privileged residents, environmental benefits only accrue to the few. Porta Nuova is a paradox of aesthetically driven green urbanism and infrastructure that claims to contribute to an innovative sustainable city while effectively wiping out a formerly public green area. Many see it as a classic example of greenwashing that incorporates, performs and reproduces spatial, social, emotional and political exclusion on different scales and undermines true environmental sustainability goals.

In response, activists and residents have created alternative greening models such as those embodied by Isola Pepe Verde as a place to interact, share and make decisions together. The fate of the project, however, remains unclear. For now it is allowed to remain, perhaps only to greenwash the image of the Milan municipality and to silence voices of dissent against Porta Nuova.

Keywords

(a) the urban development pattern of the city and neighborhood: green design and urbanism; smart city, urban redevelopment and revitalization; abundance of abandoned or poor-quality housing
(b) the urban greening of the city and/or neighborhood: Vertical Forest; green space redevelopment; fights over community gardens; community mobilization for public green spaces
(c) the inequalities at stake: green privatization, luxury housing, green privilege; lack of government response for inclusive and equitable greening

Note

1 Partisans' resistance refers to rebellious groups who opposed the Italian fascist regime during World War II. Today, partisans are a symbol of progressive revolution in most Italian *centri sociali* (social centers).

References

Anselmi, G. (2014). I Megaprogetti immobiliari nell'epoca di finanziarizzazione ed austerity: i casi di Milano e Salford. PhD Thesis, University Bicocca-Milan.

Boeri Studio (n.d.). Vertical forest. Retrieved, April 2019, from: https://www.stefanoboeriarchitetti.net/en/project/vertical-forest/

Brizioli, A. (2015). Storia dell'Isola, il quartiere storico che vogliono cancellare. *The Globalist*. [Online article]. Retrieved, December 2018, from: https://www.globalist.it/culture/2016/05/08/storia-dell-isola-il-quartiere-storico-che-vogliono-cancellare-68713.html

Comune di Milano (2019). Edifici degradati e abbandonati [Municipality official website]. Retrieved, April 2019, from: http://www.comune.milano.it/wps/portal/ist/it/servizi/territorio/Immobili_degradati

Connolly, J. (2018). From Jacobs to the Just City: A Foundation for Challenging the Green Planning Orthodoxy. *Cities* 91: 64–70.

Isola Pepe Verde (2018). Esperienza di un giardino condiviso a Milano [Printed brochure]. Isola Pepe Verde.

Legambiente (2019). Mal'aria 2019, il rapporto annuale sull'inquinamento atmosferico nelle città italiane [Blog post]. Retrieved, March 2019, from: https://www.legambiente.it/malaria-2019 il-rapporto- annuale-annuale-sullinquinamento-atmosferico-nelle-citta-italiane/

PGT (Piano di Governo del Territorio) (2018). Milano 2030: Visione, costruzione, strategie, progetto di città. Comune di Milano. [Governmental plan Milan 2030].

Porta Nuova (n.d.). Porta Nuova [Official website project, PDF brochures]. Retrieved, January 2019, from: http://www.porta-nuova.com/brochure

Signorelli, A. (2015). Quartiere Isola: la storia e l'origine del nome. *Blogo*. [Online article]. Retrieved, November 2020, from: https://www.blogo.it/post/68936/quartiere-isola-milano-storia-origine-del-nome

2

DISMANTLING THE JUST CITY

The unevenness of green experiences in Amsterdam Noord

Carmen Pérez del Pulgar

Amsterdam Noord has transformed from an industrial, working-class district into one of the trendiest and greenest parts of the city. This chapter discusses the processes leading up to this urban transformation—including policies aimed at addressing post-industrial decline, the spatial concentration of low-income dwellings and an outdated housing stock in a shrinking city—and argues that it exemplifies a case of state-led gentrification made possible by a neoliberal restructuring of the housing system. Today, Amsterdam Noord is an attractive destination for real estate capital investment, high-income residents, leisure and tourism. At the same time, the socio-economic position of many households has deteriorated over the last decade. A housing crisis and the perils of hyper-segregation, gentrification and (exclusionary) displacement are emerging. Although a relatively great deal of social equality is maintained thanks to the residuals of a strong institutional social architecture, Amsterdam's status as a just city is being challenged by this new strategy of urban renewal.

From industry and white working-class housing to ethnically mixed "problem" neighborhoods

In the early 20th century, as Amsterdam became an accessible port for steam ships, Noord become home to the municipal Schipyard Stoomvaartmaatschappij Nederland, Koninklijke Nederlandse Stoomboot-Maatschappij (K.N.S.M.) and to four private shipyards in subsequent decades. The port also housed industrial plants that produced sulfuric acid, low-voltage cables, aircrafts and refined petroleum.

The industrialization of the northern riverbank led to the gradual construction of workers' housing at the district's core. An important precedent was the 1901 Dutch housing law obliging the provision of decent housing for the

DOI: 10.4324/9781003183273-3

working classes, adopted partly as a monarchic concession after the 1848 revolts (Hochstenbach 2017). Prompted by the urgent need to house new industrial workers in Amsterdam Noord and to relocate people living in the inhabitable slums in the city center, housing associations and anarchist worker communes built seven garden villages, known in Dutch as Tuindorpen, between 1914 and 1934. These garden villages, designed as social experiments, were rent-controlled social housing in isolated village-like neighborhoods consisting of low-rise, single-family units with plenty of light, fresh air and green surroundings. With their own facilities like playgrounds, schools, bath houses and churches, garden villages evolved to have their own unique character, history and strong local identity. Some, like van der Pekbuurt (1922), were pioneering projects of mixed social housing for both workers and managers. Others, like Asterdorp[1] or Floradorp, were conceived as living schools (*woonschool*), aimed to civilize "uncivil" citizens. In exchange for housing and a stipend, residents had to stay clean, sober and do laundry, among other things. Many residents, however, did not adhere to the rules, and the stigma and marginalization associated with these garden villages prevailed. As a current resident explained,

> A hundred years ago, Amsterdam Noord was what Australia was to England. So, you sent your criminals, your misfits. Now you would say all the white trash, they sent them here. A lot of their great-grandchildren still live here, and that is really the old school white Noord.

On top of this stigma, Amsterdam Noord has long experienced isolation and political neglect. Until 2017 there was no metro line, pedestrian or bicycle connection to the city center apart from the ferryboat. Cultural facilities and other amenities were few and of poor quality. It has also persistently ranked as one of the poorest districts of the city. Among young people, the share of school drop-outs is still above average, and the proportion of people over 65 and children is the highest in Amsterdam.

Today, residents of Tuindorpen are more attached to their own garden village than to the rest of the city. Amsterdam Noord, and especially Tuindorpen residents, are often perceived by outsiders and the media as having a white, hard-working, anti-establishment identity and more recently, as supporters of right-wing populist parties (e.g. Geert Wilders' Freedom Party). Noord's identity is both ethnicity- and class-based. As a resident activist explained:

> It's very white, and it's very difficult to move here if you are a person of color or a young hipster couple with babies: you will have a tough time. It's the flame of the old Amsterdammers. If you become friends with them it's fine, but … they are the old Amsterdammers.

Despite this white working-class popular identity, Amsterdam Noord is now the district with the third highest percentage of residents with a non-western

FIGURE 2.1 Floradorp Garden Village. (Photo by: Carmen Pérez del Pulgar.)

background, mostly from Turkey and Morocco. They have made up a total of about a third of the district's population since the 1960s.

Indeed, this significant spatial concentration of ethnic minorities and unemployed populations was identified as a problem in the New Urban Renewal strategy (1997) that replaced Amsterdam's strong social tradition of planning (van Wijngaarden 2015; Fainstein 2005; Priemus and Van Kempen 1999). Since then, a set of urban renewal strategies has targeted neighborhoods with spatial concentrations of unemployed residents or ethnic minorities with the aim of improving the built environment and decreasing the spatial concentration of vulnerable residents. The 2006 Dutch Act on Extraordinary Measures for Urban Problems included a controversial article requiring households moving into a deprived area to obtain a permit, granted only if employment income, wealth or pensions, enrollment in higher education or residence in the metropolitan region for at least six continuous years could be proven. Researchers argue that this act prevents vulnerable populations from moving into deprived neighborhoods and encourages the influx of more affluent residents (van Gent, Hochstenbach, and Uitermark 2018). From 2007 until 2012, the Dutch government introduced an action plan to invest in 40 so-called *probleemwijken* (problem neighborhoods)[2] across the Netherlands, including many in Amsterdam Noord. From 2012 onwards the City of Amsterdam has continued to reinvest in these areas with its own funding under the Neighborhood Plan of Action (Ontwikelbuurten),[3] focusing on 22 neighborhoods across the city, 7 of which are in Amsterdam Noord. These interventions have focused on public space improvement, housing renovation and retrofitting of the heating infrastructure towards a carbon- and gas-free city by 2040.

Waterfront redevelopment: from contamination to green development fantasy

After decades of industrial prosperity, capital began fleeing from Amsterdam Noord in the 1970s. Amsterdam's port moved downriver, and many shipyards started filing bankruptcy. The gradual deactivation of the central and eastern harbors resulted in large swaths of industrial wasteland on both sides of the IJ river. The rest of Amsterdam was also experiencing structural economic problems, with the migration of working middle classes to the suburbs, a collapsed housing market and high levels of unemployment. In this context, Amsterdam, like many other post-industrial European cities, considered waterfront redevelopment as a strategy to revitalize the city. The current regeneration of the banks of the IJ is the city's largest and most important urban transformation in recent history.

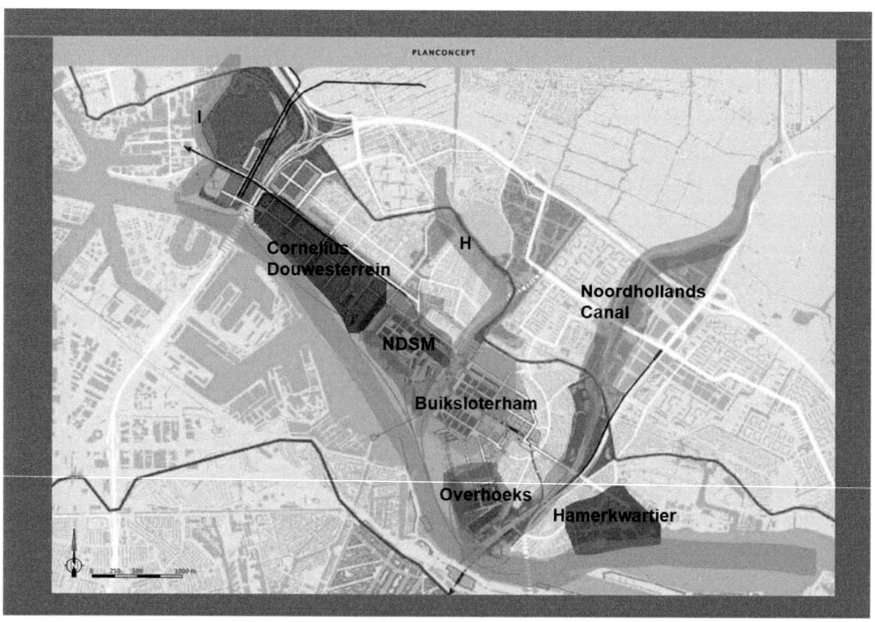

FIGURE 2.2 Masterplan for the northern bank of the IJ river. Areas of intervention. (Photo by: Carmen Pérez del Pulgar.)

Whereas the regeneration of the southern bank became tangible during the 1990s, Amsterdam Noord was still seen as "the other side" for another ten years. The neighborhood platform A.N.G.S.A.W. (Amsterdam Noord, Green City on the Water)[4] was crucial in lobbying for the redevelopment of the northern waterfront. The turning point came around 1999, when the oil company Shell moved to a new facility on the northwest end of the site, freeing 20 of the 27 hectares. Local authorities promptly acquired the heavily polluted ground and developed a Master Plan for the Northern Bank of the river IJ.[5]

The concept underlying the masterplan was the conversion of Amsterdam Noord into a creative hub, with high-quality housing, livable green public spaces and better connectivity. It was envisioned as a place of experimentation where most of the traditional social planning instruments were set aside in favor of a seemingly bottom-up, flexible and liberalized planning culture where innovation and creativity would thrive.

The specific interventions were structured around two main axes determined by the water (Figure 2.2). Alongside the horizontal axis and parallel to the waterfront, the plan identifies N.D.S.M. and Overhoeks as two central areas: the Buiksloterham transformation area and two business parks. Large green corridors were proposed for the vertical axis, connecting the IJ river to the rural hinterland alongside the most important waterways: the Noordhollands Canal and the H and I canals.

Sustainable, experimental and creative city: new centralities and transformation

The N.D.S.M.'s core was developed around the former shipyard N.D.S.M. (Figure 2.3). After falling into bankruptcy in the middle of 1980s, the contaminated site was taken over by drifters, petty thieves and artists. In 1999 the city council published a call for proposals for the development of the N.D.S.M. terrain. Kinetisch Noord was the winning project, designed by a group of N.D.S.M.-based artists and craftworkers proposing self-restored D.I.Y. affordable workspaces inside the main warehouses and the old ship slopes. The plan also included a cultural-historical impact assessment that resulted in the shipyard being granted the status of a national protected monument since 2007.

From a strategic urban governance perspective, Kinetisch Nord at N.D.S.M. was envisioned as the first flagship project of the creative city plan. The Kinetisch Nord's pioneers went through the classic phases of artist-driven gentrification: from feeling part of an urban social victory, to being used as pawns for the commercial development of the area, to becoming dispensable in making room for a wealthier creative class. As some artists and activists recall, at some point it became clear that state-led gentrification was underway: festivals, as opposed to workshops, became bigger and more commercial. The renovation of the old ship slopes suddenly had to host income-generating activities, while poorer people and social entrepreneurs were being displaced. The only remaining crane

FIGURE 2.3 Former shipyard N.D.S.M. (Photo by: Bart Stuart.)

was refurbished with government money but then privatized for a luxury hotel. M.T.V., Red Bull and many other international companies relocated to N.D.S.M.

After significant upheaval and resistance from the original warehouse occupants (mostly members of Kinetisch Nord), the city council adopted a motion to protect the old ship slope as a "cultural free space" for the remaining pioneers in 2017. During this likely temporary truce, these pioneers have launched a post-gentrification action research project called R>A>M>P[6] documenting their experience, the quality of life in the area and who the N.D.S.M. is for. As artists and activists acknowledge, "In doing so we cannot escape scrutiny ourselves ... after all, this all came about because the artists made it so attractive" (van der Lippe and Stuart 2017).

Many residents and anti-gentrification activists feel that they themselves have contributed to gentrification in N.S.D.M., arguing that the economic crisis had a peculiar two-sided effect on the redevelopment and gentrification of Amsterdam Noord. When the 2008 economic crisis stopped all construction and paused the masterplan's grandiose vision, there was suddenly a niche for bottom-up, alternative and transformative urban developments. But paradoxically, once the economy recovered, all these projects were integrated into the original ambition to create an attractive city for the creative class.

Another example that echoes this experience is the De Ceuvel sustainable workplace (Figure 2.4). Originally a contaminated old shipyard in Amsterdam

FIGURE 2.4 Sustainable workplace De Ceuvel. (Photo by: Carmen Pérez del Pulgar.)

Noord located between N.D.S.M. and Buiksloterham, the site was leased by the municipality in 2012 for ten years free of cost to residents and artists due to the absence of interested private developers or state funds. The requirement was to have a reasonable plan to clean up the brownfield site and to attract artists, which the organizers completed by building an artist village out of broken old houseboats. The site was covered with pollutant-filtering grasses that are currently decontaminating the soil. A decade later, De Ceuvel is a hub for sustainable innovation and experimentation, having attracted much attention and economic interest to the Papaverweg neighborhood. Nevertheless, vulnerable residents of the area rarely participate in De Ceuvel's workshops and activities and are often not encouraged enough to actively participate in them. Moreover, when the ten-year lease ends in 2022, De Ceuvel will no longer be able to afford the rent.

The formerly industrial area of Buiksloterham, defined as a transformation area in the masterplan, aimed to serve as a living lab for a circular, smart and bio-based city (Figure 2.2). Liberal land use policies are being tested here to possibly integrate them into new environment and planning laws. For instance, instead of planning following a strict landuse plan, in Buiksloterham only a floor space index (max volume that can be built) is regulated. Residents complain about the lack of attention to social sustainability in the developments. Only the technical aspects of the circular economy of new constructions are evaluated, but the projected uses of these do not fit local needs. Hotels, creative workspaces and high-end comfortable living are slated to coexist with scant social housing.

Attracting affluent classes: metropolitan tourism, leisure and green infrastructure

Alongside N.D.S.M., another planned central area called Overhoeks is situated on the former Shell laboratory site and named after its iconic Overhoeks Tower designed by the famous architect Arthur Staal. Envisioned as a northern extension of the inner city, the site was of utmost visual importance in the redevelopment of Amsterdam Noord (Figure 2.2). The redevelopment of the area began in 2003 with the formation of a public-private partnership with I.N.G. Real Estate Development (I.N.G. R.E.D.) as the developer. Together they commissioned Shell's Urban Design Plan (2004), which disregarded recommendations to preserve some of the existing buildings of cultural and historical value, erasing the site's industrial heritage. In general, the plan for the former Shell site sketched a high-density area with 70% of the land reserved for housing (about 2,200 mostly high-end units) and 30% for other urban functions. The high costs of the brownfield clean-up led to increased land prices of the terrains, and in turn to the shrinkage of the potential profit margins of the developers. This was then used as an argument for omitting the inclusionary zoning requirements, which in Amsterdam stipulate 40% social housing, 40% affordable housing and 20% free-market housing. Thus, only 20% of the 2,200 housing units were reserved for social housing.

In Overhoeks, a new luxury estate was built on the northern side, a mixed-use strip on the southern side and a promenade park (Oeverpark) along the IJ river to connect Overhoeks and the N.D.S.M. A nearby resident and activist described the new luxury estate as insipidly mono-functional and a sort of "patisserie urbanism" due to its monotony and socially exclusive nature. The large apartment buildings are designed as gated communities with enclosed private courtyards separated from public space by fences and accessed via private parking garages. There are other subtle aspects that exude elitism, such as marble decorations on the sidewalks or buildings named after ships linked to the Dutch colonial and slave trading past.

The mixed-use strip includes apartments, offices, a school, museums, restaurants, hotels and other tourist infrastructures such as the Holland Experience Museum. A spectacular building hosting the Eye Film Museum was planned as an anchor of urban regeneration. The old Shell canteen was transformed into a cultural leisure venue and bar, and the Overhoeks Tower was transformed into a luxury hotel with a rooftop observation deck.

The source of the Noordhollands Canal is very close to Overhoeks's new, high-end central area (Figure 2.2). The green restoration plans differ along various sections of the canal. The section between the Waterlandse Zeedijk (embankment) and the ring road is meant to have a more open and wild character, mainly intended for bike routes along the canal. The southern part between the IJ river and the Waterlandse Zeedijk embankment was transformed into the expansive metropolitan Noorderpark by merging and connecting smaller parks with four bridges (Figure 2.5). The park was designed as a strategic part of Amsterdam's green infrastructure, envisioned to serve the entire metropolitan area. In fact, the opening of the north–south underground line in 2017 featured a station at the park's entrance, making it the only metropolitan park directly accessible by metro in the city. In addition to the direct connection between the Noord and the center of Amsterdam via metro, the internal connectivity between different parts of Amsterdam Noord was greatly improved by the four new bridges embedded in the park and that connect the east and west sides of the canal previously separated by a highway and only linked via the waterfront or the ring road.

Noorderpark is adjacent to several garden villages (see Figure 2.6). The Noorderpark Foundation[7] was created to provide activities in the new park for neighboring residents and ensure that it is an attractive and safe place. Nevertheless, the park is used predominantly by commuting visitors and new affluent families that have recently moved to the area. As an employee of the Noorderpark Foundation expressed,

> It still puzzles me why for instance all these people, lower income people with small houses, why they don't use the park. There is a lot of people of Mediterranean origin here, Moroccan, Turkish … What stops them from going into the park with the families and the barbeques? We have organized a lot of things in the park, and when we asked them to cooperate, mostly they don't, because they are not interested.

FIGURE 2.5 Noorderpark. (Photo by: Carmen Pérez del Pulgar.)

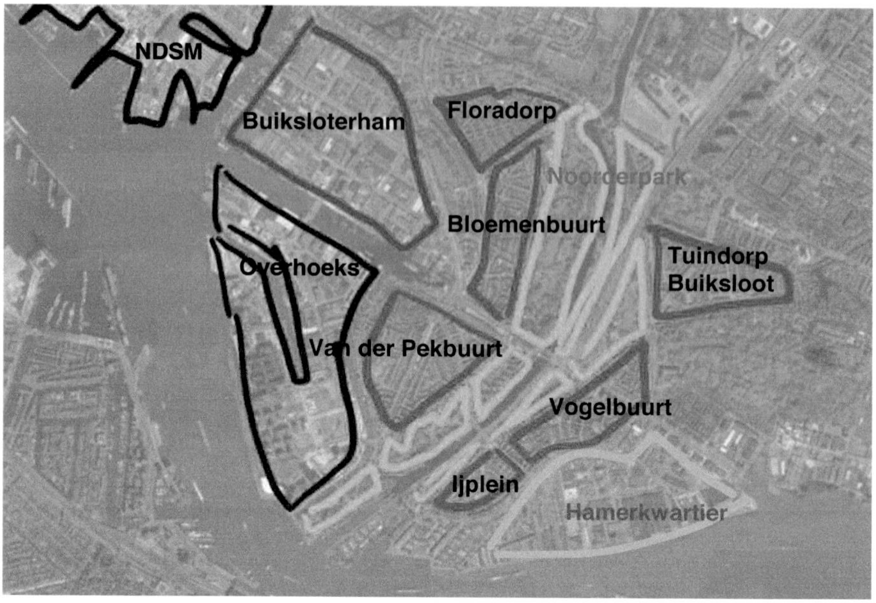

FIGURE 2.6 Noorderpark and adjacent garden villages in light gray. (Photo by: Carmen Pérez del Pulgar.)

Our interviews revealed a very coded division of space in Amsterdam Noord. The increasingly popular venues in N.D.S.M., Overhoeks and the Noorderpark mostly cater to white affluent communities. They attract large amounts of tourists, commuters from other parts of town and new, high-income and educated ethnically Dutch residents. Most community centers, such as Het Anker adjacent to the Noorderpark, are gathering points for long-term, white Amsterdam Noord residents, offering traditional Amsterdam entertainment such as Dutch music, bingo and football, where ethnic minorities and newly arrived affluent white residents don't feel like they belong. Behind the community center, however, a new swimming pool built in 2017 attracts many ethnic minority families. The Turkish and Moroccan community also frequent neighborhood's markets and shops, but rarely mix with the ethnically white and traditional Dutch spaces.

From a social city to state-led gentrification

Local residents fear the negative consequences of gentrification linked to the waterfront redevelopment and the creation of the Noorderpark. The state-led urban renewal strategy discussed earlier in this chapter was a critical step towards enabling gentrification, as was a set of institutional changes in the housing system.

Up until the 1990s, a significant share of social rental housing was provided by non-profit associations (64% of the city's housing stock in 1995) (van Gent

2013), direct subsidies to housing associations (Hochstenbach 2017), a strict and comprehensive rent regulation law and a system of public land ownership and strict landuse planning (Kadi and Ronald 2014). Yet a new neoliberal paradigm in the 1990s sought to reduce national debt, reform the housing system to promote urban competitiveness (Priemus and Van Kempen 1999) and identify and act on "problem" neighborhoods characterized by the spatial concentration of low-income, unemployed and ethnic minority populations (Priemus and Van Kempen 1999).

A combination of roll-out programs for tenure conversion and roll-back measures to promote real estate investment and homeownership led to a profound restructuring of Amsterdam's housing stock. Since 1990, (social) housing associations have been financially independent but are more regulated than other real estate developers. Since 2008, E.U. legislation on competition has forced housing associations to target social housing to economically vulnerable households (maximum household income of €36,798 annually, low for the Netherlands). E.U. regulations have also strongly limited national budget deficit levels to a maximum of 3% of G.D.P.

Since 2012, the Dutch government has reduced deficits by withdrawing money from the social rented housing sector by means of a "landlord levy" (van der Veer and Schuiling 2015). As a result, housing associations pay up to four months of their rent incomes in taxes and increasingly struggle to finance their activity and reinvestment. Furthermore, the 2015 Housing Act reduces housing association activity only to social housing provision and prohibits their efforts to provide housing for middle-income groups, project development and neighborhood maintenance (van der Veer and Schuiling 2015). In this context, housing associations are managing to survive by charging higher rents on their liberalized housing stock, the latter becoming market-rate rental housing when rent levels surpass €710/month (Hochstenbach 2017), and selling units to tenants.

As the owners of 46% of the housing stock in Amsterdam (van der Veer and Schuiling 2015), housing associations play an important role in the transformation of Amsterdam Noord. Indeed, some gentrification in the district is linked to the renovation of old social housing stock, especially the older garden villages around the Noorderpark (Figure 2.6). Housing corporations claim they cannot afford to renovate units and maintain affordable rents, which is why after renovating, they sell part of the housing stock while increasing rent on the units they keep. Therefore, although all social housing tenants are offered an alternative affordable housing option in Amsterdam or in the surrounding areas, about 20–30% of the tenants of the renovated garden villages around Noorderpark will not return after renovation. Furthermore, the criteria for selecting which residents can return after renovation are very obscure, and some activists affirm that there is clearly an ethnical bias making ethnic minorities less likely to return after renovation.

The increase in housing renovations and housing system restructuring has made investment in the private real estate market more attractive. Private housing market prices are skyrocketing in places like Amsterdam Noord, while the share of affordable housing has sharply declined and is accessed only by very vulnerable and low-income residents. As a result, exclusionary displacement has emerged as a form of hyper-segregation. "For the same house, there are residents paying 200 euros a month because they lived here for a long time and people who just bought [a home] for more than €500.000," one resident activist explained. Political subject formation, or the realization of a common political identity and possible mobilization, is seriously hindered in a context where middle low-income people are concerned about being too well-off to qualify for social housing, yet are still unable to earn enough to access the private market. While they are skeptical of investment efforts, they remain hopeful that more supply will lower housing prices.

In Amsterdam Noord, the previous social model is being fractured by the regulation of new developments. Out of a total of 9,000 new homes[8] planned in Overhoeks and N.D.S.M., 20% and 30% are designated as social housing, respectively. This share of social housing is a sharp decline in comparison with the average of 64% social housing stock in the city in 1995 (van Gent 2013) and the standing inclusionary zoning requirements of 40% social housing, 40% affordable housing and 20% free-market housing. Again, the rationale was that the type of housing stock had to be changed in order to reduce the spatial concentration of low-income people in this area (Priemus and Van Kempen 1999).

Affluent whitening and greening: exclusionary displacement and cultural dislocation

This chapter explores how a large waterfront regeneration plan coupled with a set of neighborhood improvement projects spurred by national legislative changes harnessed an influx of capital to Amsterdam Noord's deprived neighborhoods. The explicit aim was to differentiate the composition of the urban population, that is, to bring more affluent residents and prevent the growth of low-income or vulnerable populations. These interventions have been developed while the policies ensuring housing decommodification in the Netherlands were slowly dismantled.

Amsterdam Noord is undergoing significant changes. More affluent residents and young families from other parts of the city are increasingly drawn to the area, attracted by lower prices, green areas and a vibrant cultural scene. Meanwhile, Noord is also the only district in Amsterdam where the life of longtime residents has deteriorated in the last decade, and where the wealth gap between Amsterdam Noord and the rest of the city is widening. These factors have led to a segregated Amsterdam Noord where long-term, white, working-class residents, ethnic minority residents, recently arrived affluent families, tourists and city-commuters exist in separate ethnic- and class-based coded spaces. Thanks to the remnants of older institutional housing arrangements, the physical displacement of long-term residents is delayed, but social and exclusionary displacement

(Marcuse 1986) is clearly emerging. Amsterdam's "social city" model is thus slowly being dismantled in Amsterdam Noord.

Keywords

(a) the urban development pattern of the city and neighborhood: fast recent growing; recovering post-industrial; creative industry/class; touristification; income inequality; health inequality; large-scale urban renewal
(b) the urban greening of the neighborhood: waterfront redevelopment; green space; circular city
(c) the inequalities at stake: private capture of public investment; stigmatization/dismantling of social housing system; insufficient affordable housing for middle class; (green) gentrification; exclusionary displacement; lack of right to the city for low income and ethnic minority residents

Notes

1 Non-existent now.
2 https://nl.wikipedia.org/wiki/De_40_wijken_van_Vogelaar
3 https://www.amsterdam.nl/bestuur-organisatie/volg-beleid/ambities/fijne-buurten/ontwikkelbuurten/overzicht/
4 http://angsaw.nl/
5 http://www.commissiemer.nl/docs/mer/p14/p1406/1406-021masterplan_deel1.pdf
6 https://www.ndsm.nl/en/event/launch-ramp-x-helling/
7 https://noorderpark.nl/
8 Between 2010 and 2030 150,000 new houses were envisioned for the general region Randstad, 50,000 of which in Amsterdam, 9,000 in Amsterdam Noord.

References

Fainstein, Susan. 2005. "Cities and Diversity Should We Want It? Can We Plan for It?" *Urban Affairs Review* 41: 3–19.

Gent, Wouter van. 2013. "Neoliberalization, Housing Institutions and Variegated Gentrification: How the 'Third Wave' Broke in Amsterdam." *International Journal of Urban and Regional Research* 37 (2): 503–22. https://doi.org/10.1111/j.1468-2427.2012.01155.x.

Gent, Wouter van, Cody Hochstenbach, and Justus Uitermark. 2018. "Exclusion as Urban Policy: The Dutch 'Act on Extraordinary Measures for Urban Problems.'" *Urban Studies* 55 (11): 2337–53. https://doi.org/10.1177/0042098017717214.

Hochstenbach, Cody. 2017. "State-Led Gentrification and the Changing Geography of Market-Oriented Housing Policies." *Housing, Theory and Society* 34 (4): 399–419. https://doi.org/10.1080/14036096.2016.1271825.

Kadi, Justin, and Richard Ronald. 2014. "Market-Based Housing Reforms and the 'Right to the City': The Variegated Experiences of New York, Amsterdam and Tokyo." *International Journal of Housing Policy* 14 (3): 268–92. https://doi.org/10.1080/14616718.2014.928098.

Lippe, Klaar van der, and Bart Stuart. 2017. "From Labor to Leisure." R>A>M>P> IN THE SLOPE. NDSM Open Essays: 21–28.

Marcuse, Peter. 1986. "Abandonment, Gentrification, and Displacement: The Linkages in NewYork City." In *Gentrification of the City*, edited by N. Smith and P. Williams, 153–77. Allen & Unwin, Boston, MA.

Priemus, Hugo, and Ronald Van Kempen. 1999. "Restructuring Urban Neighbourhoods in the Netherlands: Four Birds with One Stone." *Journal of Housing and the Built Environment* 14 (4): 403–12. https://doi.org/10.1007/BF02496765.

Veer, Jeroen van der, and Dick Schuiling. 2015. "More than Just Housing. The Importance of Housing Associations for a Sustainable City." In *Urban Europe. Fifty Tales of the City*, edited by Virginie Mamadouh and Anne van Wageningen, 275–82. Amsterdam University Press, Amsterdam.

Wijngaarden, Arie van. 2015. "Welcome to Amsterdam! Well, Not Really. The Right to the City Requires a City in Ballance." In *Urban Europe. Fifty Tales of the City*, edited by Virginie Mamadouh and Anne van Wageningen, 267–74. Amsterdam University Press, Amsterdam.

3

A GREEN CAPITAL FOR ALL?

Austerity, inequalities and green space in Bristol

Austin Matheney, Carmen Pérez del Pulgar, and Galia Shokry

Bristol's institutional green transition was set in motion by progressive planning policies in the late 1990s that built upon a strong history of environmental activism. Though lauded for achievements such as its robust cycling infrastructure, having the lowest carbon footprint of any British city and its over 400 parks and local nature reserves, Bristol has been unable to ensure that *all* Bristolians have access to these benefits. New levels of austerity are tasking communities and the city council with discovering new paths and partnerships in their fight to build a green city that is also just, an assignment which continues to leave the most vulnerable behind.

Social deprivation in the 2015 European Green Capital

Once one of England's most important port cities throughout the Atlantic slave trade, Bristol has since aimed to reposition itself as a 'green capital' in both the UK and Europe. Heavy bombing during WWII and the downstream relocation of the city center docks in 1968 provided Bristol with an opportunity to redevelop its physical and natural environments while restructuring its economy. Since then, however, the redevelopment of vacant lots and industrial lands in line with the needs of the growing aerospace, creative media and financial sectors, as well as the proposal of new motorways, have occurred alongside persisting inequalities. Strong land use policies aimed at maintaining over 1,500 hectares of accessible green space within the city's boundaries were joined with goals of becoming a carbon-neutral city, plans for sustainable transport, green infrastructure and school programs such as the One Tree One Child program.[1] While the city's focus on the creation of new green space has since then been eclipsed by a focus on energy efficiency and carbon neutrality, Bristol's greening strategy won the city the EU's European Green Capital Award in 2015.[2]

DOI: 10.4324/9781003183273-4

Access to the progress of the city's green transformation, however, continues to be experienced unequally within a segregated city.[3] Home to over 463,000[4] people, Bristol is one of the least deprived English Core Cities but contains 41 of the 10% most deprived areas of England, including three among the top 1%, which are often adjacent to some of the country's least deprived areas.[5] As the city's population continues to grow, in large part influenced by the arrival of new residents from London and the retention of university students, a 'tale of two cities' has grown to define Bristol.[6]

This two-city divide becomes clearly evident when crossing the New Cut, a man-made channel, from the city center into South Bristol. According to the 2019 indices of deprivation, the ten most deprived neighborhoods in Bristol are all in South Bristol.[7] These high levels of social deprivation are not new. Following WWII, South Bristol and particularly the areas of Hengrove, Hartcliffe, and Withywood were the site of new council housing estates which significantly increased the population of the area. These new housing developments, however, were not met with the necessary infrastructural developments, leaving South Bristol disconnected from the rest of the city. A lack of public transportation, alongside dwindling employment opportunities, continues to play major roles in the area's socio-economic deprivation. This deprivation is also reflected in the area's green spaces. While home to an abundance of open space, including its border with the Avon Green Belt to the south, a majority of this green space is informal and of low quality. Recently, the city's attempts at bringing formal green space to the area have come into conflict with attempts to provide residents with new and affordable housing and employment opportunities.

From grassroots to institutionalized greening

Much of the greening of the city throughout the 1990s and 2000s can be attributed to the continued activism of Bristol's environmental movement. Starting with the movement's earliest actions in the 1960s, groups such as Bristol Friends of the Earth (BFoE) and Sustrans continue to impact the city and nation to this day. Throughout the late 20th century, activists fought against the construction of new motorways and increasing dependence on private automobiles, created and maintained community green spaces, pushed forward recycling efforts and demanded housing and energy justice for the most deprived wards. Put differently, Bristol's environmental movement was built upon a foundational understanding of the intimate relationship between socio-economic and environmental issues.

Between 1979 and 1986, the nonprofit Sustrans, then known as Cyclebag, constructed the 13-mile Bristol and Bath railway path and set in motion Bristol's transition to becoming Britain's first Cycling City in 2008.[8] The organization has since grown into a UK-wide charity organization with George Ferguson, the man who led their first demonstration, becoming Bristol's first elected mayor in 2012. When the city halted the collection of wastepaper recycling in 1980, members of BFoE took over and later expanded to collect cans, glass and textiles, all while employing Bristol's youth. This recycling scheme eventually led to the creation of Resources Futures, a waste consultancy organization working with clients ranging from local authorities to the national government and multinational corporations.[9] In other words, the work of Bristol's environmental movement had already begun to create a reputation of Bristol as a green and progressive city in the 1980s, a reputation which continues to attract new residents to the city.

As the environmental movement grew, however, it was met with the early years of Thatcherism, an economic recession and growing civil unrest throughout the UK. A police raid of the Black and White Café in Bristol's Inner East in 1980 triggered what is commonly known as the St Pauls Riot. In addition to the racial discrimination the Afro-Caribbean community faced through the Sus Law[10]—a stop and search tactic which unequally targeted youth of color—the riot highlighted longstanding inequalities in housing stock, education, employment, exposure to pollutants and quality green space experienced by neighborhoods of the Inner East (Townsend et al. 1985). Although an unsafe and ghettoized stigma of St Pauls persists (Slater and Anderson 2011), it has also captured the interest of property developers, and real estate prices have risen dramatically in recent years in large part due to its well-placed location, just northeast of the city center, between the central business district and two highly gentrifying/gentrified neighborhoods.[11]

Growing discontent with the progress of the previous decades culminated in the 1990s becoming recognized as a period of environmental direct action (EDA) throughout the UK (Doherty, Plows and Wall 2007). Despite the environmental movement of the previous decades having fought against the construction of new motorways, they were unsuccessful in halting the construction of the first section of the M32, which continues to split the neighborhoods of St Pauls and Easton to this day and contribute to the city center's illegal levels of air pollution. While nationwide EDA of the era primarily maintained an anti-road focus, EDA in Bristol expanded to rally against the continued loss of green space throughout the city. Direct action was taken in 1992 against the construction of a new Tesco supermarket on an open green space with activists squatting in a row of lime trees for over 90 days before being removed on live television. Despite this direct action, which was to occur again in 2011 in what came to be known as the Tesco Riots, the supermarkets were built.

As is clear, the roots of Bristol's green reputation in 2021 can be traced back for decades. However, so too can the environmental movement's inability to bring about lasting material change against corporate power in the context of an increasingly neoliberalized state. An institutional rearrangement of the region in 1996, seeing Bristol's transition from being under the administration of the Avon City Council to a unitary authority, offered the environmental movement increasing say in the (re)development of Bristol's physical and natural environments. This institutional rearrangement saw the decades-old struggles of the

FIGURE 3.1 A view of the M32 motorway. (Photo by: Galia Shokry.)

environmental movement become adopted in 1997 in Bristol's first citywide progressive planning document in over 20 years.

A green capital with little greening

After five years of debate and modifications, the 1997 Bristol Local Plan set out to protect green space, to tackle the city's transportation issues related to the lack of public transit, congestion and air pollution and to deliver new and affordable housing.[12] The plan placed emphasis on (re)developing the city center and South Bristol, aiming to (re)develop the now derelict city docks into an area of culture, heritage, business, tourism and leisure, while attention to South Bristol noted the historical deprivation of the area and its continued disconnection from the city center and northern wards.

Over the next decades, the local plan would be supplemented and replaced by numerous planning documents whose focus included access to and maintenance of quality green space (2008 Parks and Green Space Strategy), neighborhood-level supplementary planning documents (SPDs) (e.g., 2006 Planning for a Sustainable Future for St Pauls SPD10), a resilience strategy (Bristol Resilience Strategy 2016) and ending the 'tale of two cities' reputation (One City Plan 2020).

The council's approach to green space management was laid out in the 2008 Parks and Green Space Strategy, which set out quality, quantity and distance standards aimed at ensuring all Bristolians live within a 15-minute walk of a formal green space and maintaining eight square meters of green space per resident. While nearly 90% of residents already lived within 300 m of parklands and waterways, the 2008 strategy noted a significant amount of this to be poor-quality, informal green space that is underused because it is perceived as less safe and inaccessible. The One City Plan (2020) pushed the city's greening aims further by aspiring to double the city's tree coverage in 25 years. This aspiration, however, would require the planting of nearly 54 hectares of new trees every year throughout this period, as roughly 100 hectares of trees have been removed since the adoption of the 1997 Local Plan.

Collectively, the city's plans have recognized South Bristol as a key site for regeneration, due to the area's large amount of low-quality open space alongside historic deprivation. Together, these plans aimed to bring employment opportunities, higher density housing and formal green space to the area. For example, the city's largest brownfield site (76 ha) has become the site for a new park. Previously an airfield and a garbage dump, the Hengrove Park Leisure Center opened in 2012 and now hosts open space for dog walkers, children's play equipment, a skatepark, toilet facilities, climbing frames, a sports center, a bingo club, a cinema and 1,450 new homes. The regeneration of South Bristol continued with the arrival of the Filwood Green Business Park in 2015. Just across the street from the Hengrove site, the new business park brought small and medium-sized businesses to a new environmentally friendly building[13] while also delivering a new bus line, electric car charging ports and new pedestrian and cycle routes along the main road.

However, the city's plans to tackle issues related to congestion and air pollution have fallen short and continue to negatively affect residents of South Bristol. Not only are 8.5% of the deaths in the administrative area attributed to air pollution, disproportionately impacting Black, Asian and minority ethnic (BAME) populations of the Inner East, but South Bristol continues to suffer economic deprivation due to its poor connectivity to the city center. Heartbreakingly, a young girl from Hartcliffe, in the southern tip of South Bristol, who traveled to the center to discuss child-friendly cities and public transportation with the city council, revealed "it was actually the first time she had been into the city center because her parents hadn't had a car and they couldn't afford the bus fare" (member of Future Proof Parks). Further attempts to connect South Bristol to the city center via pedestrian and cycle ways have been met with disdain as they have cut through both Victoria Park and the Northern Slope Nature Reserve, with the loss of shrubbery and "through traffic" upsetting local residents.

The Northern Slopes in South Bristol, an area of deep social and ecological deprivation, has become an area where even "people who live in the area wouldn't even go … [as it is] very unkept … and there's drugs and there's litter." However, members of the Northern Slopes initiative are "trying to [not only]

FIGURE 3.2 A tightly packed St Marks Road in Easton. (Photo by: Galia Shokry.)

FIGURE 3.3 View of new housing developments in Northern Slopes. (Photo by: Carmen Pérez del Pulgar.)

reframe [themselves], but also reframe the local population." The reframing of this discourse has become a priority for the initiative as the area's reputation is being mobilized as an argument to reduce its size and to sell off land for real estate and housing developments. In hopes of initiating this transition, activities and gatherings are organized to highlight the benefits of the area which has designation as a local nature reserve.

Continued and widespread housing issues

All in all, social needs continue to plague much of Bristol, and South Bristol in particular. Influenced by each new mayor wanting to leave their mark with a new plan and the adoption of standardized policy toolkits from international policy consultants, Bristol has seemingly fallen into a loop of changing everything (in writing) so that little changes (in reality). As new developments continue to fall short in meeting the planning requirements for social housing and green spaces become increasingly exclusive and undermaintained, the council continues to fall short in reaching the goals of health and wealth equality central to their numerous plans.

Lofty aspirations for new housing units were set out for the redevelopment of the city center. The 2011 Core Strategy outlined the city center's continued growth, aiming to deliver 7,400 new homes, of which 40% were required to be affordable. Despite this requirement, as of 2017, only 11% of the 2,368 new dwellings across 11 sites were affordable, with developers citing viability and accelerated delivery as key drivers (Thorning, Balch and Essex 2019). Unfortunately, this occurrence is progressively becoming the norm, as one respondent highlights:

> there's always policies around the number of units that need to be affordable. But then what tends to happen is that the developers … will then negotiate that downwards. They often get planning permission and then start building and say "now that we're underway we've realized we can't afford to do the … the market won't allow this and that."

This process is driven by the fact "the British system is very centralized … so most of the decisions about housing … [are] made by the central government. Local government input is quite weak."

Overall, the city has seen an average of 1,341 homes built per year between 2012 and 2017, with only 12.2% being affordable. This dual state of an increasingly expensive city center, alongside a lack of affordable housing throughout the city heavily influenced by the power of the international real estate investment in the UK, has not only pushed gentrification into central neighborhoods, such as Stokes Croft and Montpellier, but has also led to the loss of green space.

FIGURE 3.4 Waterfront condos near Millennium Square and advertisement for Whapping Wharf regeneration. (Photos by: Galia Shokry.)

One such example is found in the council's plan for a new 1,435 home development at Hengrove Park. While the plan, approved in February 2020, would additionally bring a new 22-hectare park, the plan clashes with a neighborhood plan put forward by Hengrove and Whitchurch Park Neighborhood Planning Forum. This neighborhood plan called for higher density homes in order to leave room for a 30-hectare park. Despite this alternative plan being backed in a local referendum, the city council has decided to go ahead with their original plan, at odds with the local community who will be left with less green space than desired.

Similar to Whitten's (2019) study of the changing landscape of green governance in London, demographic change in Bristol is undoubtedly impacting the Bristol City Council (BCC)'s ability to deliver services to its residents. Factors including the now relatively short train commute to London, the city's claim to two world-class universities and its reputation as a green capital have led to the arrival of new residents and stressed the city's already low housing stock and affordability. And while the city council has recently reinvested in council housing and created its own council housing organization, as of January 2020 more than 13,000 families were still waiting to get a secure and affordable home.

Austerity and nonprofit stewardship within a greener Bristol

Although some may argue that the title of "green capital" is well-deserved, others disagree. Despite 98% of new business development and 94% of new homes occurring on brownfield sites and reaching the lowest carbon footprint of any British city, some say Bristol is a "self-proclaimed sustainable city," claiming "there is no real credential to support [it]." This type of argumentation not only cites the continued disparities in deprivation, continued loss and informality of green space, illegal levels of air pollution and an increasingly unaffordable housing stock but also claims Bristol's current environmental reputation as being a product of neoliberalized environmental politics, whose middle-class, creative and fun approach to greening may be driving gentrification in the city (Harper 2016). And while progress has undoubtedly been made, why do these issues persist? And more importantly, what social impacts could this persistence have as the city continues its green transformation?

As the provision of green space is still not a statutory requirement of local councils, a 66% decrease in the parks budget from 2013 to 2019 forced the BCC to entertain alternatives to funding these spaces. Reversing their position that all parks would have to become self-sustaining after April 2019, the BCC proposed revenue schemes, such as cafes, parking fees, large events, selling off of particular green space and advertising in parks as a means of funding. Public outcry, however, criticized fee-based services for making parks more exclusive while groups such as Adblock Bristol argued that the introduction of advertising in parks is antithetical to the benefits users experience in parks and green spaces. The Bristol Parks Forum, an umbrella organization representing groups directly involved in the city's parks and green spaces, was able to negotiate that the only green space sold off had little to no use for the public. While communities and the public sector were able to limit the extent to which the BCC implemented their revenue schemes, the council's budget for the maintenance of green space continues to dwindle, further pushing this task onto the public.

This transfer of responsibility is not merely a result of austerity running its course but was rather a vision outlined by the national government in their Localism Act of 2011. From a civic participation standpoint, the act aimed to relieve citizens of a sense of imposition from the central government and to bring a sense of involvement and stewardship alongside new rights and freedoms, which can be interpreted as new duties and burdens. While offering a re-assessment of local self-government, this austerity localism (Featherstone et al. 2012) has been argued to offer little in terms of lasting change as it is unrolled alongside the short-termism of austerity measures (Lowndes and Pratchett 2012). In other words, as austerity has forced the BCC to turn greening into a profitable enterprise, it has also pushed the responsibilization of parks maintenance and stewardship to local community groups.

While this task has been undertaken by the more than 70 community groups throughout the city, it has proven to be far less efficient than when undertaken by the city council. An appropriate transfer of maintenance "would need [a] huge public infrastructure or … a volunteering system that would license people to drive lawnmowers, tractors, and use chainsaws" and would begin blurring the line between civic engagement and unpaid labor. This transformation does not come without a financial and human cost to community groups additionally tasked with acquiring funding from groups such as the National Lottery Heritage Fund or relying on section 106 or community infrastructure levy (CIL) payments devolved from the city council.[14] While section 106 and CIL payments have added more than 40 million pounds to the city's revenue in the last decade, with portions of this being devolved to communities, community groups feel they have been left to "fight over the crumbs" rather than having their green spaces being maintained and affordable housing needs met.

The resulting tasks and burdens of austerity localism have highlighted who is doing and who can do this unpaid labor. Put differently, austerity localism has resulted in wealthier areas having better maintained green spaces while poorer neighborhoods are not *as* capable of undertaking the required level of volunteer work and stewardship. While far from the city's first experience with austerity, decreases in local council budgets since 2010 have been three times greater than the cutbacks of the 1970s and 1980s (Whitten 2019). Neoliberal roll-back and roll-out policies, respectively referring to "the active destruction" of the welfare state and the "purposeful construction and consolidation" of neoliberal modes of governance (Peck and Tickell 2002, 37), have increasingly led to the selling off of social housing while simultaneously prohibiting local authorities from borrowing money to build social housing or invest in maintenance, all while creating beneficial tax schemes for property investors. These negative impacts on affordability have also been met with the growing NGO-ization of urban maintenance, social support services and environmental stewardship previously filled by the public sector.

Although South Bristol, a majority white yet deprived area, is home to many community groups devoted to the maintenance of green space, the community infrastructure, like the area's public transportation, suffers from issues of cohesion. Inner East neighborhoods, especially St Pauls, continue to see the lowest levels of organizing. This is being addressed by groups such as the Green and Black Ambassadors, who conduct community research and projects to encourage stronger dialogue and engagement across BAME and environmental sustainability communities.

"Making our prosperity sustainable"

In sum, the political project of austerity pushed to new levels by the Tory government since 2010 has become a roadblock to Bristol parting with its "two city" reputation and fully embracing its place as a green capital for *all*. Despite being home to a rich infrastructure of community groups and green organizations which have pushed the city council to embrace health and wealth equality in their many plans, Bristol's green transformation continues to be experienced unequally. While those in historically affluent wards are able to withstand the effects of austerity and maintain community green spaces, and new affluent residents of the city center experience the benefits of green growth, those in historically deprived wards such as South Bristol continue to bear the brunt of the negative side-effects resulting from austerity (localism).

Even though gentrification appears to be occurring at a city-wide scale, arguably driven by the arrival of new residents and the retention of university students, the impact of the type of greening emphasized by the city council should not be underestimated. Urban greening targeting the creative classes, from the green jobs at the Filwood Green Business Park in South Bristol to the pedestrianization of the Temple Quarter which is increasingly surrounded by luxury flats and a university campus, may be making Bristol ripe for green gentrification. This is a reality that the city may not escape if the era of austerity—and thus the transfer of urban greening to the private sector—does not come to an end.

Keywords

(a) the urban development pattern of the city and neighborhood: industrial legacy; fast recent growing and green growth; creative industry/class; income inequality; health inequality
(b) the urban greening of the neighborhood: waterfront redevelopment; decontamination policies; parks; broader "European Green Capital" agenda
(c) the inequalities at stake: insufficient affordable housing for the middle class; (green) gentrification; continued exposure to contamination; budget cuts and austerity policies affecting green space; fee-based services related to green spaces

Notes

1 2008 Parks and Green Space Strategy.
2 https://ec.europa.eu/environment/europeangreencapital/winning-cities/2015-bristol/
3 https://www.bristolpost.co.uk/news/news-opinion/bristol-most-segregated-city-uk-1534458
4 https://www.bristol.gov.uk/documents/20182/32947/State+of+Bristol+-+Key+Facts+2018-19.PDF/263d5f0f-763e-9553-467d-c9704f307d7c
5 https://www.bristol.gov.uk/documents/20182/32951/Deprivation+in+Bristol+2019.pdf/ff3e5492-9849-6300-b227-1bdff2779f80
6 https://www.bristol.gov.uk/documents/20182/33904/Population+of+Bristol+September+2020.pdf/69aa0aa1-290a-ccf2-ec4f-13a7376b41a8
7 https://www.bristol.gov.uk/documents/20182/32951/Deprivation+in+Bristol+2019.pdf/ff3e5492-9849-6300-b227-1bdff2779f80
8 See more: https://www.theguardian.com/politics/2008/jun/19/transport.transport1#:~:text=Bristol%20has%20been%20named%20as,on%20the%20successful%20Paris%20scheme
9 See more: https://www.resourcefutures.co.uk/about-us/
10 https://irr.org.uk/article/fighting-sus-then-and-now/
11 See more: https://www.theguardian.com/uk-news/2016/dec/11/bristol-st-pauls-housing-hotspots-hipsters-gentrification
12 Bristol Local Plan 1997.
13 See more: https://filwoodgreen.co.uk/
14 Section 106 payments are contributions negotiated between councils and developers, while CIL apply to all development granted planning permission. These funds are then used to fund further infrastructural projects throughout the city such as schools, affordable housing, clinics and roads.

References

Doherty, Brian, Alexandra Plows, and Derek Wall. "Environmental direct action in Manchester, Oxford and North Wales: A protest event analysis." *Environmental Politics* 16, no. 5 (2007): 805–825.

Featherstone, David, Anthony Ince, Danny Mackinnon, Kendra Strauss, and Andrew Cumbers. "Progressive localism and the construction of political alternatives." *Transactions of the Institute of British Geographers* 37, no. 2 (2012): 177–182.

Harper, Earl. "Ecological gentrification, post-politics and fetishisation: Bristol, UK, as European green capital." Paper presented at *From Contested Cities to Global Urban Justice: Critical Dialogues*, Madrid, Spain, July 4–7, 2016, 1–17.

Lowndes, Vivien, and Lawrence Pratchett. "Local governance under the coalition government: Austerity, localism and the 'Big Society'." *Local Government Studies* 38, no. 1 (2012): 21–40.

Peck, Jamie, and Adam Tickell. "Neoliberalizing space." Essay. In *Spaces of Neoliberalism: Urban Restructuring in North America and Western Europe*, edited by Neil Brenner and Nik Theodore, 33–57. Malden: Blackwell, 2002.

Slater, Tom, and Ntsiki Anderson. "The reputational ghetto: Territorial stigmatisation in St Paul's, Bristol." *Transactions of the Institute of British Geographers* 37, no. 4 (2012): 530–546.

Thorning, Daniel, Christopher Balch, and Stephen Essex. "The delivery of mixed communities in the regeneration of urban waterfronts: An investigation of the comparative experience of Plymouth and Bristol." *Land Use Policy* 84 (2019): 238–251.

Townsend, Peter, Don Simpson, and Nigel Tibbs. "Inequalities in health in the city of Bristol: A preliminary review of statistical evidence." *International Journal of Health Services* 15, no. 4 (1985): 637–663.

Whitten, Meredith. "Blame it on austerity? Examining the impetus behind London's changing green space governance." *People, Place and Policy* 12, no. 3 (2019): 204–224.

4

ENACTING A RAIL-TO-PARK PROJECT IN VALENCIA PARC CENTRAL OR THE ACTUAL CONSTRUCTION OF GREEN GENTRIFICATION

Lucía Argüelles Ramos

The residents of Russafa, in Valencia, had heard about Parc Central 30 years before its construction began. A portion of the park was finally inaugurated at the end of 2018, while national, regional and city-level governments still fight over the viability of building the remaining part, which depends on the construction of a €3.1 billion 9-km tunnel across the city. In the decades waiting for the park, Russafa, a historically working-class neighborhood and multicultural hub since the 1990s, has seen a growing gentrification trend driven by speculation relating to the expected park's benefits, neighborhood revitalization and more recently, tourism and changes in housing rental markets. This chapter reviews the politics, financing, materiality and ecology in the making of the rail-to-park mega-project of Valencia Parc Central,[1] and how such processes relate to gentrification dynamics. It also examines Russafa's civic attempts to align the construction of a green space more closely with local needs.

Parc Central and Valencia's history of undemocratic and pharaonic urbanism

Valencia was governed for 24 years by the right-wing party Partido Popular, with a mayor, Rita Barberá, who was in office from 1991 to 2015. During these decades, coinciding with the Spanish economic bonanza of the 1990s and 2000s, Valencia was one of the epicenters of grandiose, undemocratic and overambitious urban interventions (Rius-Ulldemolins, Hernàndez I Martí and Torres 2016; Tarazona Vento 2017). Valencia was promoted as a touristic and sports destination, turning into a "city-spectacle" (Gaja 2007) and a capital of mega-projects intended "to produce an improved image which attracted international media attention, improved the population's self-esteem, attracted tourism and was electorally profitable for the party in office" (Tarazona Vento 2017, 74).

DOI: 10.4324/9781003183273-5

The effort to situate Valencia globally contrasted with how, on a day-to-day basis, the local government neglected to invest in neighborhood equipment and social welfare. This political environment motivated social movements to organize against undemocratic urban interventions, including urban greening projects such as the expansion of the Passeig Blasco Ibañez avenue and recurrent attempts to develop *La Horta*, a protected farmland surrounding the city (Navarro Eslava 2014; Luis del Romero Renau and Valera Lozano 2016).

Officially, the process for creating a large green area—the future Parc Central—on a space occupied by railway tracks and industrial warehouses began with the 1989 General Urban Planning Scheme. Since then, a consortium of local, regional and national governments has slowly moved towards its implementation. This public company, Valencia Parc Central S.A.[2] (V.P.C.), aims to transform the current railway infrastructure by burying the rail tracks and building a 9-km tunnel for high-speed trains, turning the existing industrial area into a park and boulevard flanked by new housing units. This rail-to-park transformation follows the current trend in the U.S. and Europe of transforming abandoned rail lines into public spaces like the much-celebrated New York High Line. In Europe, the increased popularity of high-speed rail infrastructure has led to revitalized railway stations (Garmendia, Ribalaygua and Ureña 2012), paving the way for additional regeneration opportunities and bringing visitors and business workers closer to the city center. Rail-to-park transformation has become a popular strategy in the global competition for livable cities (Garcia-Lamarca et al. 2019), which requires iconic green areas that attract both visibility and visitors.

Over the span of three decades, the park became a priority for the municipal government, which negotiated, at different levels of governance, how to make it happen. The call for project proposals was launched in 2011, the first phase of construction began in 2016, and the first area of the park (40% of the total planned, accounting for 110,000 m^2) finally opened to the public in December 2017, inaugurated by a new left-wing coalition government at the city level. The most contentious aspects of V.P.C. have been, and still are, the two conditions for completing the park: the development of housing around the revalorized park's perimeter to partially finance the project and the massive infrastructural task of burying the rail tracks. Essentially, V.P.C.'s branding of the project as having "the largest impact in terms of urban planning of the city" (Valencia Parque Central S.A. n.d.) mirrors the grandiosity and top-down planning which have characterized Valencia's recent history.

Parc Central is often portrayed in institutional discourses as the "city's green lung." Yet this narrative disregards the existence of the most popular park in Valencia, Jardins del Turia, which constitutes the largest urban green area in all of Spain. Built during the 1990s, the Jardins del Turia was the result of citizens' calls to transform the Turia riverbed into a green area instead of a highway planned by the city, once the river had been diverted for flood protection. In contrast, Parc Central did not emerge from the desire of residents: rather, it is

FIGURE 4.1 A view of Valencia's Parc Central, with the vegetable garden on the right. (Photo by: Galia Shokry.)

a top-down plan to foster investment and attract international visibility. One activist and urban planner said:

> The difference between Parc de la Rambleta, Marchalenes or Parc de la Desembocadura, or even Jardins del Turia, and Parc Central, is that no one fights for the latter. Instead, it is architects and politicians who design and plan it as a kind of New York Central Park simply because it is in the center.

The specific materiality of V.P.C., being both a park and a railway project of such large dimensions, with different stakeholders involved, provides an opportunity to study how a park is made and how this process relates to gentrification. The next sections review the political, economic and material processes behind the construction of Parc Central and its embedded gentrification effects.[3]

A political process embedded in politics of scale and tied to material infrastructure

The park essentially serves as a social and environmental justification for the much larger infrastructural project of which it forms a small, but crucial part. The project transforms the current central station Estació del Nord (currently a cul-de-sac) into a pass-through stop, in order to enable high-speed trains to arrive in the city center. The highly complex infrastructural project favors a multi-scale political process in which national (Spanish Ministry of Economic

Development, public train and infrastructure companies Renfe and Adif), regional (Valencia regional government), city-level (Valencia city's) and neighborhood-level governments (N.G.O.s and community groups involved in some consultations/opposition) all have a stake. Those different entities formed V.P.C. in 2003, the agency which took charge of the project's technical and urban development. Similar to other cases of semi-publicly managed mega-projects—a type of governance promoted as more technically efficient and collaborative (Jessop 1997)—the decision-making behind V.P.C. has in practice involved scarce participation from the general public.

Parc Central proponents recurrently use competitive urbanism arguments (Brenner 2009), suggesting that Valencia deserves the pass-through station (a highly efficient connection to the high-speed railway network) that other major Spanish and European cities already have. Such discourses often include comments about modernity and economic development assumed to be beneficial aspects of a city and expected by their citizens. However, the prioritization of a high-speed railway in the city center seems to cater to those who most value this kind of transportation, that is, visitors and businesses. In turn, part of civil society opposed burying the train tracks and considers the project a large expense and unnecessary. An alternative transit plan was designed and promoted by a number of urban planners, intellectuals and activists (Olmos and Gaja 2018), but the VPC management and the local authorities have given it little attention, privileging calls for the "modernization" of the rail system.

More recently, the Ministry of Economic Development and Renfe-Adif have advocated for a smaller scale and lower-cost project, with only a partial burying of the tracks. This proposal would require a partial redesign of part of the park and limit the housing development planned around it. However, regional and city governments oppose this alternative, calling it an orchestrated and arbitrary plan to "disinvest in the city of Valencia over the following years" (Vázquez 2016). Such politics of scale emerge out of an interest to implement a 21st-century version of modernity and strategically place Valencia as a frontrunner in the race for competitive urbanism in Spain and Europe.

Economic mechanisms to finance Parc Central: conditioning the park on land revalorization

The €3.1 billion budget for Parc Central will be partly derived from the profits gained by issuing building permits to private developers once the industrially zoned public land is turned into a residential zone. Most of this land now belongs to the public railway company Renfe-Adif. The housing and commercial development included in the plan (4,200 housing units within 8 buildings, plus commercial areas) was initially expected to pay for 15% of the project. The rest of the budget would derive from national public funding.

To manage and finance the urban development linked to the project, the city government devised an Integrated Action Plan (Plan de Actuación

Integrada—P.A.I. Parc Central) approved in 2014. Other P.A.I.s in Valencia (at least four more are projected) have been controversial and fiercely opposed by residents in nearby neighborhoods since P.A.I.s often involve the construction of a massive number of housing units without an adequate number of much-needed educational or healthcare facilities. Moreover, P.A.I.s place public urban equipment such as parks at the service of private investors by enabling them to design and implement green areas, and also to benefit from the value they add to their developments.

The dependence on private actors and markets, however, has slowed down the Parc Central project and affected its budget. One tender process designed to sell two lots for €5.8 million within the P.A.I. area has failed twice, while other lots attract little interest (Hortensia García 2020), possibly due to perceptions of high business risks in a softening real estate market in Spain. In addition, the economic crisis triggered by COVID-19 will likely have serious effects on the public budget and viability of the project. Mechanisms such as the P.A.I.s and mega-projects like V.P.C. that favor private and speculative housing development render urban development a condition for the construction of green spaces, even if the majority of the funding comes from public money. This arrangement may be key in predicting gentrification effects associated with new green spaces. In contrast, smaller parks are more often able to secure public funding with no links to new development.

A design to bring Valencia back to the international architectural arena

In 2011, Kathryn Gustafson, a Scandinavian landscape architect who gained international visibility with the Lady Di memorial in Hyde Park and the Lurie Garden at Chicago Millennium Park, won the bid for the design of Parc Central. The design of the project builds on Valencia's tradition of large-scale projects such as the Calatrava-designed Parc de les Ciencies and brings the city to the forefront of avant-garde urban greening and landscape architecture.

Gustafson's design is inspired by water and the different ecological habitats of the site: a natural reserve in a riverbed, farmland, marshland and the Mediterranean Sea. Through a minimalist design, the architect uses canals, fountains and a network of paths to connect different spaces in the park, such as an urban farm, a children's playground and a flower bed. The second phase of the park, if ever completed, will add more canals, a lake and an outdoor auditorium.

The park's appealing design and signature character seem directed at attracting international visibility and creating a new city landmark. Indeed, according to the classification of the Valencia Parks Office, Parc Central is considered a city park for tourists and visitors rather than a neighborhood park designed for daily residential uses. It is not surprising then that local users have criticized its design for favoring sporadic visits rather than day-to-day use. Residents also regret the lack of input asked of local residents, and that the park has too many paved roads

FIGURE 4.2 Parc Central's design includes canals, fountains and a lake. (Photo by: Galia Shokry.)

and is not as green as initially projected. In addition, the strict demarcation of uses does not favor social interaction. In an interview, local resident and activist of Russafa remarked that "you need an instruction manual to use it."

Another interviewee explained the difference between Parc Central and other parks:

> I think that Parc Central is very compartmentalized. It has many controlled areas. You have an outdoor gym for elderly people, and a separate area for dogs … imagine that! That's incredible. There's a parcel for everyone. In other parks these areas are not enclosed spaces inside the park, they are interwoven, so there's more of a mix of people. It's nicer. In this sense, we like the heterogeneity of El Jardín del Turia.
>
> *(Interviewee, activist and urban planner)*

In sum, the design and uses of a park shape gentrification processes, too. Favoring one type of design and user over another impacts the park's visibility and popularity, and consequently the area's future commercial and residential transformation.

The materiality of Parc Central: enmeshed ecology

The park's ecology, which includes 4,000 trees, 85,000 bushes and 70 different plant species, has been carefully crafted to be aesthetically pleasing. Aromatic plants like jasmine cover concrete structural elements, and the walls bordering the railway tracks will be covered by greenery. The trees are mostly indigenous but also come from Girona, other Spanish locations and from abroad. The successful growth of such species has required excavations of different depths and the filling of different soil types with different amounts of organic matter. An aeration tube together with 28 underground infiltration wells also creates a high-quality irrigation system that allows roots to grow faster and stronger.

This ecology requires a considerable material transformation of the park's surroundings, and an even larger one if the tracks are buried. Up until now, industrial fields with toxic hydrocarbon residues have been cleaned up, old warehouses and industrial buildings have been demolished, and dozens of centenarian trees have been cut or transplanted in order to clear space for development. The construction of gardens, irrigation systems and wells also involved massive material changes. During the first phase of the project accounting for 40% of the park's area, 500,000 m^3 of soil were moved, out of which 16,000 m^3 had to be transported (about 550 dump trucks).[4] A total of 54,000 m^3 of concrete was poured for foundations and sills and for paving 87,000 m^2 of paths. New vegetation required 152,000 m^2 of soil. In addition, the 25-floor buildings on the corners of the park will change the overall landscape and determine the incidence of sunlight on the park and surrounding blocks, and increase the traffic in the area. Lastly, the 9-km tunnel dug to bury the tracks will entail the removal and transportation of tons of soil outside the city. In short, the project's touted environmental sustainability is overshadowed by the elevated material and environmental costs that will negatively impact local residents.

The Parc Central project is also accompanied by a housing development for new upper-class residents whose planning and design mostly excluded local residents, much to their dismay. In this way, as a highly valued urban land use, not only is the park a driver of green gentrification but so is the process of park-making itself. Gentrification is constructed throughout the different layers and processes required to assemble a green area in the context of speculative urbanism, green growth and undemocratic urban interventions (Jessop 1997; Gould and Lewis 2017).

Russafa's efforts to avoid further gentrification

In Russafa, interviewees were not sure of the extent to which the unusually prolonged development of the park over the years had influenced gentrification in the neighborhood. They found it difficult to distinguish the impacts of its first announcement 30 years ago from those of the revitalization process promoted by the municipal government from 2007 to 2009. During those decades, Russafa experienced a textbook gentrification process, shifting from a "problem-hood to a trendy hood" (Romero Renau and Martín 2015). More recently, from 2014, the neighborhood's touristification and short-term rentals boom affected housing prices in unpredicted ways. It was amidst these changes, in 2011, that the Parc Central project finally kicked off.

Interviewees expressed their belief that the neighborhood was intentionally left to deteriorate after the planning of Parc Central. Residents and activists recount incidents of drug dealing, squatting and a progressive withdrawal of urban services at the end of the 1990s, when investors began buying entire blocks and pushing elderly residents out of the neighborhood. When the project was finally approved in 2011, the neighborhood's clean-up accelerated property purchases. A local Russafa resident and activist summarized the park's impact on the neighborhood: "This neighborhood is going through a rough patch. Expectations of a large green infrastructure like Parc Central have been looming for many years, as both a promise and a threat."

On the one hand, the promise is clear. Russafa was in need of more green areas, and people were looking forward to having a new park nearby. Before the first section of the park was opened in 2018, Russafa only had the Manuel Granero park, a medium-sized park frequented by the elderly and families. Despite ongoing conversations with the local government about its degraded state, plans for Parc Central quickly became an excuse to delay its repair. One interviewee explained,

> One of the consequences of Parc Central is that they won't restore our park, because we will allegedly have a super-park. But Manuel Granero is for everyone, this is our park, it's surrounded by buildings and very calm, and children are safe there. I went there with my child when he was very little, and I always felt safe because neighbours here look out for each other. That's not something that will happen in Parc Central.
>
> *(Interviewee, Russafa resident and activist)*

On the other hand, the threat is also clear. Once finished, the park and the additional developments necessary to finance it will most likely change the social and commercial make-up of neighborhood. The new high-end housing units will likely cause retail and housing values to rise. Moreover, neighbors fear that new residents will require schools, health centers, sports facilities and social centers that they have long called for and never received and which are now conditioned by Parc Central's construction. This simultaneous desire and fear of the park create an ambivalent and at times conflicting sentiment among residents.

More recently, locals have demanded the rehabilitation of several industrial warehouses within the park area adjacent to the neighborhood. In the decade prior to the park's opening, local associations from Russafa reclaimed the control of the *Naves* (warehouses) as social, cultural and sports facilities for residents. While the initial proposal by the conservative government to develop a railway museum and new facilities for Valencia University was rejected by residents, in 2015 the new government discussed the use and co-management of the Naves with local entities. The progressive Plataforma per Russafa, the main neighbors' association, organized a campaign that led to a series of workshops and a meeting with the city government to transform the Naves based on Russafa's needs (Plataforma Per Russafa 2015). Today, one of the warehouses hosts the Universidad Popular de Valencia, a historical socio-cultural center, and another will host cultural activities for social and youth organizations, though not self-managed as the residents' platform had requested. Another warehouse will be transformed into a public sports center.

A related conflict involves a dock in the warehouses area that the local government planned to redevelop into bars and restaurants for park visitors. Local residents opposed to the plan argue that it will fuel more tourism in the neighborhood. One interviewee explained:

> If the dock turns into a hotel and restaurant area, it will become more of a tourist attraction. It is in the municipality's hands to make this a welcoming square and a catalyst for neighborhood interaction. There is a social network here that needs urgent protection.
>
> *(Interviewee, Russafa resident and activist)*

By the time data analysis was completed Valencia housing prices were peaking given the high touristic pressure and the popularity of short rental apartments. Until that moment, the city had not taken specific measures at the neighborhood or local level to avoid speculation and displacement around new greening interventions, such as those described in other chapters of the book (e.g., Barcelona, Nantes or Vienna). The most effective measure to avoid further gentrification in the case of Parc Central would probably entail not developing the new high-end housing units and redirecting the park's related benefits to the surrounding neighborhoods and their residents. On the ground, only

FIGURE 4.3 One of the Naves whose control has been reclaimed by Russafa neighbors. (Photo by: Galia Shokry.)

some of the public officials interviewed acknowledged that parks can have an undesired (green) gentrification effect. City measures mostly focused on limiting the impact of short-term rentals in a context of increasing tourism: ceasing to issue new licenses within the historic area (Ciutat Vella), adding requirements to the type of apartment that can be rented (only on ground or first floors) and finally, forbidding renting individual rooms (vs. entire apartments). Another measure, the complete prohibition of touristic rentals in Ciutat Vella in 2018, brought negative consequences to other areas. That said, the Strategic Housing Plan for the city (Dec 2017) situates gentrification as a driver of the lack of affordable housing and aims at protecting and improving social housing that is already scarce in the city. In 2020, Valencia joined the network CitiesforHousing, stating an objective of improving housing affordability and security for all.

A different green future for Russafa

In the process of park-making, local, regional and national governments choose to mobilize and enact certain political, economic and social structures in order to achieve their goals. These choices are influenced by existing city imaginaries. In the rail-to-park transformation of Valencia Parc Central, two city imaginaries coexist; one that follows the paradigms of green growth, entrepreneurial urbanism and sustainable urban design, and another that champions the right to the city, social justice, housing rights and participation. As long as Parc Central is made to be part of a speculative greening process, the needs and desires of residents will remain unheard and their housing rights at stake.

Keywords

(a) the urban development pattern of the city and neighborhood: large-scale urban renewal, touristification, creative industries, entrepreneurial urbanism, sustainable urban design
(b) the urban greening of the neighborhood/city: rail-to-park transformation, public-private green partnership, iconic parks, linear green space
(c) the inequalities at stake: (green) gentrification, speculative urbanism, private capture of public investment

Notes

1 https://valenciaparquecentral.es/images/descargas/PRESENTACIONES/monografico_vpc_eng.pdf
2 https://valenciaparquecentral.es/
3 Part of the analysis and arguments found in this book chapter appeared in "Argüelles, L., Cole, H., Anguelovski, I., 2021, Rail-to-park transformations in 21st century modern cities: Green gentrification on track. Environment and Planning E.

4 http://www.valencia.es/ayuntamiento/urbanismo2.nsf/0/EFCFB6DBFC427D9CC1258266002CFF0E/$FILE/ANEJO%2029_RESUMEN%20CARACTER%C3%8DSTICAS%20GENERALES%20PROYECTO_Diligenciado.pdf?OpenElement

References

Brenner, Neil. 2009. "Cities and territorial competitiveness." In *The SAGE Handbook of European Studies*, edited by Chris Rumford, 442–463. London: Sage.

Gaja, Fernando. 2007. "El boom de València o la ciutat com a espectacle." In *Llibre verd del territori valencià*, edited by Escola Valenciana i Societat Sostenible, Congrés, Ricardo Almenar, and Federació d'Associacions per la Llengua Escola Valenciana, 201–213. València: Escola Valenciana.

Garcia-Lamarca, Melissa, Isabelle Anguelovski, Helen Cole, James J. T. Connolly, Lucía Argüelles, Francesc Baró, Stephanie Loveless, Carmen Pérez del Pulgar Frowein, and Galia Shokry. 2019. "Urban green boosterism and city affordability: For whom is the 'branded' green city?" *Urban Studies*, 58, 004209801988533. https://doi.org/10.1177/0042098019885330.

Garmendia, Maddi, Cecilia Ribalaygua, and José María Ureña. 2012. "High speed rail: Implication for cities." *Cities* 29 (December): S26–31. https://doi.org/10.1016/j.cities.2012.06.005.

Gould, Kenneth, and Tammy Lewis. 2017. *Green Gentrification: Urban Sustainability and the Struggle for Environmental Justice*. London and New York: Routledge.

Hortensia García. 2020. "Nadie quiere los solares edificables de Parc Central." *Levante-EMV*, January 16, 2020. https://www.levante-emv.com/valencia/2020/01/17/nadie-quiere-solares-edificables-parc/1966651.html.

Jessop, Bob. 1997. "The entrepreneurial city: Re-imaging localities, redesigning economic governance or restructuring capital?" In *Transforming Cities: Contested Governance and New Spatial Divisions*, edited by N. Jewson and S. MacGregor, 28–41. London and New York: Routledge.

Navarro Eslava, Luz. 2014. "Salvem El Cabanyal: Urban movements and their claim for the 'right to the city.'" *VLC Arquitectura. Research Journal* 1 (2): 47. https://doi.org/10.4995/vlc.2014.2813.

Olmos, Joan, and Fernando Gaja. 2018. "El lío ferroviario: una apuesta por la claridad y el debate." *Levante-EMV*, April 21, 2018. https://www.levante-emv.com/valencia/2018/04/21/lio-ferroviario-apuesta-claridad-debate-11994098.html

Plataforma Per Russafa. 2015. "Naves de Ribes." http://www.russafa.org/wp-content/uploads/2015/05/Les_naus_dosier.pdf.

Rius-Ulldemolins, Joaquim, Gil-Manuel Hernàndez I Martí, and Francisco Torres. 2016. "Urban development and cultural policy 'White Elephants': Barcelona and Valencia." *European Planning Studies* 24 (1): 61–75. https://doi.org/10.1080/09654313.2015.1075965.

Romero Renau, Luis del, and Antonio Valera Lozano. 2016. "From NIMBYsm to the 15M: A decade of urban conflicts in Barcelona and Valencia." *Territory, Politics, Governance* 4 (3): 375–95. https://doi.org/10.1080/21622671.2015.1042025.

Romero Renau, Luís, and Laura Lara Martín. 2015. "De Barrio-Problema a Barrio de Moda: Gentrificación Comercial En Russa-Fa, El 'Soho' Valenciano." *Anales de Geografía de La Universidad Complutense* 35 (1): 187–212. https://doi.org/10.5209/rev_AGUC.2015.v35.n1.48969.

Tarazona Vento, Amparo. 2017. "Mega-project meltdown: Post-politics, neoliberal urban regeneration and Valencia's fiscal crisis." *Urban Studies* 54 (1): 68–84. https://doi.org/10.1177/0042098015625025.

Valencia Parque Central. n.d. "Presentación." Valencia Parque Central. Accessed July 3, 2020. https://valenciaparquecentral.es/actuaciones/vision-general/presentacion.

Vázquez, Cristina. 2016. "La paisajista Kathryn Gustafson defiende su Parque Central original." *El País*, January 27, 2016, sec. Valencia. https://elpais.com/ccaa/2016/01/27/valencia/1453909325_280159.html.

PART 2

Compounded risks and impacts of urban greening in post-industrial environments

Part 2, "Compounded risks and impacts of urban greening in post-industrial environments," examines the role played by greening when juxtaposed with untreated contaminated sites and facilities. Here, socially vulnerable residents faced the dual negative dynamics of greening and unequal real estate redevelopment together with historic contamination. The continued presence of a heavily industrial landscape gives neighborhoods in Cleveland, Dallas, San Francisco, Seattle and Glasgow a distinct relationship to greening and a distinct set of social effects.

Can a historic African American post-industrial, recovering city like Cleveland address its longstanding environmental inequities without reproducing the negative effects of environmental gentrification observed in so many other cities? In the Detroit Shoreway neighborhood, despite a long history of exposure to environmental toxics, resident- and activist-driven projects may offer us some clues on how community-focused green developments might avoid the onset of green gentrification.

West Dallas is, in many ways, a classic environmental justice story—with new recent toxic twists. Environmental ills that have long contributed to health inequities among residents remain despite the opening of the Margret Hunt Hill Bridge in 2012 and its adjacent luxury retail and housing developments. Those ills may be compounded by the relocation of existing heavy industry to parts of the neighborhood not yet undergoing gentrification.

In San Francisco, Bayview-Hunters Point, despite looming environmental concerns due to remaining heavy industry and infrastructure facilities, remains a sought-after gravitational pole for the real estate sector and is one of the last remaining development frontiers in a rapidly growing city. It exemplifies how long-term environmental and health challenges fall victim to the economic

DOI: 10.4324/9781003183273-6

incentives of fast-paced luxury and green development with a consequent disregard for existing communities.

Seattle's South Park residents strive to maintain the social character of their tight-knit, largely minority community amidst city-wide gentrification pressures. Despite the legacy of extreme environmental hazards, much of this gentrification is prompted by the ongoing clean-up of the Superfund site along the Duwamish River and the new 14th Avenue bridge project, and other adjacent physical improvements.

In Glasgow's East End, land remediation on former industrial sites has produced new tensions as residents ask whether the large-scale regeneration that has brought new homes, jobs, green infrastructure and improved health outcomes will cater to the needs of long-term residents. Despite the promises made, important questions remain about whether urban greening will help remediate the enduring Glasgow Effect—low life expectancy and overall poor health of many residents.

5

IS CLEVELAND'S VISION OF A "GREEN CITY ON A BLUE LAKE" A PATH FOR SOCIAL EQUITY OR GREEN GENTRIFICATION?

Margarita Triguero-Mas and Wendy A. Kellogg

After decades of economic decline and environmental degradation, the city of Cleveland, Ohio, is experiencing a mild renaissance through greening and other revitalization efforts. Yet, by catering to and benefiting mostly wealthier, white residents, some environmental amenities seem to be perpetuating long-standing racial inequalities. Can a majority African American, post-industrial, recovering city like Cleveland address its historic environmental inequities without reproducing the negative effects of environmental gentrification[1] observed in so many other cities? In the Detroit Shoreway neighborhood, despite a long history of exposure to environmental toxics, resident- and activist-driven projects may offer some clues on how community-focused green developments might avoid the onset of green gentrification.

Cleveland's decline and reinvestment

Cleveland, with a population of 53% Black, 32% non-Hispanic white and 9% Hispanic residents, is one of the most strongly segregated cities in the U.S.[2] While the east side of Cleveland is predominantly Black, the west side of the city showcases a combination of non-Hispanic whites, Blacks and Hispanics.

In the 1950s, Cleveland reached its population peak, with over 900,000 residents when a majority of its workers (only second to Detroit in the U.S.) were employed by the transportation, machinery and iron-and-steel industries. During the 1960s, the city was fraught with civil rights conflicts, mirroring the state of unrest throughout the United States. Since then, the city has suffered drastic economic and population decline,[3] with blue-collar employment falling to only 29% of the total employment by 1986 in Cleveland and its surrounding counties.[4] This decline produced a massive population loss of 380,000 inhabitants by 2019

DOI: 10.4324/9781003183273-7

(U.S. Census Bureau 2019)—a decline illustrated by the thousands of uninhabited houses and vacant lots[5] throughout the city.

Cleveland is also characterized by environmental degradation, poverty, violence, crime and an unfair justice system, with tragic episodes like the 2014 shooting of 12-year-old African American Tamir Rice exemplifying the abuse of police force in the city.[6] From an environmental standpoint, for decades Cleveland's notoriously polluted Cuyahoga River[7] served as a sewer for industrial and other types of waste. The river and Lake Erie, declared "dead" in the 1960s from oxygen depletion, symbolized all that was wrong in a declining city such as Cleveland (Stradling and Stradling 2015). Early efforts by Cleveland Mayor Carl Stokes, the first African American mayor of a large U.S. city, focused on the right of poor residents to have access to clean water for recreation. These efforts, and the famous river fire of 1969, gave impetus to national policy on clean water.

Many environmental conditions in the city still negatively affect Cleveland's residents, however. Overall, the city has high levels of lead poisoning (with 12.83% of children under six years old presenting elevated blood lead levels in 2015[8]), infant mortality (11.98 deaths for every 1,000 live births in 2016[9]) and opioid-related deaths (Cuyahoga County Medical Examiner's Office 2018). Most of these health issues, together with smoking, obesity and low levels of Vitamin D, particularly affect low-income and Black residents. The most recent Cuyahoga County Community Food Assessment noted two startling statistics: about 50% of all Cleveland residents, and 25% of all Cuyahoga County residents, live in a food desert (Cuyahoga County Planning Commission 2017).[10] Lastly, lower-income neighborhoods and communities of color need greater access to parks and recreational spaces.

A green renaissance?

Today Cleveland is experiencing some pockets of renaissance (Detroit Shoreway Corporate Development Organization—D.S.C.D.O. 2009) despite a 35% poverty rate. Urban redevelopments have been spurred by large institutions like universities and healthcare centers, particularly around University Circle and Clark-Fulton/Brooklyn Centre, and by the work of several community development corporations (C.D.C.s) that operate in Cleveland's neighborhoods.

In the context of a legacy city like Cleveland, urban development does not seem to hold the same level of fear of increased inequality, gentrification and displacement as is held in cities with a consolidated growth trajectory like Austin or San Francisco. First, to a certain extent, some local stakeholders report that some areas of Cleveland, like the East Side, are too undesirable (in their words "too poor and too Black") to be gentrified. Second, both the municipality and several community organizations perceive reinvestment in economically distressed areas as desirable and welcomed after decades of economic decline, intergenerational concentrated poverty and socioeconomic segregation. Some even believe that gentrification, if it brings jobs, might *decrease* local urban inequalities given the core problem of widespread poverty. Lastly, reinvestment in Cleveland is not associated—at least not yet—with substantial widespread increases in housing market prices, a trend that for now diffuses fears of displacement.

There are examples, however, of possible gentrification pressures in Cleveland in some neighborhoods, moving from western Cleveland (Tremont in particular) to adjoining Ohio City, and very recently to the bordering Detroit Shoreway neighborhood,[11] largely due to efforts by community development organizations, supported by representatives to city council. The course of Cleveland's urban development in these neighborhoods is also changing with the clean-up of Lake Erie and the heavily polluted Cuyahoga River. While the river and lake were markers of environmental degradation and inequities for decades, their restoration and clean-up allowed for a concerted effort to re-brand Cleveland as "a green city on a blue lake"[12] and has significantly changed the perception residents and developers hold about the lake and the river, now seen as assets (Samanta and Kellogg 2019).

As this greening process unfolds along the river and the lake, greener or soon-to-be green ethnically diverse neighborhoods like Tremont, Ohio City and Detroit Shoreway (Detroit Shoreway Corporate Development Organization—D.S.C.D.O. 2009) are experiencing a resurgence through high-end housing developments and new environmental investments, particularly trails designed to link the neighborhoods together and to the Ohio Towpath Trail which leads to the Cuyahoga Valley National Park and to Lake Erie. Tremont hosts the South Cleveland Foundation Centennial Trail (opened in 2015), while Ohio City will receive the Irishtown Bend Cleveland Foundation Centennial Trail (awaiting a date to start construction) and the Redline Greenway (planned to be completed by 2021).[13] Detroit Shoreway will be the start of the Redline Greenway and of the Whiskey Island Connector (construction completion planned by the beginning of 2021). These new projects in Detroit Shoreway will be an addition to the Lakefront bikeway connector (completed in 2019), the improvements to Miriam Ortiz-Rush Park (inaugurated in 2019) and the dramatic green improvement that the neighborhood has experienced since Metroparks (the organization administering most of the Cuyahoga County natural areas and some city parks) began managing Edgewater Park on Lake Erie in 2013.[14] Now a new question emerges: are those green developments at risk of perpetuating historic racial inequalities in Cleveland or are they opportunities for sustainable and just urban futures?

Revitalization and greening in the Detroit Shoreway neighborhood

Detroit Shoreway is a mostly non-Hispanic white (49%) neighborhood, with a relatively high population of Blacks and Hispanics (23% each) (Cedar Lake Ventures 2018). While the racial diversity of the neighborhood is often cited as a major asset, poverty rates[15] exceed the citywide average, with 39% of residents living below the federal poverty line. Moreover, the greening-related developments of the neighborhood are far from equitably distributed across the neighborhood in terms of presence and quality.

In the northern section of Detroit Shoreway, a post-industrial area historically home to many Italian immigrants, the new Gordon Square Art District[16] has spawned an array of cultural and gastronomic offerings. The area has also recently seen new luxury housing (D.S.C.D.O. 2009) next to the restored Edgewater Park. It is important to state that Edgewater Park,[17] once riddled with crime, is now one of the city's main attractions, complete with lakeside beaches, boat ramps, a fishing pier, picnic areas and a beach house[18] with food trucks and live events. This northern area mostly attracts wealthy white couples and dissuades longtime residents due to their lower purchasing power. Moreover, its popularity is in one way or another affecting all the neighborhood residents. Many of the neighborhood residents we interviewed explained how close friends and family have not been able to find a home they can afford and how they have felt socially and physically displaced from certain areas.

> In the five years since my son bought his house [in Detroit Shoreway neighborhood], there's no house in this [Detroit Shoreway] neighborhood

FIGURE 5.1 The Edison at Gordon Square development is one of the new luxury housing developments in the northern section of Detroit Shoreway neighborhood. (Photo by: Margarita Triguero-Mas.)

FIGURE 5.2 People relaxing, napping and enjoying views of Lake Erie in the northeastern part of Edgewater Park. (Photo by: Margarita Triguero-Mas.)

> that they [my daughter and her partner] could afford. And they're not, they're not super low income ... they had to go to Cleveland Heights, you know the next, I don't know if you've been around the East Side but it's the first suburb out from Cleveland.
>
> *(Detroit Shoreway resident)*

The Detroit Shoreway neighborhood is close to downtown and motorized high-speed roads, and has sit-down restaurants and shops, all assets that characterize the attractiveness of Detroit Shoreway. However, for-profit developers highlight and use Edgewater Park as the core selling point in their real estate developments to attract residents to Detroit Shoreway. One of them highlights the green value of Battery Park for residents looking for a more dense, proximate-to-downtown "sustainable lifestyle":

> You're also seeing a lot of people who want to be close to the city, you know you can walk, it's walkable, accessible. In this specific location, you're on the water, I mean you really can't get any closer to Lake Erie, then you're also 10 to 15 minutes to downtown Cleveland ... [From] Battery Park you can walk to the Gordon Square area, you can walk to restaurants, you

> can walk to Capitol Theater, there are theaters, bars, restaurants, retail and people are really attracted to the walkability and accessibility. Those are two words that I use a lot because to be able to drive to the city in 10 minutes is a beautiful thing. To walk to Edgewater is a beautiful thing as well because there are walking paths over here, so you can actually get to Edgewater Beach.
>
> *(Developer in Detroit Shoreway neighborhood)*

The EcoVillage[19] is a different, equity-centered type of green redevelopment in the Detroit Shoreway neighborhood (D.S.C.D.O. 2009). Located in the neighborhood's southeast section, the EcoVillage is a pedestrian-friendly, Transport Orientated Development (T.O.D.) that began in 1996 (D.S.C.D.O. 2009). The EcoVillage was explicitly designed to promote social integration and sustainability (Kellogg and Keating 2011). Today, area residents have access to a 20-acre greenspace—the Michael J. Zone Recreation Centre—featuring walking paths, playgrounds, a dog park and outdoor and indoor sports fields. They also have access to a commuter train station (the West 65th–Lorain Rapid Station) and bus lines that allow them to easily reach their workplaces within 20–40 minutes.

FIGURE 5.3 New development announcement at Detroit Shoreway neighborhood. (Photo by: Margarita Triguero-Mas.)

Many new residents—mainly white and middle-income—have moved to the EcoVillage, explicitly because of their environmental values. While some security problems still exist (such as theft, drug-related and violent activities), white, Black and Hispanic populations are integrated and low- and middle-income residents coexist without any major cultural, social or physical displacement. The EcoVillage is still widely characterized as a diverse working-class and middle-income area. In view of the new type of segregation and gentrification that the development in the northern part of the neighborhood near Edgewater Park is generating, the EcoVillage stands out as a distinct example of inclusive green development.

> [The EcoVillage has made] a tiny difference [in the type of neighborhood residents]. Not a huge difference. I think the biggest [difference was] the new construction housing that was built as part of the EcoVillage [and that] definitely attracted buyers that would otherwise not live in the neighborhood. So it was a small transition. Or a small change ... because they built, as part of the EcoVillage ... maybe 20 homes over a period of 10, 12 years. [But] apart [from this, the other EcoVillage related programs] were education awareness activities, [such as] giving everybody on a certain block, replace all the light bulbs with CFLs ... So the EcoVillage is very mixed, with white, Black and Latino people.
>
> *(Detroit Shoreway resident)*

FIGURE 5.4 New housing in the EcoVillage in Detroit Shoreway neighborhood. (Photo by: Margarita Triguero-Mas.)

One of the successes of the EcoVillage project has been its emphasis on and inclusion of its least privileged residents. For example, small houses have been added to empty lots rather than transforming existing houses; the project planning has focused on envisioning the future of a community with heterogeneous educational levels, classes and ethnic backgrounds and incomes; and programs have emphasized offering residents access to capital and easy commutes to jobs (Bratman et al. 2018). The EcoVillage has also secured improvements to the transit station and the neighborhood park and strengthened weatherization projects for low-income residents. The incremental nature of the initiative has meant that even some EcoVillage residents and community organization workers are not aware of its expansion, which may partially explain why the project has not generated resentment among residents.

Cleveland's EcoVillage stands out as an exception among many ecovillages of the Global North that are often intentional communities built for and by white, college-educated, mostly middle and middle-upper middle-class residents able to afford high living costs (Accioly-Dias et al. 2017). In Cleveland, however, the Detroit Shoreway EcoVillage tells a different story, perhaps due to a slower real estate market and recovery of Cleveland after the 2008 crisis, where the number and type of new investments have remained more limited than in coastal and southern U.S. cities.

Apart from the EcoVillage and the developments close to Edgewater Park, two more areas with different characteristics coexist in the neighborhood. In both the northwest edges (where mainly Latino and white residents live) and the southwest (where Black residents are concentrated), investment has remained scarce. These two areas contain a high percentage of vacant lots, abandoned properties and more street litter than other parts of the neighborhood. In sum, while part of Detroit Shoreway is (or has been) redeveloped, some of it with a core focus on equity, the revitalization has not been homogenous.

Civic and policy efforts for an inclusive future in Cleveland

Aware of its challenges in regard to more intensified gentrification, the city of Cleveland is currently discussing racial equity needs and policies[20] to avoid displacement. Exhibitions[21] and training about structural and institutional racism, for example, have been organized in the city. As part of the Equitable Community Development Strategy of Cleveland, a study on Cleveland's property tax policy and abatement program has been requested to explore how existing policy can protect long-term residents from displacement while promoting development and taking into account differences between neighborhoods and a broad range of factors like transit and jobs.[22] Pro-equity proposals must adhere to the city's Equitable Community Development Strategy,[23] which states "the need for equitable economic development programs" given that "traditional approaches to economic development by local governments have not benefited all populations" (Councilman of Detroit Shoreway area Matt Zone 2019).[24]

Many of these municipal efforts are being further advocated for by community organizations focused on anti-displacement and housing affordability and rights. For example, since the 1987 (D.S.C.D.O. 2009) federal Low-Income Housing Tax Credit program was created, the local D.S.C.D.O. has been working to provide local affordable housing options, including 300 rental units in 16 buildings (D.S.C.D.O. 2009).[25] In fact, the D.S.C.D.O.'s history of providing a range of housing types and affordability and being an active, respected organization across the neighborhood is one of the reasons why the EcoVillage was sited in the Detroit Shoreway neighborhood (Kellogg and Keating 2011).

The most recent D.S.C.D.O. effort to provide local affordable housing in the EcoVillage is Aspen Place,[26] a development owned, built and managed by D.S.C.D.O. Inaugurated in January 2019, it offers 40 affordable rental housing units for a minimum of 30 years,[27] targeting residents with 30–60% of the area median income (A.M.I.). As a Transit Oriented Development (T.O.D.) project, Aspen Place is located next to the R.T.A. rapid public transport Lorain Station and across the street from the Zone Recreational Centre. Among other "sustainable living" amenities, Aspen Court includes indoor bike storage, gardens and reduced pricing for the R.T.A. train. Unlike many other examples in the United States, this T.O.D. is built on a social and racial equity approach as a way of "maintaining racial and economic diversity, representation of diverse people and opinions, and the power of everyday people to shape their own communities" (D.S.C.D.O. 2009).

FIGURE 5.5 A view of Aspen Place. (Photo by: Margarita Triguero-Mas.)

At the same time, several measures are being taken to ensure a more equitable distribution of green spaces in Cleveland. Supra-municipal green space management organizations like Metroparks are trying to be more present in the city, given that in the last decades suburban residents have had greater access to more and better-quality green spaces than residents from the inner city where green space and its maintenance funding is limited.

> In our last strategic plan in 2012 we made it a priority to become more involved in Cleveland and the inner ring suburbs. So more kind of the urban core of the county because we didn't have a really strong presence there prior to that … Because the way it turned out the suburbs kinda grew up around the parks that were initially rural areas as developments spread out, then those parks became surrounded by communities and so those folks are really well served, but then people that live in the older more established communities weren't really served by Metro Parks.
>
> *(Representative of Metroparks)*

Meanwhile, the Cleveland Department of Parks, Recreation and Properties is strategizing how to serve the greatest number of residents and those who need green spaces to be improved the most. Through the 2016 Cleveland Tree Plan, the city is trying to overcome the big differences in tree canopy cover between different areas of the city, seeking to reduce the urban heat island effect in the poorest neighborhoods that typically have the lowest canopy cover. This plan follows the 2013 Cleveland Climate Action Plan that engaged hundreds of residents and stakeholders. Both these plans—part of the mayor's Sustainable Cleveland 2019[28] program—focus on bringing the health benefits of environmental amenities to residents of Cleveland's neighborhoods.

Strengthening equity planning and access to quality environments in the wake of more developments

The current challenge facing Cleveland is whether recent conversations and initiatives around development and equity will bring actions quickly enough to ensure that Cleveland's vision of a "green city on a blue lake" does not perpetuate or exacerbate existing racial and social inequities. There is indeed a risk that Cleveland will follow the path of similar revival cities such as Philadelphia or Washington where the current rates of redevelopment and gentrification-driven displacement are some of the highest in the country. Moreover, associated challenges include taming green gentrification spurred by the green branding of new developments and ensuring that sustainably focused projects like the Detroit Shoreway EcoVillage do not become drivers of gentrification themselves.

Note

Some of the text in this book chapter was used as a base for "Connolly JJT, Anguelovski I. 2021. Three histories of greening and whiteness in American Cities. Front.Ecol.Evol."

Keywords

(a) the urban development pattern of the city and neighborhood industrial legacy; historical environmental degradation; historical decline; spatial racial segregation; deep poverty; nascent recovery; arts-based revitalization; livability planning
(b) the urban greening of the neighborhood/city: river contamination remediation; greening projects and development; EcoVillage; greenways; new parks
(c) the inequalities at stake: green gentrification; environmental privilege; luxury developments

Notes

1 https://www.clevescene.com/scene-and-heard/archives/2017/12/18/gentrification-what-it-means-in-the-context-of-the-rust-belt
2 https://www.politico.com/magazine/story/2015/02/cleveland-segregation-115320
3 https://www.clevescene.com/scene-and-heard/archives/2019/04/19/study-cleveland-continues-to-experience-significant-neighborhood-economic-decline-little-measurable-growth
4 https://case.edu/ech/articles/i/industry
5 https://www.clevescene.com/scene-and-heard/archives/2017/12/18/gentrification-what-it-means-in-the-context-of-the-rust-belt
6 https://crln.acrl.org/index.php/crlnews/article/view/17587/19405
7 https://www.nytimes.com/2019/06/22/climate/cleveland-fire-river-cuyahoga-1969.html
8 http://neocando.case.edu/neocando/
9 http://neocando.case.edu/neocando/
10 http://www.ccbh.net/wp-content/uploads/2017/09/2016-2017-Cuyahoga-County-Community-Food-Assessment-Report_Pt-1.pdf
11 https://www.clevescene.com/cleveland/clevelands-population-could-hinge-on-keeping-or-attracting-young-people-once-they-start-having-kids/Content?oid=19742686
12 https://www.ecowatch.com/cleveland-rocks-from-mistake-on-the-lake-to-green-city-on-a-blue-lake-1882035016.html
13 https://www.clevelandmetroparks.com and https://issuu.com/ohiocityincorporated/docs/170831_public_meeting_presentation__55e4c1eb02a061/36
14 https://friendsofedgewater.wordpress.com/2013/09/21/edgewater-park-becomes-part-of-the-metroparks-lakefront-reservation/
15 https://d3n8a8pro7vhmx.cloudfront.net/sustainablecleveland/pages/149/attachments/original/1462884616/EcoVillage_Climate_and_Sustainability.pdf?1462884616

16 http://www.gordonsquare.org/
17 https://www.cleveland.com/open/2013/02/governor_recommends_cleveland.html
18 https://www.clevelandmetroparks.com/parks/visit/parks/lakefront-reservation/edgewater-beach-house
19 https://www.dscdo.org/cleveland-ecovillage
20 http://www.clevelandnp.org/undesigntheredline/
21 Ibid.
22 https://wtam.iheart.com/content/2019-04-16-cleveland-officials-to-study-tax-policy/ and https://www.city.cleveland.oh.us/09.23.2019TaxPolicyStudy
23 https://wtam.iheart.com/content/2019-04-16-cleveland-officials-to-study-tax-policy/
24 https://www.facebook.com/gordon.detroit/posts/10214594736286244
25 https://www.dscdo.org/real-estate-development
26 https://www.cleveland.com/cuyahoga-county/2017/07/detroit-shoreway_affordable_housing_project_by_rta_station_to_get_underway_in_august.html
27 https://www.enterprisecommunity.org/news-and-events/news-releases/aspen-place-bring-affordable-homes-cleveland-ecovillage
28 https://www.sustainablecleveland.org/

References

Accioly-Dias, Maria Accioly, Carlos Frederico B. Loureiro, Leandro Chevitarese, and Cecília De Mello E de Mella Souza. 2017. «The meaning and relevance of ecovillages for the construction of sustainable societal alternatives». *Ambiente & Sociedade* 20 (3): 79–96. https://doi.org/10.1590/1809-4422asoc0083v2032017.

Bratman, Eve, Brandy Brooks, Taylor Mercedes, Josh Jancourtz, and Joel Rothschild. 2018. «Can we have communities without gentrification? Perspectives from the ecovillagers alliances». *Communities* 178: 34–37.

Cedar Lake Ventures. 2018. «The demographic statistical atlas of the United States. Race and ethnicity in Detroit Shoreway, Cleveland, Ohio». 2018. https://statisticalatlas.com/neighborhood/Ohio/Cleveland/Detroit-Shoreway/Race-and-Ethnicity.

Cuyahoga County Medical Examiner's Office. 2018. «Heorin fentanyl cocain related deaths in Cuyahoga county». 2018. http://medicalexaminer.cuyahogacounty.us/pdf_medicalexaminer/en-US/HeroinFentanylReports/100418-HeroinFentanylReport.pdf.

Cuyahoga County Planning Commission. 2017. «2016–2017 Cuyahoga county community food assessment report part 1: Current conditions». 2017. http://www.ccbh.net/wp-content/uploads/2017/09/2016-2017-Cuyahoga-County-Community-Food-Assessment-Report_Pt-1.pdf.

Detroit Shoreway Corporate Development Organization (DSCDO). 2009. «Detroit Shoreway Corporate Development Organization (DSCDO): Mission, vision, values». 2009. https://www.dscdo.org/mission-vision-values.

Kellogg, Wendy A., and W. Dennis Keating. 2011. «Cleveland's EcoVillage: Green and affordable housing through a network alliance». *Housing Policy Debate* 21 (1): 69–91. https://doi.org/10.1080/10511482.2010.533614.

Samanta, Aritree, and Wendy Kellogg. 2019. «Back to the beginning: The resurgence of the Cuyahoga River and Cleveland». In *Legacy Cities: Continuity and Change amid Decline and Revival*, edited by S. Webster and R. Tighe. Palgrave-McMillan. Pittsburgh, PA: University of Pittsburgh Press.

Stradling, David, and Richard Stradling. 2015. *Where the River Burned: Carl Stokes and the Struggle to Save Cleveland*. Ithaca, NY: Cornell University Press. Vol. 21. https://academic.oup.com/envhis/article-lookup/doi/10.1093/envhis/emw015.

U.S. Census Bureau. 2019. «QuickFacts: Cleveland City, Ohio». U.S. Census Bureau. 2019. https://www.census.gov/quickfacts/fact/table/clevelandcityohio/PST045219.

6

WEST DALLAS

The "nowhere" that became "somewhere"

Helen Cole

The long overdue clean-up of polluting industry and other neighborhood improvements in West Dallas, along with its ideal central location directly across the river from the economic center of downtown Dallas, has led to conditions ripe for gentrification. Most visible in these investments to date is the construction of the monumental Margaret Hunt Hill Bridge, opened in 2012 and touted as a catalyst for economic investment in the West Dallas area as well as a mechanism to "foster unity across the North and South divide in the city."[1] While the bridge, and subsequent luxury dining, retail and housing developments in West Dallas have served to attract visitors from the north of the city and elsewhere, the promises of unity and economic benefit of these amenities are not reaching long-time residents of West Dallas, who instead face pressures of rapidly changing real estate prices and the looming threats of disruption of their formerly isolated cultural enclaves. Furthermore, environmental hazards that have long contributed to poor health among residents remain and may be made worse by the relocation of existing heavy industry to parts of the neighborhood not yet undergoing gentrification.

Dallas' unwanted neighborhood

West Dallas consists of many small adjacent neighborhoods and is separated from the rest of the city by the Trinity River to the north and east, I-30 to the south and the West Fork of the Trinity River to the west. Prior to annexation by the city of Dallas in 1954, the area, which lacked basic infrastructure such as a sewer system and, in many homes, running water, was home to many poor white migrants from rural areas. Home to legendary criminals Bonnie and Clyde, West Dallas has also been called "the devil's backdoor" to the city. Throughout its recent history, the area has also hosted several industrial sites, including a lead

DOI: 10.4324/9781003183273-8

FIGURE 6.1 The Margaret Hunt Hill Bridge connecting West Dallas to the downtown area. (Photo by: Helen Cole.)

smelting plant, eventually owned by R.S.R. Corporation, which started operations in the mid-1930s. Many residents found work in this and other industries occupying the area. Some posit that it was the economic upturn following World War II that led many poor white residents to leave West Dallas, ultimately finding better job opportunities in the city.

Soon after annexation, the largest public housing development in the nation was built immediately across the street from the lead smelting plant and the neighboring school (Thomas Edison School), consisting of 3,500 units segregated into separate buildings for white, Black and Hispanic residents (Rodeheaver and Cutrer 1995). By the mid-1980s, over 1,300 of the original units were uninhabitable and only 1,900 families, all Black, remained, living in conditions wrought

FIGURE 6.2 A local pastor shows off his "big book," including photos of the slum conditions in his neighborhood during the 1930s. (Photo by: Helen Cole.)

with drug use and violent crime. These conditions, along with residents' claims that the housing authority had misused federal funds and purposefully prevented Black residents from moving to wealthier white parts of the city, led to the 1990 Walker vs. H.U.D. discrimination case,[2] and ultimately to the demolition of the housing project and implementation of a voucher program meant to allow public housing residents to move to housing of their choice.

Racial diversity and tensions in West Dallas

In West Dallas, boundaries between small adjacent neighborhoods also highlight racial divides in the community. By the time West Dallas' Pinkston High School, originally a Black school, was desegregated in the 1970s, many whites had left the neighborhood for wealthier neighborhoods in the northern part of the city or the suburbs, and by 1980 approximately 85% of residents identified as Black (Rodeheaver and Cutrer 1995). According to one respondent, it was the necessity to consider Hispanic residents "white" in order to desegregate the school that led to the elevation of the social position of the Mexican American community. The demolishing of public housing and mass destruction of many private homes in the early 1990s via selective enforcement of building codes, part of an effort to revitalize the city after the significant economic downturn of the 1980s, forced many Black residents out of West Dallas (Phillips 2006). While several of the area's neighborhoods, such as Muncie near the bridge and Westmoreland Heights across from the former lead smelting plant, remain predominantly African American today, West Dallas as a whole is now approximately 70% Hispanic.

FIGURE 6.3 The Trinity River as seen from below the Margaret Hunt Hill Bridge. The river separates the neighborhood from downtown Dallas, while industrial businesses line the opposite side of the neighborhood abutting highways and railway lines. (Photo by: Helen Cole.)

Tensions between racial/ethnic groups in West Dallas contribute to contrasting ideas about how best to mobilize around both existing neighborhood conditions and the increasing pressures of gentrification. Neighborhood associations for each of the small West Dallas communities tend to follow the racial/ethnic make-up of the respective neighborhoods, yet the geography of West Dallas and a sense of togetherness in the fight for justice also hold the area's diverse communities together. West Dallas' historic geographic isolation has provided the benefits of a tight-knit community, in many cases formed during childhood: "I had that bond, there was that trust, even the people that I would fight with when I was younger, they still trusted me," said one Mexican American resident. Successes of grassroots efforts led by area leaders of both races, often working together with (largely white) activists from elsewhere in the city, include the establishment of Los Barrios Unidos Community Clinic in 1972, the closure of the lead smelting plant in 1984 and the designation of a Superfund site in its place in 1995 leading to subsequent clean-up efforts.

Conflicting belongings along the Trinity River

The Trinity River, repositioned in the 1920s[3] to protect downtown from flooding, is an important feature of West Dallas but has also been the site of many instances of conflict and violence. For instance, in 1860, at the time of substantial conflict between abolitionists and slave owners, three Black men were falsely accused of starting a fire that destroyed downtown and were hanged on the banks of the Trinity River. Although a park was established near the site of the hangings bearing the name "Martyr's Park," no information is given about this history and the city has not memorialized such instances of racial violence and conflict.

Many have pleasant memories of the river as well. In recent decades, many residents recall fishing in the river and picnicking in the vast surrounding unmanicured green spaces when the river was not flooding. The wide area between the levies built to contain flooding also housed vast fields which were heavily utilized by residents for playing soccer and other sports. Although this green space and several walking and biking paths remain in the flood plain, new railings create the illusion that access is forbidden, as one life-long resident explained:

> Just the message of putting those poles there, like we're not supposed to be up here ... I mean, even though we will always know it's a river, always knew there's a flood plain, to us it's always been a park.

For this resident, there is a sense that visitors of color do not belong there.

Today, residents associate the river with the city's unfulfilled promises of planning and infrastructure. In the mid-1800s, plans to connect Dallas to the gulf, allowing for the transport of goods, eventually proved to be infeasible. Since the 1910 Kessler Plan, the original comprehensive city planning document, visions for the river have emerged and died without progress. In 2011, the city spent 4 million dollars to install artificial rapids for recreational boating which were shut down in a matter of days due to safety concerns, and several years later spent 2

FIGURE 6.4 Barriers installed along the levies create the illusion that access is forbidden. (Photo by: Helen Cole.)

million dollars to have the project partially removed.[4] This was yet another incident that respondents mentioned as an example of mismanagement of city funds, and of the river itself. Thus, it is not surprising that after decades of proposed plans for the Trinity River, many residents cited a feeling of disbelief that the current plan to create an extensive park on the river would finally be realized.

To revitalize the river, promote economic development and connect the historically segregated north and south sides, the current plan included in its first phase the construction of two monumental bridges, designed by Spanish architect Santiago Calatrava.[5] Interviewed residents unanimously agreed that it was the construction of the Margaret Hunt Hill Bridge in particular—which now provides a grand entrance to their formerly overlooked neighborhoods—which ultimately sparked both an understanding that the plans may in fact come to fruition despite widespread disbelief, and concern about the sudden changes observed with the Trinity Groves development and beyond. The bridge was referred to by some as "the bridge to nowhere" due to the relative lack of familiarity with the undervalued West Dallas area, as one city representative working in the area explained: "Yeah, and I don't think that has ever really gone over well with residents. Really think about if somebody called your home 'nowhere'."

Environmental injustice continues in the wake of gentrification

The extreme environmental degradation of West Dallas and ensuing fight to close the R.S.R. lead plant and obtain a Superfund designation have often been cited as a key example of environmental injustice (Bullard 2000; Bullard and Wright 1990). Since the opening of the bridge in 2012, dramatic changes such as the Trinity Groves development and many new luxury townhome developments remind residents that gentrification is imminent, despite lingering environmental hazards. These developments and subsequent neighborhood improvements, many of which were long fought for by residents, have also led residents to feel excluded and ignored, sometimes by their own representatives. For instance, in 2015, the area's city councilperson voted to relocate the Argos concrete plant,[6] which formerly stood at the base of the new bridge adjacent to what is now Trinity Groves, and to approve the use of 2.5 million public dollars to build a rail spur to the new location. Many residents were upset by this decision because the concrete plant was relocated not outside the neighborhood, but rather further into the center just off of the intersection of two main streets: Singleton Blvd and Westmoreland Rd, directly behind Thomas Edison School and the neighboring Superfund site. Although the new plant emits fewer pollutants due to new technology, its siting seems to ignore the wellbeing of residents. Two years later, the school was closed due to poor performance, rather than due to the unjust exposure of children to lead and other pollutants that the scenario clearly produces.[7] Despite blood lead levels (not an ideal measure for long-term exposure and impact) showed reduced exposure by 2003, high rates of asthma, diabetes

FIGURE 6.5 The former Thomas Edison School on the left, the new concrete plant on the right and the former location of the lead smelting plant, now a Superfund site, in the foreground. (Photo by: Helen Cole.)

and high blood pressure were and still are likely linked to the complex environmental exposures in the neighborhoods (Walsh, Pickens and Royster 2003).

Greening as a business investment or a resource for residents?

In recent decades, Dallas lost a few notable bids for corporate headquarters and other business deals (Boeing headquarters, the 2012 Olympic Games and, most recently, the second Amazon headquarters), ultimately inspiring city leaders to re-examine the attractiveness of their city (Graff 2008). Various proposals included the construction of extensive new parks and improving connectivity between green spaces, based on the logic that improving quality of life for residents with new greening would bolster their competitive edge by providing amenities that would also attract wealthy, white-collar workers to the city. Many of these plans have been backed by massive private donations such as the $60 million donation by the Simmons family for the construction of a new Harold Simmons Park[8] (in honor of Harold Simmons, a wealthy Dallas businessman who partially made his fortune through the lead industry[9]), slated to open in 2022 on the Trinity River alongside several West Dallas neighborhoods. This and other recent projects such as the Katy Trail (2000) and the Klyde Warren Park (2012) are touted as attractive amenities first and foremost for international investors, potential new company headquarters and regional visitors, and only secondarily for local residents.

In 2017, a new agency, the Trinity Park Conservancy, was formed to oversee the design and implementation of the current plan. Led by a well-known local architect and urban designer, this new entity has made community engagement and participation key in their process. Many residents spoke of their admiration for this new organization and its staff, despite also feeling relatively uninterested in the park itself, perhaps due to its disconnect from the concerns of many residents of color, such as the need for basic facilities like a grocery store, the presence of drug dealing and displacement due to rapid changes in their neighborhoods. Residents seemed more appreciative of improvements to existing neighborhood parks, such as a new splash pad in one small park. In a few instances, residents attributed their disengagement with the planning of Harold Simmons Park to a feeling that the "park is not for us." In other cases, residents were not aware of the plans for the new park or thought of it as another plan that would fall through, in keeping with the fate of past plans along the Trinity River.

Interviewees unanimously associated the rise of gentrification with the construction of the "bridge to nowhere." As evidenced by the relative lack of engagement and interest among residents, the question of who benefits from the new park is secondary, or tertiary, to the immediate survival and struggle of residents to maintain their vibrant communities in the face of new developments in and around the area. Respondents suggested that many of the gentrifiers were not even from Dallas, but younger residents from other cities around the country in search of work. They also claimed that ironically, many wealthy Dallasites still avoided living there due to long-held stigmas about the area. This gentrification process runs parallel to the fight for necessary resources and infrastructure

FIGURE 6.6 A new art district located inside and along local warehouses owned by West Dallas Investments LP, which also owns Trinity Groves. (Photo by: Helen Cole.)

which the neighborhoods still lack, and perhaps one reason that Dallasites may not see West Dallas as a desirable location despite the new amenities. While new construction brings improvements to infrastructure—either paid for in full by developers or through subsidies from the city—these improvements largely benefit new residents.

Leveraging community to address gentrification pressures

Several interviewees expressed hope that their children or grandchildren would one day be able to afford to live in one of the new condominiums. They expressed desire to protect their own families and neighbors from displacement and to leverage incoming economic opportunities for the benefit of their children rather than newly arriving wealthier residents. Due to limitations in political will under restrictive Texas state laws which prohibit many municipal level interventions, several West Dallas neighborhoods are looking to localized solutions such as zoning interventions used in other parts of the city to protect against displacement.

One specific public intervention being explored is the neighborhood stabilization overlay scheme which allows residents of single-family zoned neighborhoods to place restrictions on the type and size of structures built in order to preserve the architectural character of the neighborhood. One activist interviewed credited such a plan with preventing gentrification in East Dallas for over 20 years before the "economics became so mega" that the plan could no longer compete with development pressures. In East Dallas during the 1980s, residents would have never predicted the gentrification that eventually won out, simply because they perceived it as culturally impossible that wealthy residents would ever want to live in the area. Thus, the timing is perhaps different in West Dallas where it has already become clear that wealthier people are indeed interested in living in the area due to its central location and new amenities, although many feel that the new residents are predominantly new to Dallas and therefore don't hold the same disdain for the area as the city's "old rich" might. To date, La Bajada,[10] one of West Dallas' neighborhoods adjacent to the Trinity Groves development, has enacted this plan, while others are considering it.

A second scheme being discussed in West Dallas is the use of historic district designations. In nearby Tenth Street, a neighborhood in South Dallas, residents are currently fighting to keep such a designation that they argue could help protect the neighborhood from gentrification by preventing older buildings from being demolished and turned into larger condominium developments or luxury homes.[11] When the Tenth Street neighborhood, a historic freedman's town (a neighborhood or town originally settled by former slaves emancipated during or after the American Civil War) and a redlined neighborhood during the Jim Crow era, was cut off from the rest of the Oak Cliff area by the construction of I-35 in the early 1950s, its zoning was designated as industrial. Over time as historic homes and structures fell into disrepair, their zoning status as "not appropriate use" prohibited their upkeep and many were allowed to be demolished.

In 1984, the zoning was changed to "Future Growth Corridor," allowing for a mixed-use of commercial, retail, manufacturing and townhomes, but many homes were already in disrepair from the decades of neglect under industrial zoning.[12] In 2010, a new city ordinance made the destruction of older properties in disrepair even easier, and a clause that protects homes over 3,000 square feet,[13]—added at the insistence of several wealthier and predominantly white historic districts—did little to protect the small historic homes in Tenth Street (and West Dallas) from demolition. Under these policies, an over 25% reduction in the number of homes left in the neighborhood has led to an effort to remove the historic designation of the neighborhood using the argument that not enough historic structures remain to merit such a designation.

In 2018, two municipal level measures were taken: (1) a new housing policy to encourage the construction of new affordable housing, preserve existing affordable housing and provide assistance to low-income homebuyers and (2) the addition of an Office of Equity and Human Rights to oversee equity issues for the city. In addition, several local lawyers and other businesses have partnered to assist homeowners in West Dallas with obtaining clean property titles. Property titles are necessary to apply for tax incentives, sell property or apply for loans, but are often a challenge to obtain, especially in cases where property is inherited by one or multiple family members, as is common among families in West Dallas. Developers and others argue that rising property values should benefit local homeowners who could sell their property at a significant profit, but obtaining clean titles is essential to realizing any potential benefit. In any case, residents questioned the long-term economic benefit of selling their homes for a profit, as this would entail moving away from their own communities and not guarantee higher-paying employment needed to maintain the new, improved lifestyle that developers were promising.

New environmental struggles

West Dallas is a clear example of a community which, despite its own divisions, has fought for environmental justice over decades. Long-fought, successful battles include the closure of the lead plant and the ultimate establishment of a Superfund site in the mid-1990s. Yet these accomplishments have not led to systemic changes that would prevent further injustices. In fact, recent developments including the new bridge, providing direct access to the previously isolated area from downtown Dallas, and subsequent commercial development and new luxury housing that significantly contrasts with the character of West Dallas neighborhoods have led to the continued unjust exposure of long-term residents to environmental hazards such as air pollution and pollutants. Now the plan for an expansive regional park along the waterfront jeopardizes the connection felt by long-term residents to the riverbank's untamed greenspace and threatens local communities with the economic burdens of rising property taxes and rents and cultural changes that implicate the erasure of rich local cultures.

Keywords

(a) the urban development pattern of the city and neighborhood: fast recent growing; no spatial boundaries to growth; lack of public infrastructure in historically marginalized areas; commercial redevelopments
(b) the urban greening of the city and/or neighborhood: new mega-park; green resilient infrastructure; new iconic bridge
(c) the inequalities at stake: continued exposure to contamination; green gentrification; luxury redevelopment; transportation infrastructure and new connections to downtown; lack of equity policies

Notes

1 As stated on the Trinity River Project website overview of the Margaret Hunt Hill Bridge: https://www.trinityrivercorridor.com/transportation/margaret-hunt-hill-bridge
2 For details on the case and a description of conditions, see the case report for Walker vs. the U.S. Department of Housing and Urban Development here: https://casetext.com/case/walker-v-us-dept-of-housing-and-urban-dev
3 The tale of the Trinity River can be found here: https://www.dmagazine.com/publications/d-magazine/2017/march/wild-dallas-trinity-river-guide-dallas/taming-the-trinity/
4 https://www.dallasnews.com/news/2017/06/14/dallas-will-spend-2-million-to-remove-4-million-whitewater-feature-from-trinity-river/
5 https://www.designboom.com/architecture/santiago-calatrava-margaret-mcdermott-bridge-ih30-dallas-texas-09-07-2015/
6 https://www.dallasobserver.com/news/despite-protests-council-oks-cash-for-concrete-plant-to-move-near-school-7726782
7 http://stories.kera.org/saving-schools/2018/05/07/dallas-school-to-close-even-though-district-tried-to-save-it/
8 https://trinityparkconservancy.org/the-park/
9 https://www.dmagazine.com/publications/d-magazine/2010/february/harold-simmons-is-dallas-most-evil-genius/
10 The ordinance documenting the neighborhood stabilization overlay of La Bajada can be found here: https://dallascityhall.com/departments/sustainabledevelopment/DCH%20documents/pdf/nso_LaBajada-Z112-220.pdf
11 https://www.aiadallas.org/v/columns-detail/The-Road-to-Disinvestment-How-Highways-Divided-the-City-and-Destroyed-Neighborhoods/pt/
12 https://candysdirt.com/2019/05/30/positive-news-for-once-in-tenth-street-historic-district/
13 http://dallas-tx.elaws.us/code/dallas/51a-4.501/

References

Bullard, Robert D. (Robert Doyle). 2000. *Dumping in Dixie: Race, Class, and Environmental Quality*. Boulder, CO: Westview Press.

Bullard, Robert D. and Beverly H. Wright. 1990. "The Quest for Environmental Equity: Mobilizing the African-American Community for Social Change." *Society and Natural Resources* 3(4):301–11.

Graff, Harvey J. 2008. *The Dallas Myth: The Making and Unmaking of an American City*. Minneapolis, MN and London: University of Minnesota Press.

Phillips, Michael. 2006. *White Metropolis: Race, Ethnicity, and Religion in Dallas, 1841–2001.* Austin, TX: University of Texas Press.

Rodeheaver, Daniel G. and Jennifer G. Cutrer. 1995. "Environmental Equity and the Case of West Dallas." In *An Aging Population, An Aging Planet, and a Sustainable Future,* 203–33, edited by S. R. Ingman, X. Pei, C. Ekstrom, H. Friedsam, and K. Bartlett. Denton, TX: University of North Texas Press.

Walsh, Brad, Sue Pickens, and Marcene Royster. 2003. *Final Report, Dallas Lead Project.* Dallas, TX. Retrieved from https://www.dshs.state.tx.us/epitox/tehiprojects.shtm (accessed July 30, 2021).

7

LAND REMEDIATION IN GLASGOW'S EAST END

A "sustainability fix" for whose benefit?

Melissa García-Lamarca and Neil Gray

Following an industrial boom from the mid-to-late 19th century, Glasgow's East End underwent exceptional levels of industrial decline. By the 1960s, it suffered from wholesale abandonment and devaluation, visible through widespread swathes of vacant and derelict land and decrepit building structures. After several unsuccessful regeneration attempts over the decades, in 2007 Glasgow City Council (G.C.C.) won the bid to host the 2014 Commonwealth Games in the East End. In 2008, the same area was subject to the largest regeneration project in Scotland—Clyde Gateway[1]—rooted in sustainability discourses and the provision of new green and blue infrastructure. Clyde Gateway has invested hundreds of millions of public funds across 840 hectares of land, 350 hectares of which was defined as surplus, vacant, derelict, polluted or in need of substantial infrastructural investment. Here we explore whether substantial benefits from regeneration are in fact trickling down to the local community through the measures being implemented, or whether the "sustainability fix" merely operates to legitimize and accommodate the contradictory impulses of profit-making urbanism and environmentalism (While, Jonas and Gibbs 2004; Temenos and McCann 2012). In essence we ask: who are the new businesses, jobs, homes and green-blue infrastructure really for, and at what cost?

From urban decline to new urban frontier?

Glasgow's East End was once a major industrial powerhouse, but the "unplanned, unregulated operation of free enterprise" tended inexorably to industrial contraction, capital flight, mass unemployment and accelerated urban decline (Middleton 1987, 13). Extensive industrial and chemical industries brought great wealth to owners in the "Second City of Empire," yet workers suffered from stark exploitation, toxic working conditions, cyclical unemployment and insalubrious

DOI: 10.4324/9781003183273-9

living conditions. A legacy of deteriorating industrial infrastructure, contaminated vacant and derelict land and under-maintained slum housing became an intractable barrier to new investment. By 1957, several areas in the East End became part of the city's Comprehensive Development Area (C.D.A.) scheme established to address slum housing and over-population issues and which led to the demolition of 95,000 homes citywide by 1975. But a dawning recognition of the enormous socio-economic, environmental and psychological costs of the C.D.A. scheme––in conjunction with overspill planning to fund new residential settlements called New Towns––led to its curtailment in the early 1970s (Pacione 1995). Once tightly woven together in factory and housing settlements, inner-city urban communities and workplaces in the East End were left abandoned amid multiple brownfield gap sites as people flowed outwards to new towns and suburbs (McDonald 1987).

By the mid-1970s, Glasgow's East End was vilified as "the most striking example of metropolitan decline in the United Kingdom" (Wannop and Leclerc 1987, 70). In response, the multi-agency Glasgow Eastern Area Renewal (G.E.A.R.) project (1976–1987) was established, a precedent to other influential U.K. inner-city urban renewal projects. In what became the G.E.A.R. area, corresponding closely to the current Clyde Gateway project area we examine in this chapter, the population contracted from 145,000 in 1951 to 41,000 in 1981, with G.E.A.R. doing little to arrest this decline (Middleton 1987). Between 1976 and 1984, overall unemployment in the G.E.A.R. area almost doubled, with male unemployment in Bridgeton and Dalmarnock higher than anywhere else in the region at the end of the project (McArthur 1987). G.E.A.R. had more success regarding housing: 4,000 tenements were rehabilitated, 8,000 interwar houses modernized and 2,000 council and 2,000 private houses constructed. Yet, these gains must be placed in context, since the C.D.A. demolition program contributed to a reduction of houses in the G.E.A.R. area from 28,500 in 1971 to 15,049 in 1981. Overall, the legacy of G.E.A.R. was "an unstructured patchwork of housing, industrial, and recreational activities" amid vast swathes of vacant and derelict land (Reed 1999, 212). Major pump-priming public funding attracted scarce private investment, foreshadowing Clyde Gateway's regeneration project.

By 2006 in the administrative ward of Calton in Eastern Glasgow, set to be the principal site of Commonwealth Games 2014[2] and the Clyde Gateway regeneration activity, 99.4% of the population lived within 500 m of a vacant or derelict site, compared to a Scottish average of 26.6% (Gray 2008). As a result of this and other factors, East End residents have notoriously poor health conditions and low life expectancies. Today, there is still a 15-year gap in male life expectancy between the wealthiest West End neighborhoods and the most deprived East End neighborhoods (Nixon 2016; Cowley, Kiely and Collins 2016), with public health researchers coining the term the "Glasgow effect" to indicate excess mortality in Glasgow compared to that of other similar post-industrial cities in the U.K. The combined factors of deindustrialization, mass unemployment, slum housing, urban restructuring and overspill policy make the East End a prime example of their effects. Add to this the reality of very poor access to fresh and healthy food, something that has only started to change in the past few years with the operation of an open-air market in central Calton a few days a week. A Calton local explained how the deep stigmas that accompany multiple forms of deprivation in the East End have been internalized by many residents over generations:

> The more that you hear from external sources how deprived and poor and how we're all undereducated … you know the health stats are so bad and all of that might be true, but if that's all that people see and all that gets any publicity then I think that has a massive impact on well-being.

Such mental and physical health problems have been exacerbated by numerous stigmatizing discourses around the East End, especially in the period preceding the Games and the Clyde Gateway project. Conservative politicians, policy think-tanks and the media have insistently conflated the local population with environmental characterizations of the East End as redundant, decayed and worse (Gray and Mooney 2011). Such discourses have material effects as well as symbolic ones. Accusations of decay, blight and obsolescence have been mobilized to legitimize publicly funded yet privatized regeneration plans in the area (ibid.) and related exceptionality measures in planning (Gray and Porter 2015).

FIGURE 7.1 Derelict land in the East End: view from the Athlete's Village toward Dalmarnock train station. (Photo by: Melissa García-Lamarca.)

The East End has thus been characterized as a depopulated and de-civilized "new urban frontier" necessitating extensive regeneration and justifying substantial public funding for essentially private economic and real estate development (Gray 2008; Gray and Mooney 2011).

Enter Clyde Gateway, whose original objectives in 2008 included the creation of 21,000 new jobs, 10,000 new housing units, 20,000 population increase, 400,000 square meters of business space and £1.5 billion of private-sector investment. The remediation of vacant and derelict land is central to these ambitious plans, with a (green) "infrastructure first" approach to create the conditions for profitable investment. By 2015, Clyde Gateway had remediated 208 hectares of vacant and derelict land out of a 350-hectare target. In the next two sections we explore key sites of remediation and regeneration, examining the costs and socio-environmental outcomes thus far.

Remediating land: for whom and to what end?

Athletes' Games Village: one of GCC's first major interventions in the East End, in partnership with Clyde Gateway and the Commonwealth Games 2014 Organizing Committee, was the 700-unit Athletes' Village[3] in Dalmarnock along the banks of the River Clyde. A second residential phase was planned but is yet to be developed. The site's environmental credentials include sustainable drainage systems (S.u.D.S.) infrastructure, a new district heating system and eco-housing. On top of a costly and controversial process of deal-making and land assembly, estimated at around £30 million in total (Gray and Porter 2015, 2017), the public cost of remediation on the 35-hectare site was substantial, including £7.7 million funding from G.C.C. via the Scottish Government's Vacant and Derelict Land Fund (V.D.L.F.) scheme. City Legacy Consortium, the public-private partnership who won the bid to construct the Athletes' Village, were then gifted the site at nil cost by its G.C.C. partner.

A prominent legacy of socio-environmental injustice in the Village area is the uneven implementation of the exceptional compulsory purchase powers granted in Section 42 of the Commonwealth Games Bill.[4] Private developers were exempted from compulsory purchase powers, but they were applied in a blanket fashion on long-term residents and local shopkeepers. One family in private housing was subjected to violent eviction[5] by more than 100 police officers in a high-profile dawn raid after mounting a much-publicized campaign along with the independent Glasgow Games Monitor 2014[6] to resist displacement and receive a decent settlement (Gray and Porter 2015, 2017). After the Games events, the housing nestled along the River Clyde was retrofitted and sold as private market housing (300 units), and the remaining interior-located housing became social housing (400 units). While in the contemporary UK housing development context this is a reasonable ratio, the site previously comprised 1,589 homes, almost exclusively social rented, in structurally sound tenement buildings. In short, the regeneration process resulted in a significant net loss of public/social housing. Yet, the housing development is considered to be of architectural merit, despite reportedly poorer maintenance of the social housing and occasional social tensions across social and private housing tenures (Kidd and Kearns 2018).

Riverside Dalmarnock[7]: Clyde Gateway spent nearly £10 million in the early 2010s to remediate a heavily contaminated former power plant site adjacent to the Athletes' Village and build S.u.D.S. infrastructure. This figure, however, does not take into account the extensive public funding used to buy, remediate and sell the site, a process that typifies the byzantine role of public funding in the development of the East End's land and housing market. Clyde Gateway sold the 8.9-hectare site to Link Group for £5.7 million in 2015 after extensive and costly remediation. Yet, the purchase of this land occurred through a complex and heavily publicly subsidized process, with the transaction only being possible after the original owner (Murray Estates) was paid 12 times the amount for which they had bought the land only 5 years earlier.

Link Group,[8] one of the biggest social landlords in Scotland, and private developer Laurel Homes[9] are building 562 homes on the site by 2026 with an allegedly "neutral tenure mix": 206 private homes to be delivered by Laurel

FIGURE 7.2 Privatized public green space in Athlete's Village. The fence bordering the path alongside the Clyde River was meant to be removed after the Commonwealth Games, but residents refuse to do so. (Photo by: Melissa García-Lamarca.)

Homes and 356 affordable homes to be delivered by Link Group. The latter includes 50 New Supply Shared Equity[10] homes (grants provided by the Scottish Government to help registered social landlords build new homes for sale, and to help low- to medium-income households to purchase them); 88 mid-market rent[11] homes (created by the Scottish Government in the mid-2010s as an affordable alternative to the private rented sector, with rents set in between social housing and private market rates); and 218 homes for affordable rent.[12] Despite the immense amount of public money poured into the site, the total is skewed toward private or intermediate (privatized) housing: 338 homes (including New Supply Shared Equity and mid-market rent homes) as opposed to 218 homes for affordable rent. In Scotland, affordable rent is not clearly defined––it generally means up to 80% of market rent––but appears to provide more latitude in terms of price-setting than the distinct category of "social rent" in social housing, which is more expensive than local authority (public) housing. Since the local community is deeply impoverished in relative terms, most cannot afford private homes at Riverside Dalmarnock, where sales prices range from £145,000 to £195,000.

Shawfield Business District: the 64-hectare Shawfield site is key to meeting Clyde Gateway's targets on remediation, business space and jobs. However, the site was once home to the notorious J&J Whites Chemical Works that generated extensive, highly toxic soluble chromium (VI) contamination.[13] An employee from Clyde Gateway told us that the site will never be "clean enough for residential [occupation] so it's got consent for a million square feet of commercial office space." The first 11-hectare phase of land remediation and infrastructural development benefited from over £27 million in public European, national and local authority funding, including the £4.8 million Shawfield/Dalmarnock "smart bridge" that connects––via S.u.D.S. greenway environmental infrastructure––the business district with the Dalmarnock railway station, which was itself refurbished with £11 million in public money. Despite this significant outlay, 53 hectares of deeply contaminated land still require remediation, at a projected cost of £68 million. Thus, the estimated public cost of land assembly and remediation amounts to over £92 million before building construction is even discussed.

To date, one building has been developed on site by Clyde Gateway. The Red Tree Magenta building suite, topped by a green roof garden terrace, was at full occupancy within days of its official opening in 2019. Shawfield business district is targeting commercial development in the form of the finance, insurance and real estate (F.I.R.E.) sectors and the creative industries, seeking to convince businesses to relocate to the East End with attractive rents in comparison to the city center, and a cocktail of incentives and funding support.[14] However, given long-term issues with training, education and unemployment of the local population, it seems very unlikely that such businesses will provide them jobs beyond low-paid cleaning, retail and other service-based opportunities.

In summary, the key green-blue infrastructure developments facilitated by Clyde Gateway involve very substantial state funding to de-risk inward

FIGURE 7.3 The bridge over the Clyde River to the former site of the J&J White's Chemical Works. The Red Tree Magenta is in the distance. (Photo by: Melissa García-Lamarca.)

investment, correct market failure and underwrite essential site preparation and remediation costs so that private developers do not have to do so. The developments are primarily office/business spaces and private housing, with new employment opportunities created mainly for the business and so-called creative classes. Meanwhile, 80% of all housing constructed so far is designated for private sale or involves intermediate forms of tenure. No concrete figures have ever been given for social housing despite the evident need of the multiply deprived social demography of the area. It is difficult to see how such developments will address well-documented issues of social marginalization, ill health and poverty in the local community.

Some environmental exemplars in Glasgow's East End

More positively, the Clyde Gateway initiative includes community benefits clauses in many of its development tenders, and one of two praiseworthy projects in the area funded by a clause is the Baltic Street Adventure Playground.[15] Award-winning architects Assemble[16] designed this small site with local children in a model participant-led intervention. It is truly a child-led space, created for 6–12-year-old kids who are supported and enabled to act on their right to play. Addressing the local lack of play spaces (due to historic under-funding and under-maintenance), adult supervision and healthy food, the play area includes spaces where children learn how to grow, harvest and cook food. Propagate, a local collective that provides food education at the site, explained that they were "helping kids, parents and carers to grow, cook and eat vegetables, and understand where their food comes from … and about the wider kind of sustainability issues that come with that as well." For some children, the Baltic Street Adventure Playground is their first experience preparing and eating non-processed food. The space is free access, and children can participate in day-to-day management and site development. All sorts of everyday materials are used to create the play spaces, which are the opposite of manicured playgrounds. The success of the adventure playground has attracted children from wealthier parts of Glasgow, but it is created and largely used by local children. Its development in some respects compensates for the removal of an informal football pitch across the street that was turned into an enormous car park for the Commonwealth Games, now tarmacked and largely abandoned.

Cuningar Loop Woodland Park[17] is another recently created local green space benefiting local and broader Glaswegian communities. On a larger scale of 15 hectares, the park follows a "social forestry" approach, aiming to increase access to forests and wild spaces for urban communities across Scotland. Scottish Forestry partnered with Clyde Gateway, two local councils, the National Health Service and Creative Scotland to create the Woodland Park, which was a vacant and contaminated edgelands site used as a landfill for the rubble from the nearby Gorbals slum clearances in the 1960s. The majority of park funding––£4 million of the £9 million budget––was spent on site remediation, an "invisible" cost given that the apparently wooded and green site was in reality an unused informal landfill. Scottish Forestry employs a dedicated engagement officer who works with minority groups to encourage health and well-being activities, and around £600,000 has been spent on natural play zones, boulders and bike paths aimed at attracting marginalized families who may not normally visit the woodlands. Their approach is one of "health by stealth," encouraging widespread use of the park and meeting socio-environmental justice concerns. A bridge funded largely by G.C.C. was key in connecting this South Lanarkshire site to the Athletes' Village on the Glasgow side of the river. However, broader park access needs to be improved because the other main entrance/exit is poorly signposted and some distance from the park's main amenities.

Due to the long-term fragmentation and dispersal of the community and the resulting processes of political decomposition, community mobilization in the East End has been relatively marginal. This may be one reason why such areas are chosen for large-scale regeneration, given that populations in wealthier areas have more social capital to resist unwanted processes of urban transformation.

However, the Calton Heritage and Learning Center's Green Volunteers program is one example of a project that has been developed as a way to counteract the troubling realities that the combination of large amounts of derelict land, drug-selling and drug use bring to the neighborhood. As explained by the center's director:

> We were seeing needles being thrown in people's gardens so they were lying under children's play toys and we started removing them and then from there we started teaming up with community safety to get gloves and picker uppers and bags and then we started doing that … The police told us that anti-social behavior was going down because people could see into the gardens now, so it wasn't such a safe place to hide so things like people using drugs in your front garden, people having sex and paying for sex in your garden stopped. There wasn't lots of this going on but enough to create a horrible feeling.

Funded through Commonwealth Games legacy and Lottery funding, the intention of the Green Volunteers is not to replace publicly provided services but rather to help improve local health and well-being:

> Our idea is that the Green Volunteers get to find out through gardening that not only do they improve the environment and the way people feel about where they live, but we get to know where the most isolated people are, we take them soup in the winter, we scrape away the ice from their front door … we'll go to the shops for them whether it be the post office or pick up a prescription or whatever it might be and then we just keep in touch with them and let them know there's people in the community that care, that know their name, that know they're there and that they can call upon.

Toward environmental justice in the East End?

While we recognize that the hundreds of millions of public funds spent decontaminating, remediating and providing green and blue infrastructure in the East End provide genuine environmental benefits, in this chapter we call into question who will reap them: local communities subjected to the nefarious effects of toxic environmental conditions for the last 50 years, or newcomers contributing to (green) gentrification in the East End? Notwithstanding honorable exceptions, contradictions and tensions abound in the Clyde Gateway's "sustainability fix." These include, for example, concern for community and environmental benefits while land remediation largely attracts the business/creative classes; the mobilization of public funding for unaffordable housing; and the most attractive locations within development sites—–e.g., next to the River Clyde—given to the private sector. Most developments do little to address local employment and housing needs and are relatively marginal in comparison to an overall business-oriented approach that leaves genuine socio-economic and environmental justice in the East End reliant on the long-proven fallacy of trickle-down benefits.

In order for this to change, what is required is political will, a regulatory environment geared toward genuine equitable distribution and the participation of local people not just post factum but in the very planning and production of the urban environment. The Glasgow Games Monitor 2014 is one group that has attempted to raise awareness and foster such involvement but, unfortunately, the fragmentation of the area and its population through previous rounds of disinvestment, abandonment and stigmatization means that there has been limited community engagement so far. Positive lessons can be gleaned from exemplary green leisure and play spaces, such as the Baltic Street Adventure Playground and the Cuningar Loop Woodland Park, which fulfill local needs. While these processes and spaces have at least somewhat mediated the inequitable consequences of publicly funded private development in the area, much more is required.

Keywords

(a) the urban development pattern of the city and neighborhood: recovering post-industrial; income inequality; enduring health inequality; vacant space; large-scale urban regeneration; creative class
(b) the urban greening of the neighborhood: land remediation; green infrastructure; green space
(c) the inequalities at stake: insufficient affordable housing; private capture of public investment; employment and economic development needs; (green) gentrification

Notes

1 http://www.clydegateway.com/
2 http://www.glasgow2014.com/
3 http://c-c-g.co.uk/project/commonwealth-games-athletes-village/
4 https://www.parliament.scot/parliamentarybusiness/Bills/16291.aspx
5 https://www.bbc.com/news/uk-scotland-glasgow-west-11435253
6 See https://gamesmonitor2014.org/
7 https://www.riversidedalmarnock.co.uk/
8 https://linkhousing.org.uk/what-we-do/about-link/
9 https://www.laurelhomes.co.uk/
10 https://www.mygov.scot/new-supply-shared-equity-scheme/how-it-works/
11 https://www.gov.scot/publications/rent-affordability-affordable-housing-sector-literature-review/pages/6/
12 https://linkhousing.org.uk/what-we-do/more-than-a-landlord/development/where-we-are-building/riverside-dalmarnock-glasgow/
13 https://www.glasgowtimes.co.uk/news/17473890.polmadie-burn-poisonous-time-bomb-leaves-glasgow-families-fearing-theyll-be-sold-down-the-river/
14 https://www.magentaglasgow.com/help-support/
15 http://www.balticstreetadventureplay.co.uk/
16 https://assemblestudio.co.uk/projects/baltic-street-adventure-playground
17 https://forestryandland.gov.scot/visit/cuningar-loop

References

Cowley, Joe, Kiely, John, and Collins, Dave. 2016. "Unravelling the Glasgow Effect: The Relationship between Accumulative Bio-Psychosocial Stress, Stress Reactivity and Scotland's Health Problems." *Preventive Medicine Reports* 4: 370–375.

Gray, Neil. 2008. "Clyde Gateway: A New Urban Frontier." *Variant* 33 (Winter): 8–12.

Gray, Neil, and Mooney, Gerry. 2011. "Glasgow's New Urban Frontier: 'Civilising' the Population of 'Glasgow East'." *City* 15, no. 1: 4–24.

Gray, Neil, and Porter, Libby. 2015. "By any Means Necessary: Urban Regeneration and the 'State of Exception' in Glasgow's Commonwealth Games 2014." *Antipode* 47, no. 2: 380–400.

Gray, Neil, and Porter, Libby. 2017. "The Right to the City and Its Limits: Contested Property Claims, Urban Exceptionality, and the Fight for Relational Space in Glasgow's Commonwealth Games 2014." In *Contested Property Claims: What Disagreement Tells Us about Ownership*, edited by Maja Hojer Bruun, Patrick J. L. Cockburn, Bjarke Skærlund Risager, and Mikkel Thorup, 23–38. London: Routledge.

Kidd, Maureen, and Kearns, Ade. 2018. *Village Life: The Early Experience of Living in the Commonwealth Games Athletes' Village Development, Glasgow*. Glasgow: Glasgow Centre for Population Health. Accessed June 5, 2020. https://www.gowellonline.com/assets/0000/3958/Village_life_-_the_early_experience_of_living_in_the_Commonwealth_Games_Athletes____Village_development.pdf.

McArthur, Andrew A. 1987. "Jobs and Income." In *Regenerating the Inner City: Glasgow's Experience*, edited by David Donnison and Alan C. Middleton, 72–92. London: Routledge.

McDonald, Sheila. 1987. "Some Environmental Considerations." In *Regenerating the Inner City: Glasgow's Experience*, edited by David Donnison and Alan C. Middleton, 136–150. London: Routledge.

Middleton, Alan. 1987. "Glasgow and Its East End." In *Regenerating the Inner City: Glasgow's Experience*, edited by David Donnison and Alan C. Middleton, 3–32. London: Routledge.

Nixon, Ruairidh. 2016. *Comparison of Aspects of Glasgow's 56 Neighbourhoods*. Glasgow: Glasgow Centre for Population Health. Accessed May 26, 2020. https://www.gcph.co.uk/assets/0000/5492/Comparisons_of_aspects_of_Glasgows_56_neighbourhoods.pdf.

Pacione, Michael. 1995. *Glasgow: The Socio-Spatial Development of the City*. Chichester: John Wiley and Sons.

Reed, Peter. 1999. "Beyond the Centre." In *Glasgow: The Forming of the City*, edited by Peter Reed, 202–222. Edinburgh: Edinburgh University Press.

Temenos, Cristina, and McCann, Eugene. 2012. "The Local Politics of Policy Mobility: Learning, Persuasion, and the Production of a Municipal Sustainability Fix." *Environment and Planning A: Economy and Space* 44, no. 6: 1389–1406.

Wannop, Urlan, and Leclerc, Roger. 1987. "Urban Renewal and the Origins of GEAR." In *Regenerating the Inner City: Glasgow's Experience*, edited by David Donnison and Alan C. Middleton, 61–71. London: Routledge.

While, Aidan, Jonas, Andrew E. G., and Gibbs, David. 2004. "The Environment and the Entrepreneurial City: Searching for the Urban 'Sustainability Fix' in Manchester and Leeds." *International Journal of Urban and Regional Research* 28, no. 3: 549–569.

8

A COMMUNITY FIGHTS FOR ITS HEALTH WHILE BATTLING IMPENDING GENTRIFICATION

Bayview-Hunters Point, San Francisco

James J.T. Connolly

Gentrification is a hard thing to see in real time, but for several years there has been a place in San Francisco where you can watch it happen. The Third Street T streetcar that crawls slowly down the southeastern side of the city, parallel to the eastern waterfront, was the first piece of transit infrastructure built in San Francisco in five decades when phase one of the line opened in 2007. Toward the city center, the streetcar follows the new frontier of gentrification, and once outside it moves along a rapidly changing industrial area that abuts Bayview-Hunters Point (B.V.H.P.), an ethnically mixed working-class neighborhood with the last historic concentration of African Americans in San Francisco. B.V.H.P. is the end of the line for the new streetcar's first phase and a good point to reflect on the rising pressures for change coming down the line. For now, the gentrified upheaval that is so visible further up the line is only a shallow presence in Bayview-Hunters Point. Yet many residents and community activists perceive that a fast-moving gentrification frontier leaves them with the conundrum of either staying in a contaminated, post-industrial and climate-exposed environment or advocating for their health but being at risk of displacement. They are facing what is described as "complex environmental riskscapes" and "overlapping environmental injustices" (Cole et al. accepted).

A history of local organizing in a context of environmental contamination, distrust and neglect

B.V.H.P. is located in the southeastern part of San Francisco, which has been historically populated by low-income communities of color and other ethnic minorities. It is still home to heavy industry facilities, some of which are defunct and some of which continue to operate. These include a former P.G.&E. power plant, the Southeast Sewage Treatment plant and the closed Naval Shipyard, as

DOI: 10.4324/9781003183273-10

FIGURE 8.1 The Third Street streetcar pulls into Bayview near Williams Avenue, ca. 2017. (Photo by: Pi.1415926535, CC BY-SA 3.0 via Wikimedia Commons.)

well as several scrapyards, diesel freight transportation and two freeways. There is a widely perceived condition of long-term neglect of the area from the responsible federal and local agencies and representatives. As one resident succinctly put it, "people assigned to protect us have either bailed out or sold us out." A director of one of the agencies regarded with distrust by community members agreed, "I would say that the City largely turned its back [on areas like Bayview]."

As a result of this neglect and years of contaminating industrial activity, Bayview-Hunters Point residents have been exposed to ongoing and disproportionately high health risks such as asthma and other respiratory infections, cancer, low birth rates, and premature diabetes as well as psychological and mental health issues (Dillon 2014; Liu and Shatford 2019). In addition, residents have long complained of toxic smells from the sewage treatment plant and the area has been a notorious target for illegal dumping by contractors unloading their waste from other parts of the city. In recent years it was not uncommon for B.V.H.P. residents to witness many streets full of trash, breathe noxious air and be exposed to contaminated waterways and soil (Metzger and Lendvay 2006). Given that the area has for many decades been predominantly non-white, B.V.H.P. is a prototypical instance of what is often referred to as environmental racism in American cities (Schwartz 1997). At the same time, home is home, and the connection to place runs deep for neighborhood residents: "It's a beautiful place if people cared about us, but those two little letters—IF—mean a lot ... if you care about us, help us. If you don't care about us, leave us alone."

In response to decades of neglect from city, state and federal agencies as well as private entities, a number of B.V.H.P. community organizations have stepped forward to defend their own health through various activities. The range of actions underway demonstrates a robust community-built support system: from community gardens led by groups like Literacy for Environmental Justice aiming to educate people on the benefits of urban greening and healthy nutrition; to youth support programs led by groups like Hunter's Point Family; to the individual activism of people like Marie Harrison, who fought to clean up toxic conditions in her neighborhood and served on community-led advisory boards for air quality and data collection; to the political organizing of groups like Greenaction for Health and Environmental Justice, which organizes tours and collaborates with a nationwide environmental justice response network (the I.V.A.N. network) enabling communities to report polluting activities. These are only a few examples among many; there is no shortage of locally engaged activism in B.V.H.P.

For a long time, activists and residents in Bayview-Hunters Point felt they were left to go it alone in a corner of the city overlooked by public officials and those with investment capital. Speaking to residents, it becomes clear that this sense of being on their own is pervasive, built over decades of fighting

FIGURE 8.2 Looking over the Bayview neighborhood toward downtown San Francisco from the entrance of Bayview Park, ca. 2013. (Photo by: Tim Adams, CC BY 3.0 via Wikimedia Commons.)

FIGURE 8.3 Volunteer day at one of Bayview's community gardens. (Photo by: James J.T. Connolly.)

tooth-and-nail for basic needs. It shapes how they perceive what is happening now in their area. As a wave of new development threatens to engulf the whole of the eastern waterfront in San Francisco, many battle-weary citizens of B.V.H.P. are suspicious of new interests pointed in their direction, prompting questions like "what is in it for us?," "why now?" and "what will become of us?" These questions are directed toward external and internal interests and supplanting older concerns about the local environment. As the gentrification pressure mounts, the issues most affecting the day-to-day lives of residents shift toward housing affordability. One local organizer observed,

> Now everyone is trying to stabilize housing. If you don't have a place to stay, you don't care that the sea level is going to rise, you don't care that your air is bad, you don't care about the quality of the environment.

The perception that demographic shifts generate a more acute threat than environmental conditions is reasonable given recent census data. While in 1980 72% of residents self-identified as African American, today only 27% identify as Black (from any background). This decline was described by the San Francisco Bayview Newspaper as "the smallest Black population and the fastest exodus of any big city in America" (Epps 2020). In contrast, according to the 2009–2013 American Community Survey, 19.5% of all neighborhood residents are now white. As an example of the housing boom characterizing the area, recently homes have gone for sale for more than \$800,000 to \$1 million for a two-bedroom, and prices are expected to rise steeply—well above the financial capacity of existing residents. This shift has created a new sense of acute insecurity, wherein the benefits of the environmental clean-ups in their neighborhoods are lost on existing residents of color if they become displaced due to housing costs.

Who can afford to live in San Francisco's last development frontier?

Development and gentrification pressures are nothing new to San Francisco neighborhoods. Especially since nearby Silicon Valley became the center of the high technology universe in the 1980s, historically lower-income and diverse San Francisco neighborhoods (including the well-known case of the historically Latinx Mission District, which underwent intense gentrification in the 1990s and early 2000s) have been overrun with investors and higher-income residents (Mirabal 2009). This pressure has only intensified as more tech firms flooded the city and their workers showed a sustained preference for city living. Grassroots organizations in areas like Bayview-Hunters Point took note. In the 1990s and 2000s these neighborhoods were still seen as mostly industrial and unappealing, but it became clear that if they waited too long to tackle gentrification pressures like they had in the Mission District, it would be too late.

Indeed, there is a sense that much has changed since the early 2000s when Bayview-Hunters Point was off of the gentrification map for the city. Now, neighborhood residents, public officials and developers interviewed for this chapter were acutely aware that the neighborhood is seen as perhaps the last development frontier in San Francisco. One major development site underway alongside Hunters Point is the India Basin mixed-use project, which features 15.5 acres of open space and is touted as "a pedestrian-first, human-scaled, bicycle-friendly, residential village where neighborhood amenities are within a short walk."[1] Also in the early phase of development, Candlestick Point in Bayview is slated to be a new small town within the city with over 7,000 new residential units, 750,000 square feet of office space, a hotel and 300,000 square feet of retail space abutting large green spaces on the waterfront.

FIGURE 8.4 The scars of a rapid and brutal gentrification process which evicted thousands of residents and uprooted families with generational roots in the area remain in the Mission District. This mural, ca. 2019, memorializes the evicted residents who were displaced from the city. The Mission District gentrification process remains a key reference point for many local organizers. (Photo by: James J.T. Connolly.)

Diverging community voices in the context of an environmental clean-up scandal

A third large development site at the Hunters Point Naval Shipyard is a major preoccupation for those familiar with the traumatic displacement that occurred in the Mission District and the extreme environmental hazards inflicted on nearby residents in B.V.H.P. Given the extreme levels of contamination found at the Naval Shipyard, the site was designated as part of the federal Superfund program and was targeted for clean-up in 1989. The U.S. government hired the private company Tetra Tech to lead and monitor its remediation, which, in a high-profile example of the distrustful dynamic that characterizes much of the redevelopment efforts in B.V.H.P., was eventually sued for reporting falsified results of the actual percentage of radioactive emissions on the site. The resulting scandal[2] was a public expression of the growing skepticism that energized the now robust local environmental activism in the area, which organized to push back against efforts to sweep under the rug decades of environmental neglect. As one local activist reflected on the level of distrust surrounding the clean-up process at the Naval Shipyard, she highlighted the need for continuous vigilance from community and environmental activist organizations, recalling that

> nobody would give an audience to the whistle-blowers when they started reporting a few years ago and then it came to community action ... and now because of that action, because of these tiny but mighty groups of people coming forward with truth [there is a move toward proper clean-up at the site].

In 2004, Parcel A (part of the Naval Shipyard) was deemed suitable for reuse and was transferred to the City of San Francisco for housing development.[3] The main shipyard developer, Lennar Urban, capped the parcel in order to proceed with construction, a move later challenged in court as insufficient by those who purchased homes nearby. In 2020, Lennar settled out of court with residents, eager to move onto later phases of development at the now sold out (despite environmental concerns) shipyard site. The lingering concerns were expressed by the head of one local group, who commented,

> And so, my concern is like, well the cap is underground. It's got houses on it. Right? Like what happens with sea level rise here, right? Does it rise up? Do we start having, you know, we're not modeling. What is it? How does nuclear radiation proliferate into people's basements, where children will play a lot, right?

This set of conditions that characterizes development at Parcel A highlights what most longtime residents are feeling—that environmental clean-up, greening and new development are co-occurring at a rapid rate and generating what almost feels like a fog of war given the demand to act in the face of incomplete information. That mix is generating a rising tide of gentrification pressure as well, and the associated profits to be had by developers and property owners are seen by many as the underlying motivation for some to push through or look the other way on conditions that would otherwise be unacceptable. This ratcheting up of green redevelopment in the context of San Francisco's expensive housing market has combined with a history of local neglect and distrust in B.V.H.P. and a city-wide history of brutal displacement of lower-income people of color to generate a fractious and tense environment for local organizers.

FIGURE 8.5 New mixed-income housing development at the former Naval Shipyard site. A total of 500 acres in the Hunters Point and Candlestick area is the focus of the new redevelopment plan for the San Francisco southeastern waterfront. The project has been marketed for its green and sustainable design, offering up to 13,000 new homes, 32% of which are meant for affordable housing (https://sfocii.org/hunters-point-shipyard-and-candlestick-point). (Photo by: James J.T. Connolly.)

The fractiousness is fueled on the one hand by top-down conditions for engagement established by developers and regulatory agencies with typical pro-growth agendas. Several local organizers critical of the community engagement process as part of the redevelopment of the Naval Shipyard pointed out that, on the advisory committee, only 1 out of 11 people is from B.V.H.P. and requests for more community seats were denied. This was seen by one longtime resident as a "systematic, consistent, plotted-out process to remove transparency," despite allowing a veneer of transparency to be laid over the planning of the project. Like the growth machines of old, wherein local developers, politicians and media conspire to make cities platforms for growth, the media are also viewed as "unreliable partners with unreliable narratives" by those who seek greater transparency in the development process. In short, the redevelopment of the Naval Shipyard made it clear to many that, while a nod would be easily given to a thin style of neighborhood participation in planning, a real voice in the process would be much harder to come by.

On the other hand, and perhaps the most challenging aspect of the fractiousness that has emerged around the Naval Shipyard project, is the extent to which suspicions are directed internally upon those who live in the neighborhood and make claims of representativeness. Several activists described disagreements over who actually represents Bayview-Hunters Point and the motivations of individuals that have engaged in planning and development processes as formal representatives. One went so far as to argue that "there are lots of community organization 'leaders' in the neighborhood and lots of external leaders … that have hurt the community in order to get paid." Another organizer characterized the motivations that generate these tensions within the larger context of economic injustice: "When we start succeeding we move out because there is no support for us there … you name an injustice and I will say that it is because we do not have economic clout."

The overall view that systemic under-representation and lack of economic clout largely determine which terms of redevelopment are negotiable by local residents is underpinned—for Black residents—by an engrained awareness of historic racism. One local activist who grew up in B.V.H.P. recalled her early experiences in school, remembering how there was no homework for Black kids in her class because teachers assumed that they had no one at home to supervise them. Others reflected on the neglect of the neighborhood as a reflection of racism on the part of city administrators.

This sense of racist and economic structural exclusion pointed several residents toward a feeling that their starting point in negotiations would mean that they would be forced to accept the terms given in exchange for basic needs that have been demanded for decades, but not granted. The redevelopment of the degraded and unsafe Alice Griffith Homes public housing was seen as a case in point, wherein the overall cost versus benefit of the project was hard for some local organizers to discern (Chapple and Elias 2018). The homes, they were told, would be redeveloped as mixed-income housing with space for existing tenants contingent upon the Naval Shipyard going forward in order to leverage funds from the new development.[4] This needed redevelopment pitted local residents who wanted to slow down the Shipyard due to environmental concerns against those who wanted to speed up the Alice Griffith Homes. The net benefit for the neighborhood, as a result, was greatly muted in the eyes of some. The extent to which voice in the larger development storyline for the neighborhood would have to be sacrificed meant that the win at Alice Griffith was cold comfort for some organizers.

Others view the source of tensions and fractiousness in B.V.H.P. as rooted in the area's demographic shifts. From this perspective, Bayview-Hunters Point is no longer one community, but rather five or six communities speaking in separate voices. There are both longtime and new Black, Asian and Latinx residents, as well as newly arrived white residents. For some, mirroring the critique of identity politics more broadly, this identity-based fractiousness is the key dimension of the neighborhood being leveraged by those seeking to push development through by pitting certain groups against each another in order to diminish their capacity to act as a unified front. One young white resident, self-described as a gentrifier, pointed toward the issue of density as a possible opportunity to overcome this fractiousness. Longtime organizers and newly arrived gentrifiers, he argued, might share concerns about a push toward high-density development in B.V.H.P. This view perhaps presages a later phase of post-gentrification politics for B.V.H.P., one that in other contexts such as Austin or Atlanta has been largely about preserving single-family property values for those who are there after the initial gentrification wave passes.

Arriving at a conundrum: health or gentrification?

B.V.H.P. residents have been calling for the neighborhood to be cleaned up and greened for decades, a call which mostly fell on deaf ears. The deep suspicion among Bayview-Hunters Point residents about the motivations behind clean-up and greening efforts has led some to resist greening on the grounds that it may open the gates for a Trojan Horse of gentrification. As one citywide greening group said,

> We have encountered some real resistance to [greening] in this part of town … If we look at a map of where we have been active over the last four years, you know, this is the one area where we have been unsuccessful … and this is primarily about social justice concerns.

People of color living near recently greened areas expressed feeling highly vulnerable to displacement in high-growth cities, even if those are still contaminated and industrial areas. For these residents, high-profile green projects like those in Bayview-Hunters Point and the associated branding of the area seem to be a precursor of shifts toward a whiter population.

One other manifestation of the concern over looming gentrification in B.V.H.P. reported by local activists is the fear that residents have in reporting and advocating for a healthy environment. Fearful of the implications for both renters and homeowners to be able to stay in their current homes, people are reticent to report poor environmental conditions. One activist commented, "One of the reasons why a lot of people don't come forward is 'cause they don't want to lose their house even though it's contaminated and they're not getting the proper maintenance from the landlord or management." In other words, people are concerned that the more attention they draw to their specific conditions, the greater the risk of being asked to leave in order to perform clean-up or incur costs they cannot afford.

Overall, many B.V.H.P. residents are happy to see new restaurants, shops and services being established in locations that were previously abandoned, and overjoyed at the thought of amazing new green spaces near their homes.

However, as much as new amenities are desired, a home that they can stay in is a necessity. In the end, there is a great deal of insecurity over whether greening will benefit current or new residents. The green spaces attached to large-scale developments were described by some as "window dressing." This circumstance leaves many feeling obliged to choose between living in an environment where they risk their health due to contamination or risk displacement due to gentrification.

The city planning agency in San Francisco has tried to learn from past lessons and internalize these concerns, a process which led to the creation of the community stabilization initiative. This initiative works at the neighborhood level to develop customized plans that seek to account for specific pressures and to apply tenant protections and housing production goals to vulnerable areas. So far, B.V.H.P. has engaged with this initiative mostly through cultural preservation programs that are still nascent. Cultural heritage is certainly an important aspect of addressing potential displacement, but much more remains to be done.

The discourse of current environmental justice activism and the frame of policy responses

In the end, the citywide push for growth has left one B.V.H.P. activist to conclude, almost in desperation, that "planning has to find a motive other than increasing the general fund." The first priority remains to ensure that heavy industries are held accountable for harmful emissions and that there is a systematic and comprehensible amount of data collected to monitor air quality and toxicity. Furthermore, it is apparent that there is a thriving community of agencies and organizations operating on the ground, but if they are to address the second priority of intervening in processes of displacement, they must have access to funding and greater political voice. On a wider policy level, the simplistic frame wherein a choice must be made between environmental justice and economic growth needs to be dismantled in places like Bayview-Hunters Point. This is one of the great challenges faced by the planning profession.

Keywords

(a) the urban development pattern of the city and neighborhood (i.e., fast recent growing; recovering; long-term stable growth; industrial; tech-driven, etc.): long-term growth; tech industry; spatial racial segregation; waterfront redevelopment
(b) the urban greening of the neighborhood/city (i.e., new parks and gardens; green resilient infrastructure; food production): environmental and ecosystem preservation; neighborhood greening; waterfront promenade; greenway; state and federal parks; community gardens

(c) the inequalities at stake (i.e., green gentrification and/or displacement; deep inequities in access to green; climate resilience/unequal climate protection from greening; continued exposure to contamination): enduring segregation; environmental privilege; green gentrification; luxury developments; climate health risks; continued exposure to contamination

Notes

1 Design | BUILD SF (bldsf.com).
2 https://www.sfchronicle.com/bayarea/article/U-S-sues-Tetra-Tech-over-Hunters-Point-shipyard-13536013.php
3 https://sfocii.org/hunters-point-shipyard-and-candlestick-point
4 Alice Griffith housing project is on thin ice—the *San Francisco Examiner* (sfexaminer.com).

References

Chapple, Karen, and Renee Roy Elias. "Analyzing investment flows in comprehensive community revitalization: The case of Bayview Hunters Point, San Francisco." *Journal of Urban Affairs* 40, no. 4 (2018): 494–517.

Cole, Helen, Isabelle Anguelovski, James J. T. Connolly, Melissa Garcia Lamarca, Carmen Pérez del Pulgar, Galia Shokry, and Margarita Triguero-Mas. "Adapting the environmental risk transition theory for urban health inequities: Examining complex environmental riskscapes and overlapping injustices in seven Global North cities." *Social Science & Medicine* 277 (2021): 113907.

Dillon, Lindsey. "Race, waste, and space: Brownfield redevelopment and environmental justice at the Hunters Point Shipyard." *Antipode* 46, no. 5 (2014): 1205–1221.

Epps, Kevin. "San Francisco don't like Black people." *SF Bayview*. February 11. (2020) Accessed online October 12, 2020. Available at: https://sfbayview.com/2020/02/san-francisco-dont-like-black-people/

Liu, XiaoHang, and Sally Shatford. "Affordable housing and residential segregation in San Francisco: A spatial analysis between 2000 and 2015." *Applied Housing Research Initiative, Working Paper* 2019-2 (2019): 1–14.

Metzger, Eliot S., and John M. Lendvay. "Commentary: Seeking environmental justice through public participation: A community-based water quality assessment in Bayview Hunters Point." *Environmental Practice* 8, no. 2 (2006): 104–114.

Mirabal, Nancy Raquel. "Geographies of displacement: Latina/os, oral history, and the politics of gentrification in San Francisco's Mission District." *The Public Historian* 31, no. 2 (2009): 7–31.

Schwartz, Diane. "Environmental racism: Using legal and social means to achieve environmental justice." *Journal of Environmental Law and Litigation* 12 (1997): 409.

9

RESISTING GREEN GENTRIFICATION

Seattle's South Park neighborhood struggles for environmental justice

Helen Cole and Troy D. Abel

In Seattle's South Park neighborhood, residents strive to maintain the character of their tight-knit, largely minority community amidst citywide gentrification pressures. After decades of neglect and environmental struggles, the city continues to clean up the Superfund site along the adjacent Duwamish River while finally implementing several infrastructure, park and public space improvements. In addition, in 2014 the new South Park Bridge connecting South Park to other parts of the city was completed, replacing the dilapidated bridge condemned in 2010. Those physical improvements have facilitated rapid change in the neighborhood at a time when few affordable neighborhoods remain in Seattle and South Park attracts city-dwellers seeking a small-town feel. Vulnerable residents are having to balance priorities with limited and minority-driven organizing resources, often feeling forced to prioritize affordable housing over the threat of extreme environmental hazards that have long plagued the mixed industrial and residential neighborhood.

A history of environmental injustice in South Park

Home to the Microsoft and Amazon headquarters, Seattle is at the forefront of the tech industry and a major driver of growth and development in the city. Given that many of the area's new employees are in high-paying tech jobs (paying an average of $130,915 per year for software developers, for example, and tech jobs account for 42% of the office jobs in the city),[1] real estate development has largely targeted luxury housing and high-end restaurants catering to this new workforce. These changes have spread well beyond the boundaries of the tech company neighborhoods, affecting the cost of living in every corner of the small city, including homes in South Park, a long-underinvested neighborhood in South Seattle.

DOI: 10.4324/9781003183273-11

South Park, situated on the western bank of the Duwamish River, is separated from Downtown Seattle by an industrial zone and lags far behind other neighborhoods in the city in terms of infrastructure, quality and quantity of park space and air and water pollution resulting from the polluting industries on the river (Abel and White, 2011). It is also one of the few neighborhoods in Seattle with a strong presence of people of color, primarily Latinx, along with smaller Vietnamese and East African communities. Only one-third of South Park residents are non-Hispanic white compared to a citywide average of approximately 66%.[2]

South Park lies southeast of the Duwamish River and in the traditional territory of the Duwamish Tribe,[3] one of the tribes represented by Chief Seattle, an original signatory of the 1855 Treaty of Point Elliott[4] land use agreement between local indigenous tribes and the United States government. After the arrival of white settlers to Seattle in the mid-1800s, South Park became a thriving farming community and home to many immigrant populations—at first Italians, soon followed by Japanese. By the late 1800s, industrial development began in neighboring Georgetown on the other side of the river. Between 1913 and 1920, the Duwamish River was rechanneled, converting the once meandering stream into a straight channel and making way for industry along the newly reengineered waterfront (Cummings, 2020). Over the next few decades, the Boeing airplane company and many other industries sprouted up along the river. This development also brought hundreds of workers, transforming South Park from a small farming community to a heavily industry one (Zahler, Marti and Thomsen, 2006). Similar to many other neighborhoods experiencing environmental injustices, these changes led to the unequal exposure of residents to

FIGURE 9.1 Industrial businesses line the Duwamish River, as seen from Duwamish Waterfront Park in South Park. (Photo by: Helen Cole.)

industry-related pollution. South Park's transition to hosting many industries led the city of Seattle to attempt to change the neighborhood zoning to "industrial" in the mid-1960s, but protests staged by over 4,000 residents resulted in preserving its low-density residential zoning, thereby maintaining residents' right to maintain homes despite rapid industrialization.

The siting of industries in South Park and the rest of the Duwamish River Valley has also grown in part due to the expansion of the tech industry in Seattle since the 1990s. As tech firms have opened up in other former industrial areas such as Ballard and South Lake Union[5] (farther north in Seattle), industries being pushed out of other parts of Seattle have further concentrated in the Duwamish Valley (Abel, White and Clauson, 2015). Industrialization over the past 100 years has thus led to high levels of air pollution and environmental degradation in South Park and nearby neighborhoods, resulting in 2001 in the declaration of the five-mile segment of the Duwamish River passing through South Park as a Superfund site by the Environmental Protection Agency (E.P.A.). Clean-up is particularly contested due to the quantity and diversity of industries in the area and to the difficulty of identifying the extent of responsibility for each party, making it difficult to follow the E.P.A.'s policy of charging responsible parties with the burden of cleaning up Superfund sites.

In 2013 the Duwamish River Cleanup Coalition (D.R.C.C.),[6] a nonprofit organization originally founded in 2001 to advise the E.P.A. in clean-up efforts along the river, along with Just Health Action,[7] documented those convergent stressors on local residents' health. Their study showed that South Park has the

FIGURE 9.2 South Lake Union, the home of the Amazon headquarters, a former industrial area, was once proposed as a large central park which was contested by residents fearing the displacement of local businesses. (Photo by: Helen Cole.)

FIGURE 9.3 An unpaved road in an industrial-zoned section of South Park. (Photo by: Helen Cole.)

highest or near highest exposure to a number of pollutants compared to other neighborhoods in Seattle, higher rates of poverty, more foreign-born residents and residents of color and lower rates of health insurance. Overall, the study found that life expectancy in the South Park and Georgetown area is a full 13 years lower than the life expectancy of residents in wealthier parts of the city.[8]

Resident-led greening and overdue municipal investments

In addition to clean-up efforts overseen by the D.R.C.C., many greening efforts in South Park in recent years have been community-led. For instance, several small neighborhood parks such as Cesar Chavez Park and the River City SkatePark (both owned or stewarded by SeaMar, a nonprofit organization and landowner in the neighborhood) have been designed and implemented by the community. The private ownership of the River City SkatePark not only means that they maintain the right to use the park for graffiti art—which would be prohibited if the city were to take over its maintenance—but also that the burden of maintaining the space falls on the community. Many of these small community-initiated parks are overdue for refurbishment and residents feel either overextended by their leadership and activism roles, or have moved away from the neighborhood. The city now faces the challenge of eliciting residents' input, and balancing the varying and often evolving needs and demands of the increasingly diverse neighborhood.

In addition, recent interest from the city to improve conditions in South Park has resulted in the implementation of many concurrent projects, either new or

FIGURE 9.4 South Park Community Center and surrounding field, soon to be refurbished in part by a 4 million dollar donation from a private school in the wealthy Capitol Hill neighborhood in exchange for the use of the field at specific hours (https://parkways.seattle.gov/2019/05/13/south-park-redevelopment-project-moves-forward-with-seattle-academy-partnership/). (Photo by: Helen Cole.)

delayed. These include a safe streets project for improved pedestrian and traffic infrastructure, the renovation of smaller green or open spaces such as Duwamish Waterway Park[9] and the field alongside the South Park Community Center, three public utilities projects for infrastructure improvements long overdue (including the South Park pump station, a stormwater treatment facility and a stormwater collection system for roadways)[10] and "special emphasis"[11] police patrols—a decision announced in May 2019 to the surprise of residents. In addition, as mentioned above, plans are being finalized for a plaza at the base of the newly rebuilt 14th Ave bridge,[12] the result of years of community advocacy for more open space that resulted in the land purchase by the city parks department via parks levy funding. Many of these also focus on climate resilience, as the neighborhood is also at an elevated risk for flooding due to its low-lying waterfront position, particularly noticeable when a combination of a high tide and stormwater leads to backups in the drainage system.

Race, ethnicity and difficult representation for environmental justice in Seattle

Unlike many U.S. cities where racial and ethnic minorities make up the majority of urban residents, Seattle is a solid minority-minority city in which at least

65.6% of the city's population identifies as non-Hispanic white. In addition to the already relatively small minority population, due to widespread gentrification and displacement and a quickly rising cost of living, many communities of color have been displaced to areas in the south or outside of the city. Historical neighborhoods of color include the Central District, the original center of the Black community in Seattle, which was 73% Black in 1970. Today in the Central District, after a steady decline over the past 50 years, few residents of color remain (only about 14% of the neighborhood identifies as Black).[13] South Park is another of the few neighborhoods historically home to communities of color in the city, although with fewer historic ties to any specific ethnic or racial community than the Central District, making it also one of Seattle's most diverse neighborhoods.

The dearth of racial and ethnic minorities in Seattle places strain on the possibility of gathering mass power to fight for environmental justice in neighborhoods such as South Park (Svarstad et al., 2011). However, contributing to efforts of procedural justice, Seattle has for several years placed substantial emphasis on social equity in planning efforts. One such attempt is through the Equity and Environment Agenda[14] which includes an Environmental Justice Committee meant to "strengthen community partnerships with City departments" and "connect community-based solutions into government." Furthermore, the Racial and Social Justice Initiative[15] office, with its own staff and a core team of various city departments, works across the city to address institutional racism. Acknowledging their own limitations in attempting to address equity with a predominantly white staff, many city departments mentioned efforts to hire people of color, particularly to work on projects with a specific equity focus. However, these new hires often previously worked in Seattle's nonprofit and advocacy organizations fighting for social and/or environmental justice, meaning these municipal recruitment efforts have also led to an appropriation of leaders from the nonprofit sector. Due to the scarcity of people of color in the city, the strategy of achieving environmental justice through representation and participation is particularly challenging.

In formal city-led planning processes, particularly for implementing large infrastructure projects and participatory planning, efforts at interdepartmental collaboration may also help achieve procedural justice. Those often include representatives from multiple departments, such as Seattle Utilities, Parks and Recreation, Office of Planning and Community Development. While this strategy seems to acknowledge the cross-relevance of many of the projects implemented in South Park (and others) and the needs of residents, some respondents regretted that this approach has not led, in their views, to improvements in equity, and may in some cases slow down implementation by adding inefficient bureaucratic processes. Many respondents also highlighted the additional challenge of cross-department involvement specifically in South Park in terms of green space development given that multiple departments in the city and county own land for potential open space (Seattle Department of Transportation, Kings County, the Port of Seattle, Seattle Parks and Recreation and private entities such as SeaMar). In response, Outside Citywide,[16] a relatively new office housed within the Office of Planning and Community Development, has been created to specifically align work between departments working on public open space. In fact, one of this new department's first areas of focus was to hone in on South Park and create a holistic approach to addressing issues in the neighborhood.

Addressing community priorities with finite resources

South Park's environmental challenges as well as its relative isolation have to some extent protected the neighborhood from rising prices despite widespread citywide gentrification in Seattle. However, the opening of the new South Park Bridge in 2014 led to the turnover of long-standing businesses along 16th Ave, which themselves were threatened by financial hardship during the four-year closure of the bridge.[17] This immediate crisis, and the relative lack of ability on the part of the city to address it in a timely manner, has instigated tension in the neighborhood around housing versus greening, perceived by many in the community to be in direct competition with each other for finite resources. An example of this tension is the resident opposition to the plaza project on the part of several residents, some arguing that the land would be better used to meet their immediate and urgent housing needs in view of increasing housing costs and displacement threats. While they applaud the city for finally addressing older demands—including more public/open space, infrastructure improvements and the clean-up of the river—they do not see any real ability to address affordability. Furthermore, they fear that concerns for housing rights will receive the same "late treatment" as the environmental problems residents fought against for decades. As one activist explained:

> When the bridge was closed and once the city finally paid enough attention to the neighborhood, the residents really started saying, well it's great that now you're paying attention but at the same time we don't want you to, we don't want this to create more displacement.

In fast-growing cities like Seattle, where few affordable neighborhoods remain, gentrification is inevitable in those last frontier neighborhoods. In fact, many South Park residents—some of whom were "gentrifiers" themselves due to their social profile—said in interviews that they had moved there for its tight-knit community and small-town feel in spite of its environmental challenges. Those who have the resources to move elsewhere in the city cited these same environmental problems as the potential reason they may choose to leave the neighborhood. "I was well aware of the pollution, but you know it's one thing to be well aware of it, it's another thing to be living in it, I can tell you that much," one recently arrived homeowner explained. Several homeowners in the neighborhood also mentioned the impacts of climate change on the neighborhood as a looming threat to the already environmentally vulnerable

FIGURE 9.5 New housing in South Park across from the South Park branch library, which opened in 2006. (Photo by: Helen Cole.)

neighborhood. The city's small-scale investments to improve these environmental problems may not spur gentrification, but they may speed up the process by prompting homeowners and investors to accept these challenges with hopes of a better future.

Safeguarding housing rights and accessing resources in a diverse neighborhood

As mentioned before, the major concern is that affordable housing is not apparently being addressed in a timely and meaningful way. Seattle's new Mandatory Affordable Housing (M.H.A.) plan,[18] adopted in March 2019, requires new developments to either include 5–11% affordable units, where a higher percentage applies in neighborhoods with higher housing costs, or to contribute to a citywide fund for affordable housing. This may have a long-term effect on the availability of housing, but many are doubtful. Due to the high price of land all around the city, developers report that it is not economically feasible to pencil in affordable units in their projects. One architect reported that out of 16 multifamily projects he'd worked on in the past few months, only 1 opted to take the option of including affordable housing in the design, which resulted in only 1 affordable unit. The others opted to contribute to the fund, but the fund is not tied to the neighborhood itself and it doesn't produce housing as immediately as the present development of market-price housing, leaving residents skeptical about the ability of this new policy to meet residents' needs. As luxury housing goes up around the city, homelessness is also a major concern, with a census of over 11,000 homeless individuals in 2019,[19] indicative of the acuteness of the affordable housing shortage in Seattle.

In addition to the lack of affordable housing, access to existing housing and other resources available to residents is not equitable, such as digital access to housing opportunities. For example, many new units are advertised only online and are often taken before local residents find out they exist. This digital divide is also a barrier to accessing other neighborhood resources. For instance, the parks department uses an online system to formally reserve use of the field at the Community Center, which blocks those who don't have consistent internet access and goes against the more spontaneous way in which the community prefers to use public space.[20] There is a growing trend in providing information and access to services and resources online, in line with smart cities initiatives, and with the tech-centricism of Seattle specifically.

Divisions in accessing resources also occur due to the existence of closed informal networks through which information and resources are spread. For example, respondents report that landowners who own a few units or houses tend to rent these units at relatively affordable prices, but often seek people with similar interests and from similar social circles (they also tend to be white and not well-connected to Latinx or other minority groups), relying on word of mouth to find new tenants, as one landlord explained. This practice further limits access to diversity, leading to parallel communities that seldom interact with each other. Meanwhile, efforts at inclusion are often limited to producing materials in Spanish or offering interpreters for public events, addressing only some of the more obvious demands regarding access to information, but not more complex informal mechanisms of exclusion.

FIGURE 9.6 Mayor Durkin addresses the South Park community as housing activists appear with signs demanding affordable housing. (Photo by: Helen Cole.)

In South Park, a small number of residents comprise several neighborhood organizations and groups which plan neighborhood community activities. Many residents serving on the boards of these organizations feel over-extended with little time to manage their many commitments. Furthermore, many groups are predominantly white. In one specific case, the South Park Area Redevelopment Committee (S.P.A.R.C.) controls many resources in the community due to its status as a 501-c3 nonprofit organization. Formed in the 1970s, S.P.A.R.C. has long served as the fiscal sponsor for community projects and organizations without this formal status. It operates, among other resources, the Neighborhood Center—often the only building in the neighborhood with space to host community meetings and events—and the offices of small community services nonprofits such as the South Park Information and Resource Center (S.P.I.A.R.C., which provides health, leadership and economic development services predominantly in Spanish), a food bank and a senior program. This control of resources is often a source of tension, given that S.P.A.R.C. does not occupy the space that it manages, and particularly since the small leadership is entirely comprised of white residents after the only Latino member was hired by a city office. Thus, the struggle for resources in South Park, and the division and tension between racial and ethnic groups, is also demonstrated within the nonprofit and charity community.

Green accelerators of gentrification in the last affordable neighborhood of Seattle

A recent *New York Times* Travel section article was headlined, "For a Seattle Enclave, Isolation May Be Its Salvation." South Park does sit relatively isolated from the rest of the city, separated from central neighborhoods by an industrial district and a river. However, after a four-year hiatus, the opening of a replacement bridge in 2014 added additional access and connectivity, and in turn, investment opportunity in the neighborhood. Aside from the relative geographic isolation of the area, lasting environmental concerns such as a polluted river, acoustic contamination from planes[21] and air pollution are also cited as reasons for the relative affordability of the area and slowing development. As many homeowners proclaimed: "If I move away from here, it will be because of the air pollution." The promise of projects and programs that address the neighborhood's lack of green space, high levels of pollution and climate-adaptive infrastructure is not the main driver of gentrification in this charming, tight-knit community. Rather, these improvements, now finally coming to fruition after decades, may accelerate the process in what is already destined to be the "last gentrified" neighborhood amidst broad gentrification and displacement processes sweeping the entire city of Seattle.

Keywords

(a) the urban development pattern of the city and neighborhood: tech-driven development; recent fast-growing city

(b) the urban greening of the neighborhood: neighborhood park improvements; climate resilience planning; environmental remediation; community-led greening
(c) the inequalities at stake: legacy of environmental justice struggles; continued exposure to contamination; displacement and citywide fast-paced gentrification; poor political representation of historically marginalized groups and conflicts over access to resources

Notes

1 https://www.geekwire.com/2018/seattle-nations-fastest-growing-tech-hub-study-says-bay-area-engineering-centers-continue-expand/
2 http://seattlecitygis.maps.arcgis.com/apps/MapSeries/index.html?appid=bf93420ee86147e9ba6de9cadecfc57e
3 https://www.duwamishtribe.org/history
4 https://goia.wa.gov/tribal-government/treaty-point-elliott-1855
5 In fact, in the mid-1990s, a large central park area was proposed for the then-industrial South Lake Union area. However, this idea was shot down by voters in 1994 and 1995. Ironically, voters were concerned that such a park would be too expensive and that it would spur development in the area which would drive out smaller local businesses. Paul Allen had contributed a loan of 20 million dollars to purchase land for the project and after its defeat, he took this land back, later developing it as the tech hub it is now. See: https://crosscut.com/2015/12/south-lake-union-could-have-been-seattles-central-park
6 http://duwamishcleanup.org/
7 http://justhealthaction.org/about/
8 The report is available online: http://justhealthaction.org/wp-content/uploads/2013/03/Duwamish-Valley-Cumulative-Health-Impacts-Analysis-Seattle-WA.pdf
9 https://www.seattle.gov/parks/find/parks/duwamish-waterway-park
10 These three projects are highlighted on the Seattle Public Utility's webpage: https://www.seattle.gov/utilities/neighborhood-projects/current-projects
11 https://www.seattletimes.com/seattle-news/crime/seattle-police-to-conduct-emphasis-patrols-in-7-neighborhoods-ahead-of-summer/
12 https://www.seattle.gov/parks/about-us/current-projects/south-park-plaza-landbanked-site-park-development
13 https://www.seattletimes.com/seattle-news/data/historically-black-central-district-could-be-less-than-10-black-in-a-decade/
14 https://www.seattle.gov/environment/equity-and-environment/equity-and-environment-initiative
15 https://www.seattle.gov/rsji/about
16 https://www.seattle.gov/opcd/outsidecitywide
17 Original plans to replace the bridge would have called for the new bridge to be built while the old one was still in use but concerns about the structure of the old bridge led to its sudden closure, which left the neighborhood relatively isolated while the new bridge was being built. At this time, the Office of Economic Development worked with local leaders and business owners to develop an assistance plan for local businesses which were threatened with reduced business, but many were frustrated by the limitations in how the funding, which came from the federal Community Development Block Grant program, could be used.
18 https://www.seattle.gov/hala/about/mandatory-housing-affordability-(mha)
19 http://allhomekc.org/wp-content/uploads/2019/05/2019-Report_KingCounty_FINAL.pdf

20 One respondent reported that this tendency by residents to use the space informally rather than requesting official reservations online led to records indicating low usage of the space, when in fact it was highly utilized by the community. Currently this space (see Figure 9.4) is being remodeled and the private school providing funding has reserved the right to exclusively use this public space during certain hours, a decision which was backed by the citation of low usage.

21 Several respondents pointed to this noise as a driver of decreased quality of life in the neighborhood. Several mentioned that the flight path directly above the neighborhood was intended to be used only during specific weather conditions but with time, they say, it seems to be in constant use. See: https://crosscut.com/2013/01/faa-greener-skies-quieter-skies-beacon-hill-jet-no

References

Abel, T., White, J. and Clauson, S. 2015. "Risky business: Sustainability and industrial land use across Seattle's gentrifying riskscape." *Sustainability* 7, no. 11: 15718–15753. http://doi:10.3390/su71115718.

Abel, T. D. and White, J. 2011. "Skewed riskscapes and gentrified inequities: Environmental exposure disparities in Seattle, Washington." *American Journal of Public Health* 101, Supplement 1: S246. http://doi:10.2105/AJPH.2011.300174.

Cummings, B. J. 2020. *The River that Made Seattle: A Human and Natural History of the Duwamish*. Seattle, WA: University of Washington Press.

Zahler, A., Marti, A. and Thomsen, G. 2006. *Seattle's South Park*. Charleston, SC/Chicago, IL/Portsmouth, NH/San Francisco, CA: Arcadia Publishing.

PART 3

(Re)creating unjust racialized landscapes in the green city?

Part 3, "(Re)creating unjust racialized landscapes in the green city?," analyzes how greening agendas create, re-create and sustain unjust racialized landscapes in historically segregated sites. It articulates the whiteness of greening and of its (new) users vis-à-vis historical legacies of segregation and discrimination. While greening is not solely linked with white culture, Atlanta, Washington, Austin and East Boston show how greening paves the way for white privilege.

In Atlanta, one of the most segregated and race- and class-divided cities in the U.S., the Beltline greenbelt, which includes a greenway planned to encircle the city and a proposed expansion of the rail system, is no exception to these divides. The trend of green gentrification along its path has made it difficult for long-term low-income Black residents, many of them multi-generational residents, to stay in their formerly neglected neighborhoods.

Washington, D.C., also a long-segregated city and now the most rapidly gentrifying metro area in the U.S., is scheduled to inaugurate the 11th Street Bridge Park in 2021. In Anacostia, Southeast D.C., where the bridge will be built, the profound racialized inequality is allowing greening, sustainability and corporate financial capture to advance. Despite the inclusivity and boldness of the accompanying Equitable Development Plan and the hope for reconciliation of communities long divided by the river, green transformations reveal the absence of an abolitionist green practice, one that recognizes the centrality of land recognition, redistribution and control for residents.

In Austin, greening is illustrated by smart growth and densification plans in East Austin, along with targeted new investments in green projects. This recent practice has in theory made the city more compact and sustainable while preserving the environmentally fragile areas of West Austin. Yet, it has also contributed to socio-environmental conflicts and major issues of displacement and gentrification affecting racialized communities. In the meantime, the city still

DOI: 10.4324/9781003183273-12

suffers from a lack of affordable housing that a history of racist land use policy further aggravates.

In East Boston, green assets in this coastal community have enhanced its environmental and economic value and built new climate protection while creating housing vulnerability and social exclusion for residents of Latin American and Italian descent. However, combined with large-scale real estate investments along the neighborhood's waterfront, the neighborhood's vulnerability has been further compounded. The transformation of environmental and coastal elements (both as assets and risks) is a new source of displacement and climate gentrification in East Boston.

10

RESHAPING LEGACIES OF GREEN AND TRANSIT JUSTICE THROUGH THE ATLANTA BELTLINE

Helen Cole and Daniel Immergluck

In Atlanta, one of the most segregated cities in the U.S. with a long history of concentrated urban poverty and a host of race and class divisions, the Beltline is transforming neighborhoods along its path. Atlanta's race and class divides can be seen clearly by examining the history of the city's public transit system, and the Beltline project, which includes a greenway and a proposed expansion to the rail system, is no exception. Alongside the promise of new environmental and transit amenities to long-neglected neighborhoods, it has manifested clear examples of green gentrification as property prices rise exponentially along its path, making it difficult for long-term low-income Black residents, many of them multi-generational residents, to stay in their formerly neglected neighborhoods.[1]

Atlanta's legacy of segregation

During the Jim Crow era of "separate but equal" policies, racial segregation was no exception in Atlanta and became more intense following the 1906 Atlanta Race Riots in which 25 Black residents were killed and many downtown Black-owned businesses were destroyed at the hands of a mob of middle-class white residents protesting that their social status and jobs were threatened by the growing successful Black community (Taylor and Hill 2013). Following the riot, an exclusively Black business district was established along "Sweet" Auburn Avenue, which included Black-owned financial institutions such as life insurance and banks, as well as churches and night clubs. John Wesley Dobbs, a notable activist and political leader who served as the informal mayor of Sweet Auburn, famously referred to the street as "the richest Negro street in the world" (Pomerantz 1997). The resulting geographical concentration of the Black community worsened class tensions as members of the well-educated Black elite began to gain political power over the following decades.

DOI: 10.4324/9781003183273-13

In his book, *Regime Politics: Governing Atlanta, 1946–1988*, political scientist Clarence Stone describes the evolution of politics in Atlanta as the rising political power of the Black elite joined forces with the existing white economic power structure forming a biracial coalition and leading to substantial physical redevelopment and a business focus in city, a focus which tolerated widespread poverty and inequality. Soon after the assassination of Dr. Martin Luther King, Jr., Atlanta's "favorite son," Mayor Jackson, elected in 1973 with the support of Atlanta's Black communities and a new wave of white progressives, led the formation of the city's Neighborhood Planning Units, which were designed to provide the city's Black communities more say in city planning at the neighborhood level.

Although undergoing rapid demographic change, Atlanta still remains a majority-Black city that ranks only second in level of segregation in the U.S. (Silver 2015), with further segregation by class among Black residents. Segregation patterns can also be seen in many aspects of city infrastructure such as the Metro Atlanta Regional Transit Authority (M.A.R.T.A.; colloquially referred to by some as the "Moving African Americans Rapidly Through Atlanta"). The success of the M.A.R.T.A. system, born in 1971 out of a need for public transportation in a car-dominated city, was limited by a fear of granting relatively easy access to poor and Black communities to sprawling suburban metropolitan areas. After several rounds of negotiations, the final plan was passed in large part due to the support of Black voters, the same force that shut down an earlier version of the proposed system due to its lack of service to Black neighborhoods (Anon 2017; Johnson Gaither et al. 2016). Transformations to the city's urban form were also influenced by the 1996 Summer Olympics, which saw the construction of Centennial Olympic Park, a large central park owned and operated by a private entity, and other facilities. Private individuals and foundations have also had a sizeable impact on the city. For instance, Arthur Blank, owner of the professional American football team, the Atlanta Falcons, and Office Depot, built a new stadium for the Falcons, for which over $700 million in tax dollars were contributed,[2] in the center of the city.

As in many American cities, the physical and social divisions of the city are marked by highways and M.A.R.T.A. transit lines, dividing areas of opportunity from those without. In the city's South and Southwest, lower-class Black neighborhoods suffer unjustly from the consequences of many environmental ills. In addition to lacking transportation infrastructure, services and jobs, flooding is a major problem that causes mold and other problems in public and low-income housing facilities in the area. These same neighborhoods were prime victims of predatory lending and subsequently saw high levels of eviction during the 2008 financial crisis (Lartey 2018) which, along with decades of neglect, have left South and Southwest Atlanta scattered with vacant homes. Neighborhoods like Peoplestown and Pittsburgh in South Atlanta, sitting directly along the path of the future Beltline, have seen dramatic increases in rent prices and property taxes, further displacing long-term residents who have weathered these storms.

The Beltline and Atlanta parks and green space management

Despite Atlanta's layered racial and class inequities, many of the city's segregated, predominantly Black neighborhoods have relatively high levels of greenness both in terms of tree canopy and public parks (Weber et al. 2017). However, the city lacks connectivity between neighborhoods and equitable access to green spaces. The Atlanta Beltline and its network of parks aim to address both challenges, highlighting on its website the potential to "break down barriers and connect people, communities, and cultures that have been intentionally segregated for generations."[3] The Beltline itself was inspired by a 1999 master's thesis by planning student Ryan Gravel[4] envisioning a more livable and accessible city through the repurposing of abandoned rail lines encircling the city into a greenbelt meant for recreational purposes and accompanying new transit lines connecting with existing M.A.R.T.A. lines. The abandoned rail lines are remnants of a time in which Atlanta was a center of freight shipping, dominated by competing railway companies that left non-intersecting, loop-like rail lines surrounding the city as freight transportation largely fell out of fashion. Gravel himself claims that, although his work addressed such issues as feasibility, his plan was purely intended as an academic exercise.

The project came to light and grew in popularity, however, when in 2000, Gravel and two colleagues, having grown increasingly frustrated with the city's transit infrastructure, mailed copies of the Beltline proposal to several government and business leaders, including an influential city council person, Cathy Woolard, who found the idea interesting and took it on. However, it was not until 2005 that the Atlanta Beltline Partnership was established to oversee fundraising and awareness of the project. In 2006, Atlanta Beltline Inc[5] was formed by Invest Atlanta (the official city economic development authority) to oversee the planning, development and execution of the Beltline. Although both entities work closely with the city of Atlanta, they are private, nonprofit organizations. In addition, an extensive network of nonprofit organizations is involved in the project, primarily through acquiring land for the project, which has grown to be one of the biggest challenges of completing the full loop of the Beltline, particularly given the rapidly rising property prices along its path and the difficulty in navigating bureaucratic city processes for procuring land quickly. As the Beltline construction broke ground in 2010, along with the greenbelt and transit lines, the plan grew to incorporate a system of interconnected parks,[6] currently at various stages of development.

Fears of dramatic real estate construction and housing price increases along the Beltline

As the Beltline has been implemented in stages, starting with the Eastside, some have seen development in adjacent neighborhoods as a warning sign of what may come to the west, south and north ends. Along the Eastside of the Beltline, dramatic economic changes have already taken place, and many businesses along the corridor have "turned" to face the Beltline, as explained by a local architect: "the Beltline is this fascinating study in fronts versus backs … where you have retailers whose front door was on Ponce de Leon Place, [and] now their front door's on the Beltline, their entire floor plan inverted." One of the most well-known and emblematic new developments is the Ponce City Market, a collection of high-end restaurants and shops (with a rooftop amusement park) located in a refurbished building purchased from the city of Atlanta in 2011 for $27 million on the Eastside Beltline and developed by Jamestown, an investment company also responsible for the Chelsea Market on New York City's Highline.

FIGURE 10.1 Ponce City Market, developed by Jamestown real estate, offers an eclectic mix of high-end shopping, dining and entertainment, conveniently accessed via the Eastside Beltline. (Photo by: Helen Cole.)

Ponce City Market also abuts the Old 4th Ward, a historic Black neighborhood where the Historic 4th Ward Park was one of the first in the network of parks along the Beltline to be refurbished. This project is seen by many as a forewarning of displacement, particularly by residents of other Black neighborhoods in South and Southwest Atlanta, which have suffered similar histories of disinvestment, flooding and other environmental challenges over decades. In the case of the Old 4th Ward, the new park features a 2-acre lake which doubles as a stormwater basin and connects the neighborhood to the Beltline. It is now surrounded by luxury townhouses and the neighborhood has been seen as a quintessential example of gentrification and cultural erasure. As one local activist noted, "In 2009 when the Fourth Ward went off market, I realized, 'Oh I'm about to be displaced in my own city.'"

Meanwhile, in the South Atlanta neighborhoods of Peoplestown and Pittsburgh, which await a future segment of the Beltline, and in West Atlanta, home to the most recent stretch of the Beltline, residents see the new infrastructure as a major cultural and economic threat. "The Beltline is nothing but a sidewalk to displacement," one activist stated. Thus far, the west section of the Beltline is far less utilized, perhaps demonstrating its ill fit with community needs and preferences. A string of craft beer breweries and other "hipster" businesses are opening in formerly industrial buildings along its path, signaling demographic and cultural shifts in the neighborhood. Meanwhile, West Atlanta, which lies within the catchment area of Proctor Creek, still experiences regular flooding, which activists such as the West Atlanta Watershed Alliance have long drawn attention to with little or no response from the city. In that sense, the Beltline plays a strong role in greening, beautifying and changing the imaginary and character of West and South Atlanta, while leaving other environmental and

FIGURE 10.2 Walkers enjoy the Eastside Beltline. (Photo by: Helen Cole.)

FIGURE 10.3 The less frequented Westside Beltline trail weaves through a warehouse district, where space has now been converted into a kombucha bar and brewery, an artisanal gelato shop and more. (Photo by: Helen Cole.)

FIGURE 10.4 A vacant property in Pittsburgh covered in signs from developers stating "we buy houses." (Photo by: Helen Cole.)

climate risks untouched. Black residents face multiple socio-ecological risks of displacement, both from the Beltline and from unmitigated flooding.

In Peoplestown and Pittsburgh, the Beltline is slated to connect many neighborhoods of varying racial and socioeconomic characteristics; thus, it struggles to adapt to the demands and interests of residents in each neighborhood. Pittsburgh is physically separated from other areas by a system of interstates and train tracks, on which stalling freight trains can separate the neighborhood for hours, placing a major strain on anyone trying to reach jobs, or other opportunities outside of a neighborhood where there are few services and jobs, and a large percentage of the housing is vacant. Despite these disadvantages, residents fear displacement and associate rising prices of living and the presence of new investors who are primarily "flipping" properties in the neighborhood to substantial financial gain, with the imminent arrival of the Beltline. Such fears are confirmed by recent research documenting the rapid change of property prices since the announcement of the Beltline (Immergluck 2009; Immergluck and Balan 2018).

Struggle to include anti-displacement protections

Fears of displacement along the Beltline are connected with the failed promise of affordable housing, although the City of Atlanta and Atlanta Beltline Inc acknowledge the lack of affordable housing and the need to accelerate construction. New stock of affordable housing is set to be funded in part through the Tax Allocation District, an agreement which called for the allocation of tax revenue generated from so-called "underutilized" properties near the Beltline to the Beltline project for the next 25 years. The lack of progress on affordable housing and other equity issues around the planning and implementation of the Beltline received negative publicity in 2016 when the "father" of the Beltline, Ryan Gravel, and Nathaniel Smith, the founder and executive director of the social equity nonprofit Partnership for Southern Equity, resigned from the Beltline Partnership board.[7]

In addition to their resignation, another turning point in the focus of Atlanta Beltline Inc came in 2017 when residents on the Westside protested an offensive art installation[8] meant to "activate" the Westside Beltline. The exhibit featured a set of life-sized photographs of Black men in prison uniforms posing with dogs. Residents were deeply upset about the marketization of Black suffering to brand their neighborhood as cool and trendy. Their reaction prompted the Beltline to redesign the process by which they choose and install art along the corridor, adopting a more participatory project and a community advisory board. In addition, a Chief Equity Officer was hired as a full-time staff member, which was intended to bring equity to the center of strategic planning.

Efforts to address the affordable housing crisis have fallen short in areas around the Beltline and the city as a whole. First, affordable housing targets are often based on units being affordable to households with less than 80% of the metropolitan median income, which amounts to a fairly high level of $66,000 for a family of four in 2020. This does little for people living in many of the neighborhoods on the Beltline's path, which are among the poorest in Atlanta and have incomes far below this amount. In short, they are simply too poor to afford it. Because many residents near the Beltline have very low incomes, affordable housing programs should be targeted at much lower income levels, such as below 50% of the area's median income, with some additional targeting at below 30% of area median income. The path along the north–south M.A.R.T.A. line, which intersects the future Beltline twice, connects Buckhead, a neighborhood with a median household income of $104,518, to the West End, the poorest M.A.R.T.A.-adjacent neighborhood with a median household income of only $19,447 (Commission 2016). The gap between those living in the poorest neighborhoods and those living in the wealthiest neighborhoods is so large that basing the affordability of housing on a level such as $66,000 does not provide affordability to many low-income families in the city.

Second, policies to create new so-called affordable housing do not protect those currently living in gentrifying neighborhoods. Atlanta's history as one of the nation's first cities to see a free Black community settle and thrive in the years after the civil war led to many Black families actually owning their own homes—at least before the recession of 2008—making home ownership a source of pride and a legacy of Atlanta's Black community. Yet the rate at which prices are rising along the Beltline means that property prices and taxes for these residents (which are effectively paid by renters as well as homeowners) may exceed their ability to maintain their homes or afford their rent. Ironically, these are the same families that stayed in the city center, maintaining the "urban form" that others claim has been lost through decades of suburbanization.

A future transportation corridor and broader resilience initiative

Controversy around the Beltline continues in 2020. As envisioned by Ryan Gravel, the Beltline's strength is not only as a greenway but also as a transportation corridor and a resilience and sustainability initiative. The already constructed sections include ample physical space to implement the project's original transportation plan, which would bring rail to the route and connect to the existing M.A.R.T.A. trains that serve the city. Gravel and others argue that the proposed Beltline transit line would respond to predicted growth, shifting toward a more forward-thinking mindset in transportation planning. However, the vision of the Beltline as a great connector is not shared by all.

One lifelong resident of Peoplestown said:

> The Beltline is not a green project in my opinion, not at all … the Beltline is the tool that they're using to displace people and basically build a solid base for the white populace around the nation to have somewhere to be comfortable in. That's really what I've gotten out of the Beltline. I'm from here so like those of us that are from here, we refer to it as a de facto boundary.

Thus, many see the growth referred to by supporters of the Beltline plan as a euphemism for gentrification processes that are pushing out lower-income and multi-generational Black residents.

In addition to the loss of support due to gentrification observed along its path, and its role as a de facto boundary in an already highly segregated city, several logistical problems would need to be overcome to fulfill the Beltline M.A.R.T.A. expansion plan. First, although the Beltline intersects with M.A.R.T.A. lines on all four segments, this does not always occur at all existing M.A.R.T.A. stations, which would thus require additional infrastructure funding to add stations or the disruption of transit trips in order to connect. Second, the existing public transit system, as in many U.S. cities, has fallen behind other hard infrastructure expansion, leaving many parts of the city (including some of the sections with most employment hubs, namely the Emory University/Centers for Disease Control and Prevention hub to the northeast of the city) with no direct access to the transit system, which the proposed Beltline does little to alleviate. Third, many of the city's poorer residents do not live within walking distance of a M.A.R.T.A. station or the proposed Beltline transit. As the Beltline continues to cause gentrification nearby, this situation will worsen. Many poorer residents are heavily dependent on M.A.R.T.A.'s bus system, which is significantly unreliable and has been losing ridership (partly as a result of such unreliability). Transit justice demands a substantial investment in better bus service, especially in neighborhoods with high shares of transit-dependent residents.

Finally, Atlanta's transit development pattern has been shaped over decades by historic segregation, racial and class-based political decisions and public opinion (Anon 2017), in particular the reaction of the city's wealthy white communities after the initial construction of the M.A.R.T.A. system in the 1970s. Back then, it was Black leaders that insisted that the east–west line be built first, following the substantial support of Black voters who approved the project in the first place. By the time the north–south line connecting some of the city's wealthier white neighborhoods with downtown was complete, many white residents refused to ride the train because it was perceived as "Black."

As neighborhoods near the Beltline gentrify, efforts to create transit on the Beltline—rather than along an artery through the southwest part of the city, or connecting the existing M.A.R.T.A. rail to the job-rich Emory/C.D.C. corridor to which many healthcare workers commute—are viewed by some as reinforcing transit injustices, with higher-quality and higher-cost rail designed to serve the new urban elite while working-class communities of color are left with inferior, unreliable and poorly funded bus services. Meanwhile, arguments are complicated by the changing suburban demographics, such as the nearby Gwinnett County and other suburban areas that are experiencing a growth in poverty as poor, Black and immigrant communities are increasingly forced out of the city center.

The massive Atlanta Beltline project continues amidst a long history of racial and class segregation as well as emerging social tensions around gentrification and environmental justice, begging the question: what will the Beltline's combined green and transit infrastructure be an access spine for? And for whom? Where will lower-income Black residents live and how will they circulate through the metropolitan area if they have to live further away from the core central districts and move to car-dependent counties far from jobs, services and existing social networks?

Keywords

(a) the urban development pattern of the city and neighborhood: race and class segregation; income inequality; recent fast-growing city; recreation and travel industry
(b) the urban greening of the neighborhood/city: greenbelt; greenways; green space connectivity; climate resilience planning
(c) the inequalities at stake: transportation equity; green gentrification; luxury developments; affordable housing crisis; high-end commercial development

Notes

1 Some of the text in this book chapter has appeared in Argüelles L, **Cole HVS**, Anguelovski I. 2021. Rail-to-park transformations in 21st century modern cities: Green gentrification on track. *Environment and Planning E- Nature and Space.*
2 https://www.theguardian.com/sport/2017/sep/29/why-are-georgia-taxpayers-paying-700m-for-a-new-nfl-stadium
3 From the Atlanta Beltline, Inc website: https://beltline.org/the-project/project-goals/equity-and-inclusion/ (accessed April 8, 2020)
4 Mr. Gravel's thesis can now be accessed via the Atlanta Beltline, Inc. website: http://beltline.org/wp-content/uploads/2012/04/Ryan-Gravel-Thesis-1999.pdf
5 https://beltline.org/the-project/project-goals/
6 Current and future parks are listed here: https://beltline.org/the-project/project-goals/project-goal-parks/
7 For example, see the Atlanta Journal Constitution article by Leon Stafford from 2016: https://www.ajc.com/news/local-govt--politics/beltline-creator-gravel-resigns-from-partnership/xw6pzH118XduyzFlYwJNvN/
8 See https://newsone.com/3745104/atlanta-beltline-apologizes-for-racist-art-black-inmates/

References

Anon. 2017. *Opportunity Deferred: Race, Transportation, and the Future of Metropolitan Atlanta.* Atlanta, GA. Retrieved from https://serve-learn-sustain.gatech.edu/sites/default/files/documents/2017_pse-opportunity_deferred.pdf.

Anon. 2016. *Monday Mapday: Income Divide.* Atlanta, GA. Retrieved from https://33n.atlantaregional.com/monday-mapday/monday-mapday-income-divide.

Immergluck, Dan. 2009. "Large Redevelopment Initiatives, Housing Values and Gentrification: The Case of the Atlanta Beltline." *Urban Studies* 46(8):1723–45. Retrieved April 8, 2020, from http://journals.sagepub.com/doi/10.1177/0042098009105500.

Immergluck, Dan and Tharunya Balan. 2018. "Sustainable for Whom? Green Urban Development, Environmental Gentrification, and the Atlanta Beltline." *Urban Geography* 39(4):546–62.

Johnson Gaither, Cassandra, David Himmelfarb, Sarah Hitchner, John Schelhas, J. Marshall Shepherd, and Binita K.C. 2016. "'Where the Sidewalk Ends': Sustainable Mobility in Atlanta's Cascade Community." *City and Society* 28(2):174–97.

Lartey, Jamiles. 2018. "Nowhere for People to Go: Who Will Survive the Gentrification of Atlanta?" *The Guardian*. Retrieved April 8, 2020, from https://www.theguardian.com/cities/2018/oct/23/nowhere-for-people-to-go-who-will-survive-the-gentrification-of-atlanta.

Pomerantz, Gary M. 1997. *Where Peachtree Meets Sweet Auburn: A Saga of Race and Family*. New York: Penguin Books.

Silver, Nate. 2015. "The Most Diverse Cities Are Often the Most Segregated | FiveThirtyEight." Retrieved April 8, 2020, from https://fivethirtyeight.com/features/the-most-diverse-cities-are-often-the-most-segregated/.

Taylor, Henry Louis, and Walter Hill. 2013. *Historical Roots of the Urban Crisis: African Americans in the Industrial City, 1900–1950*.

Weber, Sarah, B. Bynum Boley, Nathan Palardy, and Cassandra Johnson Gaither. 2017. "The Impact of Urban Greenways on Residential Concerns: Findings from the Atlanta BeltLine Trail." *Landscape and Urban Planning* 167:147–56.

11

A NEW SHADE OF GREEN

From historic environmental inequalities over green amenities to exclusive green growth in Austin

James J.T. Connolly and Mateus Lira

Austin is traditionally known for being Texas' most green and ecological city, with 15% of the land within city limits dedicated to parks and other green spaces and an abundance of regional nature reserves. The city was founded as a capital with priority given to preserving the natural beauty of the Texas Hill Country over development, making Austin a rare exception to the common model of American industrial urbanism (Busch 2017). As a consequence of a long history of spatial segregation along racial lines, access to the healthy environments and natural green amenities with which the city is endowed has never been equitable. This trend has continued, as the city itself is becoming a privileged space due to expanded gentrification in once-marginalized neighborhoods such as East Austin. It is this history of racial divide and recurring environmental privilege that has shaped local activists' responses to green inequalities.

The overall municipal response to the challenge of balancing growth, equity and environment has been to densify Austin. For some, this approach reduces sprawl, resulting in a more compact and sustainable city and, at the same time, increases the existing housing supply, which in theory would make housing more affordable. This efficiency approach, however, has not diminished the socio-environmental conflicts and major issues of displacement, gentrification and lack of affordable housing that Austin faces today. As a result, Austin residents remain deeply divided over the question of densification, and this division is at the heart of efforts to make a greener and more socially just Austin.

A history of preserving the environment in West Austin at new social costs

Austin's urban development has been historically guided by a good offer of green spaces and preservation areas, but also by high levels of segregation and

DOI: 10.4324/9781003183273-14

exclusion. From the 1928 masterplan that designated a "Negro District" in the eastern side of Austin, Black residents were forced to move to East Austin, the part of the city with the worst infrastructure and proximity to polluting industries, by a government who threatened to cut off basic city services to anyone who would not relocate. These explicit racist policies and norms produced deeply segregated and long-lasting residential patterns and inequalities, especially so in regard to access to city services, quality of schools and environmental amenities for African American and Latinx residents. In the early 1990s, those residents were still living close to six tank farms emitting contaminating gasoline vapors. While minorities in East Austin were exposed to poor environmental and health conditions, white residents of West Austin could enjoy living in the greenest part of the city. In response, residents mobilized around the grassroots environmental justice groups such as People Organized in Defense of Earth and Her Resources (P.O.D.E.R.), which resulted in the successful removal of polluting industrial plants in the area, including the closure of the Holly Street Power Plant, first partially in 1998 and fully then by 2005.

In the meantime, Austin solidified its reputation as a green and "livable" city. In August of 1992, voters overwhelmingly passed by referendum the Save Our Springs Ordinance, which limited development rights over the Edwards Aquifer in southwest Austin. The vote followed two years of intense organizing that galvanized a community of local environmental activists who had been gathering

FIGURE 11.1 The Holly Street Power Plant (ca. 1980) on the banks of Town Lake was a major landmark in the East Austin neighborhood for decades until it was removed in the early 2000s. (Photo by: Joe Mabel, CC BY 3.0 via Wikimedia Commons.)

support for water quality-based land use regulations since the 1970s into the Save Our Springs (S.O.S.) Alliance. The organizers behind S.O.S. went on in the next decades to become a powerful force for shaping development based on water quality protections, endangered species preservation and air quality concerns. Politically, the S.O.S. ordinance established mainstream environmental preservation groups rooted in mostly white, relatively wealthy and westside neighborhoods as the key opposition to unfettered growth with the capacity to sway electoral politics and garner the economic resources needed to challenge developers in court.

In 1997, this growing presence led to S.O.S. environmentalists taking the majority of the seats on the city council despite challenges by traditional pro-development candidates, who normally won these seats. This victory ushered in what was referred to as the "Green Council." During the Green Council era of the early 2000s, two essential and ultimately contradictory moves were made. First, the new council immediately began working with environmental justice advocates who lived and worked in East Austin to empower ongoing neighborhood planning processes. Through the development and passage of an East Austin Overlay District that resulted, it became more difficult to build anything in the area that did not align with existing uses or the community-generated

FIGURE 11.2 The Barton Creek Greenbelt (ca. 2010) was a major focus of preservation efforts in West Austin throughout the 1990s and early 2000s. The greenbelt now serves as a major recreational area and feeds the Barton Springs swimming pool, one of the city's most-known amenities. It runs through several southwest Austin neighborhoods. (Photo by: LoneStarMike, CC BY-SA 3.0 via Wikimedia Commons.)

neighborhood plan. The intent of this move was to bring some of the land use protections readily afforded to West Austin neighborhoods to the east side. In the end, the overlay was mostly used to exclude new polluting industries from locating in East Austin and support the push to clean up existing polluters.

Soon after, however, the Green Council made another move that undercut the land use controls given to East Austin residents through the overlay. The Council passed a citywide Smart Growth plan that designated most of the east side of Austin as the main "desired development zone" for the city. That the Smart Growth plan identified East Austin in particular as a key receiving point for new residential growth reflected a deal made behind closed doors between West Austin S.O.S. environmentalists linked to the Green Council and development interests who had been ratcheting up pressure in an open war over land use for the prior five years. This deal, essentially, designated West Austin as a high-priority area for ecological preservation and East Austin as a high-priority area for new environmental clean-up and greening initiatives accompanied by new higher density real estate developments.

The end result was that environmental preservation enjoyed for decades in the whiter areas of Austin and environmental clean-up that was happening in the non-white areas became packaged together by public officials and private developers as a marketing tool for the city to attract even more new investment—a form of green city branding (García-Lamarca et al. 2021). This more effective branding tool further fueled the conversion beginning in the 1990s of the once university- and government-centered town into a hotspot of tech industry companies, with wealthy employees willing to pay high prices to access Austin's green and hip lifestyle. By the 2010s, Austin was for a while the fastest-growing large or mid-sized U.S. city. The pressure of this level of growth coupled with Smart Growth plans led to a subsequent flood of development in East Austin.

A green and livable lifestyle for whom?

As the effects of this shift became clear, the already fragile collaboration between mainstream environmental activists and E.J. groups became more and more strained. Between 2000 and 2010, while Austin's population increased by 20%, the city lost 5.4% of its African American population (Tang and Ren 2014), with the top stated reason among those who left (54%) being unaffordability (Tang and Falola 2017). Here, increasing property taxes and cost of living during the 2010s forced residents of its historically stable working-class neighborhoods out to lower-cost and more dispersed communities. Increasing affordability concerns in the eastern part of the city are also linked to targeted greening initiatives, a critical example of sustainability fix for some (Tretter 2013; Long 2016). Specifically, a $40 million municipal bond helped to purchase new land for green space that was strategically placed to encourage development in the east and accelerate the desirability of those neighborhoods for new investment.

As reflected by one high-level city agency member, "We were trying to buy land out there ahead of development and hopefully draw development out there with parkland. And it was hugely successful."

It was clear by the end of the 2010s that there was an established precedent wherein the mainstream politics of environmental preservation left East Austin deeply vulnerable to gentrification and residential displacement (in addition to increased flood and contaminant vulnerabilities from new development), while West Austin neighborhoods became intensely expensive without widespread social and cultural displacement. This circumstance fractured, often along generational lines, the once nearly unified environmental activist block that ushered in the Green Council 20 years earlier. Many were openly questioning the role of environmental activism in fueling the processes that generate social injustices in their city.

This series of events points toward a challenging circumstance for high-growth cities like Austin. The S.O.S. ordinance was successful at achieving its goal, but the current critiques highlight the narrowness of the goal itself. Water quality and local ecological resources were preserved in a manner that certainly would not have occurred if developers were allowed to move forward as they wished in the 1990s. To this day, many developers see the environmental controls like the S.O.S. ordinance as an outsized impediment for new real estate in

FIGURE 11.3 A typical new condo development—this one in East Riverside, ca. 2019—in the newly densifying East Austin. (Photo by: Sk5893, CC BY-SA 4.0 via Wikimedia Commons.)

FIGURE 11.4 The extent to which geography shapes a social divide in Austin is sometimes a palpable aspect of the local conversation. (Photo by: Roland Tanglao, CC BY 2.0 via Wikimedia Commons.)

the city, and have continued the unabated push toward what they see as a greater reconciliation between their interests and the environmental community. On the one hand, then, the goal of reining in the worst effects of unfettered development was accomplished, but really only on the west side.

Meanwhile, some housing rights groups explicitly argue that gentrification and new developments in East Austin are happening due to environmental protection in the West and because of the lack of the east side's political agency in the municipal government. However, others also argue that even if more buildings could be built on the west side, the entire city would see an increase in development, without reducing the construction of new developments in East Austin. So, another consequence of the land use politics of the past two decades is that East Austin has become one of the most rapidly gentrifying neighborhoods in the country, displacing many low-income people of color (Maciag 2015). Inequality worsened citywide as a result, so that by 2017 Austin had the fourth largest wage gap between the highest-income and middle-class households in the country (Wile 2017).

What is clear is that the historic power imbalance along racial lines, contrasting claims of environmental movements and differing social configurations between East and West Austin generated development asymmetries. Most fundamentally, these asymmetries resulted in intense development pressure in East Austin because of the presence of previously industrial and cheap land. East

Austin residents logically sought to clean up and shut down polluting facilities, but were left with the subsequent struggle of living in a pricier area and associated green gentrification concerns. In response, a common reflection around procedural injustice started taking shape from those who negotiated the 1990s Smart Growth agenda: African American and Latino/a east side residents "were not at the table" at that time. Since then, some vocal environmental activists who were original supporters of the Smart Growth initiative apologized for the impact of gentrification on East Austin residents.[1] Many are now recognizing that the commitment to building Austin's green credentials in terms of preservation on the west side and new greening on the east side came at the expense of gentrification, displacement and continued environmental injustice for minority residents. It also eventually reflected how strong greening was mirrored by a weak housing policy and an ahistorical lens on land use in the city. It seems that efforts to address the challenging circumstances in Austin are now turning in this direction.

Conflicting debates over tools to increase affordability

In Austin, the socio-environmental conflict between real estate developers, environmentalists and displaced residents is woven together with the struggle for affordable housing. Responding to the urgent challenge of securing greater affordability throughout the city, densification and increased housing construction are seen as possible solutions by some environmental activist groups, municipal agencies and developers. In the early 2000s, some local environmental groups began to see densification as the necessary link between historic concerns for ecological preservation and a broader focus on urban sustainability. From this perspective, increased density promotes an alternative to urban sprawl and reduction in car dependency. Contrary to the S.O.S. history, this approach politically aligns some environmentalists with infill developers who complain about facing too many restrictions, not enough governmental subsidies to build affordable housing and anti-density positions from local residents.

Meanwhile, there is another side of the local environmental position, which sees density as a trojan horse for growth that respects neither ecological preservation nor affordability. The argument from this perspective does not see density as a solution to affordability because of the broader challenge faced by the Austin region's decades-old push for unfettered growth on the back of green lifestyle boosterism. Even the recent push to introduce cost savings that might generate greater affordability is not seen as productive. One activist reports:

> In many places you can reduce the cost of construction but it doesn't necessarily bring the price down. In a market-driven economy, price is set by the market. So if you increase the demand by inviting all sorts of rich people to come to Austin, Texas, then the price point goes up no matter how much you save on permitting.

In their views, many so-called affordable housing units are just smaller units at slightly less expensive prices, and the newly denser areas of Austin will never be affordable. As a result, such activists report that many low-income residents are faced with having to leave the city.

> It's funny people talk about "our poverty rate is going down!" and it's like "yeah, but that's basically just because poor people can't afford to live in the city, you know the poverty rate in the surrounding areas has gone way up."

This activist was pointing this out despite the new dense construction she observed in lower-income areas.

As Austin's original Latinx—especially low-income—residents are increasingly exposed to rising housing prices, unemployment and displacement, in the mid-2010s, issues of equity and affordability became more central to the municipality's agenda. So far, the municipality's response has been to declare an official concern regarding housing affordability in the immediate surroundings of new implemented greening initiatives or to incorporate an equity lens in the implementation of new green infrastructure. However, most municipal departments of the city operate in a segmented or sectoral way, making the design and implementation of green equity measures particularly challenging. One community organizer described this initiative relative to another example of an effort to address these issues:

> The city ... put together a Gentrification Taskforce like in 2002 ... that did a whole series of studies and came out with recommendations, and where did it go? It went on the shelf too, just like our environmental justice went on there before them.

Overall, providing and securing affordable housing in Austin is not easy. Land for new developments is relatively scarce, especially so in the preserved areas of West Austin. In addition, there is a lack of subsidies or incentives for developers to build affordable housing, which leads them to mostly target the new influx of affluent and white residents who can afford market-price housing. As well, governmental controls on housing process are extremely limited. For instance, inclusionary zoning, a planning mechanism which increases the offer of affordable housing by providing incentives and regulations to developers, has been restricted by Texas state law since 2005, precluding cities from putting mandatory affordability or rent control requirements on developers. Further, even the miniscule provisions offered by the city for public housing and public financing of housing tend to be met with political contestation. While much of the subsidized housing in Austin is generated by federal tax credits, the production of this type of housing has also been legally limited by the state legislature, and the expense of construction makes it difficult to accomplish. The extremely constrained state and city resources available are in part reflective of the Texan desire

FIGURE 11.5 Many new homes have been built in East Austin by incoming residents that employ green building standards. Often, these new homes mark the slow transition of the neighborhood through gentrification. (Photo by: Larry D. Moore, CC BY-SA 4.0 via Wikimedia Commons.)

to generate a business-friendly and low-tax environment, but they also make it quite challenging to intervene in the historical process that wraps greening into inequitable land controls.

Among equitable and green initiatives, Austin's program for developing more green buildings has been highlighted by institutions such as Foundation Communities, Austin Energy and the Office of Sustainability as a solution to provide housing that would be both ecological and affordable. Meanwhile, developers have been using the green building agenda to move ahead of environmentalists in their battle over sustainable land use practices, arguing for example that green roofs will compensate for the environmental impact of new constructions. However, some activists and community members perceive these new green developments as a reinforcement of environmental privilege, since green buildings are not perceived as affordable, but rather as a luxury.

Beyond density: alternatives for social justice, equity and affordability

Ongoing debates on new zoning and land use regulations, most notably those that led to the demise of a new land development code termed CodeNext, have made disagreements over how to introduce affordability and density throughout the city a front-and-center aspect of local politics. One of the main concerns is whether new land use initiatives will force greater density in West Austin, which has stood staunchly against such a push on environmental and neighborhood preservation grounds. Now, there is a rumbling within longtime environmental justice communities about whether such a push is desirable. Perhaps, they reason, the only pragmatic solution for neighborhoods like East Austin is to rapidly grab onto the types of local controls that have protected West Austin throughout the development boom years.

Most recently, a newcomer organization led by young people of color, the Austin Justice Coalition, has entered the density debate to argue against land use policies that keep blindly pushing for targeted areas to be the center of increased growth. In their view, simply allowing more density in marginalized areas could result in the erasure of Latinx and other minority communities. The Coalition also rejects assertions that market behaviors will produce affordability in Austin—density alone, they reason, will not do the job. Rather, it pushes for new policies that overtly acknowledge the historic racial inequities embedded in land use and development in the city. In that context, joint proposals by city councilors and the Austin Justice Coalition for an "equity overlay" to the land development code are gaining support and building on the University of Texas's Uprooted Project. Their proposal plans for additional affordability and multi-family protections in gentrification-vulnerable areas targeted for increased density. In sum, their advocacy pushes density initiatives away from a narrow environmental or economic justification and toward direct discussions around the manifestations and drivers of social justice.

As affordability continues to be one of Austin's most crucial challenges, the question remains: is density enough? Is the city really solving its socio-environmental conflicts and reducing gentrification just by building more? Most likely not. What is particularly missing from the political context is a willingness to seriously expand support for alternative housing models and collective ownership. While community land trusts and cooperatives exist in Austin, such anti-speculation and anti-displacement alternatives have not been seriously incentivized through citywide policy. If Austin is to protect its residents from gentrification and displacement, remaining green but at the same time equitable, it needs not only more buildings but also fair housing policies that accompany the green boosterism of the past decades. In other words, instead of targeting developers' profit-making or production of units for high-income newcomers, Austin's urban development needs to make its historically marginalized and low-income residents a priority.

As a partial response to the political challenges of unequal green growth policies, the City of Austin openly inaugurated an equity office in 2016 in order to evaluate how existing policies impact racial equity and which policies can be counter-formulated. Most recently, this office's equity evaluation is turning toward the land use and greening policies that helped generate a city that was designated the most economically segregated in the country in 2015 (Florida and Mellander 2015). The effort is being informed by prominent experts in the city and there is hope that if the office is able to reach across agencies, it may change the historical pattern of ineffectiveness around these issues.

The challenge of equitable development

In its green equity pursuit, Austin faces serious institutional hurdles and illustrates the challenges for affordability created in the absence of the underlying conditions for strong participatory process, for state and federal support and for a unified push from the local population toward equitable growth outcomes. In this context, Austin is left with cross-agency coordination as a potential tool for managing existing green tech growth dynamics. It remains to be seen if the strong relationship between greening and unaffordability coupled with a historic neglect for equity that made Austin an extreme case of unequal green growth can be undone with this tool. If Austin wants to remain both green *and* equitable, it cannot rely on the same system that put social and environmental goals at odds and generated deep social divides.

Keywords

(a) the urban development pattern of the city and neighborhood: spatial racial segregation; long-term growth; tech industry; densification; Smart Growth; sustainability fix
(b) the urban greening of the neighborhood/city: environmental and ecosystem preservation; neighborhood greening; livable city planning; green buildings
(c) the inequalities at stake: enduring segregation; environmental privilege; green gentrification; increased flooding risks for vulnerable groups; luxury developments; affordable housing crisis; low taxation; unaddressed (re)-zoning needs

Note

1 See https://www.mystatesman.com/news/opinion/phillips-how-environmental-deal-led-austin-gentrification/iVkoOTmU0tHZcxCKMg4j5O/

References

Busch, Andrew M. 2017. *City in a Garden: Environmental Transformations and Racial Justice in Twentieth-Century Austin, Texas.* Chapel HIll, NC: University of North Carolina Press.

Florida, Richard, and Charlotta Mellander. 2015. *Segregated City: The Geography of Economic Segregation in America's Metros.* Report. Toronto, ON: Martin Prosperity Institute.

García-Lamarca, Melissa, Isabelle Anguelovski, Helen Cole, James JT Connolly, Lucía Argüelles, Francesc Baró, Carmen Perez del Pulgar, and Galia Shokry. 2021. "Urban green boosterism and city affordability: For whom is the 'branded' green city?" *Urban Studies* 58(1): 90–112.

Long, Joshua. 2016. "Constructing the narrative of the sustainability fix: Sustainability, social justice and representation in Austin, TX." *Urban Studies* 53(1): 149–172.

Maciag, Mike. 2015. "Gentrification in America report." *Governing*, February.

Tang, Eric, and Ren Chunhui. 2014. *Outlier: The Case of Austin's Declining African-American Population*. Report. Austin, TX: The Institute for Urban Policy Research and Analysis at the University of Texas at Austin.

Tang, Eric, and Bisola Falola. 2017. *Those Who Left: Austin's Declining African American Population*. Report. Austin, TX: The Institute for Urban Policy Research and Analysis at the University of Texas at Austin.

Tretter, Eliot M. 2013. "Contesting sustainability: 'SMART Growth' and the redevelopment of Austin's eastside." *International Journal of Urban and Regional Research* 37(1): 297–310.

Wile, R. 2017. "These are the 10 cities where inequality is increasing the fastest." *Time*, 26 April.

12

THE RACIAL INEQUITIES OF GREEN GENTRIFICATION IN WASHINGTON, D.C.

Isabelle Anguelovski, Malini Ranganathan, and Derek Hyra

Washington, D.C., is one of the most intensively gentrifying cities in the United States (Hyra 2015; Hyra and Prince 2015; Prince 2016; Richardson, Mitchell, and Franco 2019), but gentrification and displacement are not homogenous across its landscape. This chapter considers the process of green gentrification, focusing on the transformations of the Anacostia River. Many residents remember the Anacostia River as a highly polluted, neglected and segregating river. More recently, however, the river's partial clean-up and revitalization have turned it into a new green icon of the revived U.S. capital. While many see it as a symbol and opportunity for racial reconciliation in the highly segregated "Chocolate City," (Asch and Musgrove, 2017) it also increasingly embodies the traits of a gentrifying and displacing riverfront.[1]

Neighborhoods in southeast Washington along the Anacostia River have come to represent two extremes and frontiers: on the one hand, a green and gentrification frontier in the wealthy Navy Yard on the western shore, and on the other, an open frontier on the east shore (Williams 2015; Ash and Musgrove, 2017). In this chapter, we examine urban redevelopment and greening trends in the east frontier and zoom in on the 11th Street Bridge Park construction project and its surrounding area. Our analysis demonstrates that while this project claims to be equity-driven, it currently fails to fundamentally address the city's historic racial inequality and the erosion of Black land ownership. Furthermore, as the intervention is driven by external corporate and philanthropic interventions, it risks neglecting or undermining longstanding racial justice efforts at the grassroots level and overlooks local calls and mobilization for more abolitionist and emancipatory urban greening practice.

Contrasting legacies east and west of the Anacostia River

The area east of the Anacostia River, which includes the broader neighborhood of far southeast/southwest, encompassing Historic Anacostia, has traditionally

DOI: 10.4324/9781003183273-15

been an area with the largest concentration of African American residents (still 95%[2] up to this date, and home to almost 50% of D.C.'s total Black residents. See Zippel, 2016). These demographics are explained by generations of African American land- and homeowners settling in the area. Indeed, from a slave plantation during most of the 18th and 19th centuries, the community of Barry Farm became the first African American settlement of freedmen in post-civil war Washington. Such freedmen worked in the Navy Yard or on Capitol Hill and returned at night across the river to build their homes. These hard-fought achievements, also reflected in the growth of local Black businesses, were followed by post-World War II urban renewal policies in Southwest D.C. (Gillette 1995) that evicted African American residents and rehoused them into concentrated public housing units, many of which were built in Anacostia (such as the 444-unit development known as Barry Farm[3]).

In the 1970s and 1980s, many subsidized rental units for working-class residents were also developed in the southeast. As expected, after decades of housing segregation policies and practices and urban renewal interventions, poverty rates east of the river soared. During this time, Black, formerly middle-class and disinvested residential areas such as Mayfair Mansions in the northeast also came to be associated with outdoor drug markets, tense police relations and patrolling by Black community groups, including those associated with the Nation of Islam (Page 1989). More recently, the financial crisis of the late 2000s and ensuing unequal recovery contributed to widening racial inequality in the D.C. region (Hyra 2017). This is the context in which aggressive new real estate and commercial investments have taken place, largely from the 2010s onwards (Hyra, Fulilove, Moulden and Silva 2020).

By contrast, directly west of the Anacostia River, Navy Yard (together with near southeast) is a post-military and post-industrial neighborhood cleaned up and redeveloped in 2020 to host new townhomes, bike lanes, shops, restaurants,

FIGURE 12.1 Historic Anacostia (left) vs. remaining buildings and demolition of Barry Farms public housing development (right). (Photos by: Isabelle Anguelovski.)

wineries, a Whole Foods Market, restored waterfronts, bike and walking nature trails and other so-called accessible and attractive green spaces. Coordinated by the Capitol Riverfront Business Improvement District, the Navy Yard transformation is presented by D.C. officials, planners and business leaders as a successful example of clean-up, rebranding, economic development and demographic growth. According to its supporters,[4] the Navy Yard demonstrates best practices around what an environmentally sustainable, compact, transit-oriented development can produce—a new green asset that increases livability in an economically booming and attractive Washington. Notably, Navy Yard has also seen the most dramatic influx of affluent white residents and households earning \$100,000 or more (+29% and +18% between 2006–2010 and 2011–2015 respectively) (Rabinowitz 2017).

In short, as of 2020, the west and east sides of the Anacostia River could not be more different. This chapter focuses specifically on how new green developments in particular are set to impact the east side of the Anacostia River. In 2014, the global New York City architecture firm O.M.A. won the competition for the 11th Street Bridge Park (henceforth "Bridge Park"), a \$50–60 million planned development to enhance environmental and recreational access between the Navy Yard neighborhood and Anacostia, while further contributing to the clean-up and revitalization of the two shorelines and parks. Both metaphorically and literally, the Bridge Park aims to connect the segregated east to the west banks of the river. The project brings new attention to formerly disinvested areas east of the river, and associates public (i.e., Anacostia Waterfront Initiative), private (i.e., Capitol Riverfront B.I.D. members) and nonprofit (i.e., Anacostia Watershed Society) efforts that began in the 2000s to make the Anacostia a renewed environmental amenity. There are also plans to enhance the environmental and recreational offerings of the Anacostia Park. Note, however, that the remediation of heavy contamination elsewhere in the region, including the former trash incinerator at Kenilworth Park in Ward 7 and the toxicity near Barry Farms housing in Ward 8, are not part of these environmental plans. This choice

FIGURE 12.2 The 2010 Navy Yard Park and adjacent new public space. (Photos by: authors.)

demonstrates that not all environmental issues are receiving the same treatment east of the river, with some projects being glossier than others and privileging green growth and sustainability fixes (Long 2016; Loughran 2014).

The futuristic 11th Street Bridge Park project

Supported by a comprehensive 2018 Equitable Development Plan[5] (E.D.P.) and headed by the nonprofit Building Bridges Across the River conceived by the D.C. developer W.C. Smith to support the revitalization of southeast D.C., the Bridge Park has been touted as possibly the best model of green, inclusive, healthy and housing rights-centered infrastructure planning in the U.S. (Avni 2018). It did not arise directly from residents' demands but is rather an externally conceived, nonprofit-driven project. However, the nonprofit leaders of the Bridge Park have explicitly framed the project as community-centered, unlike other projects like the now infamous New York High Line, the Chicago 606 or the Atlanta Beltline, which have had negative economic and racial impacts. For instance, Bridge Park explicitly put African American arts and culture at the center of the vision for the new space.

Based on a transparent, comprehensive community-driven planning process, the E.D.P. includes workforce training and development for local residents, support to new or existing small businesses, a community land trust (C.L.T.) for the construction of new affordable housing, technical help and training towards homeownership opportunities for existing renters (using what are known as the Tenant Opportunity to Purchase Act (T.O.P.A.), and the District Opportunity to Purchase Act (D.O.P.A.), D.C. legislations) and partnerships with local lenders and nonprofit developers such as Manna, Inc. and L.I.S.C. towards housing

FIGURE 12.3 Rendering of the 11th Street Bridge Park. (Source: OMA+OLIN.)

down-payment assistance. Through Manna, Inc., the equity plan is also creating a home-buyer's club to support residents in becoming homeowners. All of these tools are envisioned as buffers and mechanisms to promote economic stability, wealth creation and permanent affordability. Just in the last three to four years, Building Bridges Across the River has secured $54 million of funding towards these programs and others linked to environmental education, the arts and culture—even before the start of construction, scheduled to begin in 2021 and finish by 2023.

One of the core anti-displacement elements of the plan is the Douglass community land trust, which will develop 250–350 units to preserve affordable rental housing for residents in the 30–50% area median income range. As one of the leaders of the 11th Street project explains,

> when we're talking about housing, it's about working with some of these housing developers, where if they have some kind of equity gap, or they are looking to provide some affordable housing, the community land trust can help to add some equity to lessen the gap of what's affordable and what's not affordable. With the community land trust model, a housing subsidy continues to stay in the home, so that those homes are permanently affordable. Because one thing we know is for community land trusts to be truly successful it can't be done without the city's support, whatever the municipality is.

The Bridge Park is not the only new green redevelopment east of the river. In 2006, the local government had already announced the demolition and redevelopment of the Barry Farm public housing complex into a mixed-use, mixed-income, transit-oriented, open space, new public infrastructure and recreation center development (Turè 2017). Adjacent to it is a new pedestrian and transit-oriented mixed-use redevelopment on Poplar Point which will include a new 70-acre park, solar panels and bike lanes. The new development is touted by the Deputy Mayor for Planning and Equity development as "one of the last great urban redevelopment opportunities on the whole East Coast of the United States." However, equity considerations have been largely overlooked as part of the planning process. Only 10% of the 700–800 new units are reserved for affordable housing. In the words of Ari Theresa, the attorney of Barry Farm residents, Red Brick Partners has failed to account for issues of gentrification included in the D.C. comprehensive plan and preserve the neighborhood (Zhan 2019). Interestingly, brochures about Poplar Point already advertise the future 11th Street Bridge Park as a new recreational amenity for residents. While Poplar Point is the largest development on the waterfront, inland in Anacostia the St. Elizabeth's hospital conversion will anchor the new headquarters of the Department of Homeland Security (D.H.S.) and welcome multi-billion-dollar investments in related industries, contractors, services, housing, small business developments and beyond (Cowell and Mayer 2015).

Racial justice concerns

The above discussion should make it clear that the Bridge Project is making the investment path into the Anacostia neighborhood more direct and attractive for new real estate developments that have already reshaped the broader political economy of the city. One current developer justifies his "first mover" investment strategy in Anacostia by pointing at the project (and the private investors that support it) as the "first entry point" to Anacostia, offering the needed credibility that legitimizes investments. In his own words, "The strategy of our company is to go in early to emerging neighborhoods that … are ready for redevelopment and to buy property and redevelop it and re-tenant it." But such large-scale developments require deeper conversations about political, cultural and economic displacement (Hyra 2015) and about who will be able to own land and homes and rent at affordable rates.

For good reason, critics have called the Bridge Park project the "bridge to gentrification," arguing that it offers a direct physical path for new gateway projects into the Anacostia neighborhood, including the equity-oriented M.L.K. Gateway project and the Reunion Square development project, funded through a $60.8 million Tax Increment Financing (T.I.F.) tax scheme widely used in D.C. to fund redevelopment projects in "emerging" neighborhoods.[6] Additionally, Anacostia's attractiveness has been spurred more recently by the opening of a

FIGURE 12.4 New real estate project rebranding Anacostia. (Photo by: Isabelle Anguelovski.)

new Busboys and Poets café, a local hip restaurant and book store chain showcasing the work of Black artists and intellectuals in the District and beyond, by the recent Historic Anacostia preservation trail and by the renewed attention to the once home of Frederick Douglass, a former escaped slave and historic American abolitionist. In this redevelopment context, a broader question arises: whose Blackness is being valued, for which purpose and for whose benefits in the green and redeveloped Anacostia?

On the one hand, the strategy used by nonprofits such as Building Bridges Across the River to bring new financial resources to Anacostia by using the bridge project's aesthetic, environmental and recreational attractiveness is a remarkable shift. Anacostia is a community of color that has suffered from disinvestment, and the redevelopment effort aims to benefit longtime residents through support for C.L.T.s and affordable housing measures in a neighborhood where only 20% of residents own homes and residents are particularly sensitive to rent increases. On the other hand, local activists and residents point to the limitation of the community land trust model in building financial equity and, in turn, socio-economic mobility. They recognize that equity challenges run deep. In a country where the racial wealth gap is enormous—median white families are projected to own 86 times more wealth than Black families over the next five years (Nieves, Collins and Hoxie 2017)—land is seen as key in addressing this gap.

To this end, some nonprofits such as Organizing Neighborhood Equity (O.N.E. D.C.) are mobilizing to secure additional land and resources for Black families. As one organizer explained, "You need a different plan for people to control the land. We tie the environment to housing and land." Some housing and economic development activists also argue that the $50–60 million that the green bridge will cost could be easily reverted into funding for community-initiated green projects and enterprises and other minority-driven and Anacostia-based developments and buildings, while limited funding could be allocated to the renovation of the existing Anacostia Park and Trail east of the river.

Activists and residents fear that the 11th Street Bridge Park, with its acquired financial value, will activate the commodification of greenness and diversity together, enabling the profit-driven consumption of race and ethnicity through "cool," comfortable, multi-cultural events in new green areas and their surrounding retails spaces (even if those embody Black achievement) (Summers 2019). Residents note that such green projects are envisioned and driven by white—albeit progressive—leaders, financed by global banking institutions such as J.P.Morgan Chase, and designed by international New York City-based firms (O.M.A.), revealing who holds the economic power in the southeast and who might make the next move in Anacostia.

The fact remains that no matter how much equity planning is embedded into urban greening interventions, high-value green real estate projects might be directly paving the way for a new dispossession and displacement frontier. Such fears are justified in light of one developer's vision to make the east of the river a "destination for residents from across the city." As he further explains, the more his firm "can own in the neighborhood, the more successful the firm will be." Local fears of broader neighborhood loss already materialized in 2016 when a developer's "space finding" bus tour through Anacostia showcased potential sites to developers seeking ground zero spaces or land to buy. The tour bus was kicked out by protestors, who denounced how developers had just discovered Anacostia as an investment site, exhibiting what Safransky calls "green settler colonialism" (Safransky 2014) and threatening residents with racialized dispossession. Anacostia residents risk suffering what Oren Yiftachel calls (Yiftachel 2020) a "permanent state of displaceability," or erasure, in the quest for more profit. In Anacostia, house prices are already going up quickly—median sale prices have multiplied by 2.5 between 2014 and 2018[7]—signifying that, except for wealthier African American residents who might be able to pay premium prices for homes and increased property values, the majority of African Americans will likely lose out, and will ultimately be displaced

Towards abolitionist imperatives for greening and sustainability

In the push to green formerly disinvested areas and carve them up for new market opportunities, Blackness is being rebranded to become palatable to white consumers who enjoy Black heritage, arts and music as "cool" without necessarily interacting with Black people and older residents of the area (Summers 2019). This commodified Blackness has been deployed by developers for the purpose of pushing the green development agenda in the southeast, whereas the Blackness that actually built the southeast and its historic relationship with the river—and made its neighborhoods thrive despite racialized land appropriation, abandonment and conversion—seems erased. Such an erasure undermines the ability of the community to thrive. As a longtime resident says about the future Bridge Park, "I will be good enough to serve you slurpies and hotdogs at the river festival, but not good enough to live there." Good, green intentions are not good enough to reconcile and empower.

Furthermore, some activists and residents see these gentrified green spaces as a new plantation for day laborer residents who cannot access or feel uncomfortable in these places but must work in or near them in order not to be displaced. Here, grassroots movements such as O.N.E. D.C. or EmpowerD.C. are playing an important role in empowering residents to reaffirm their roots and identity. Others have even articulated a reparation claim for gentrified Anacostia residents, arguing that

> Barry Farm residents were injured because they were getting pushed out, they were injured because they were living in horrible conditions and the idea is to get them some kind of remedy. The demolition went forward. People were displaced. Barry Farm just was like the most egregious example of everything wrong with zoning.

Ultimately, the Bridge Park and other green improvements east of the river invisibilize existing modes of place-making, everyday resistance and bottom-up networks of resilience, support and care that are so fundamental to the sustainability of the city. This is particularly true for low-income residents and those living in affordable or public housing complexes who fear they will experience social or economic barriers or discomfort in attending recreational green events and lose a sense of belonging. Outsider-imposed plans in the Anacostia region might undermine the dignity, humanity and resilience of long-term vulnerable residents. What might also become erased through such a grand project is collective memory and history with regard to the river. Building on McKittrick's work (McKittrick 2011), we see that new green projects fail to recognize how Black residents practice place-making outside of what McKittrick sees as normalized, accepted and official cartographies of power.

An alternative to top-down greening efforts would be an abolitionist ethics of environmental justice (Ranganathan and Bratman 2019) and urban greening (Anguelovski et al. 2020). As Ranganathan and Bratman (2019) have argued, longtime African American residents of Wards 7 and 8 have developed social and neighborhood networks rooted in an ethics of care and healing. During times of weather emergency, for instance, neighbors bring food for the elderly and offer childcare support—a pattern that critical urban and environmental scholars have argued is essential for building bottom-up resilience (Klinenberg 2002). As one interviewee stated,

> People offer each other all sorts of in-kind support in times of crisis; they look after each other's kids, share rides, check in on each other. Now, how do we strengthen those bonds? That's what we need to focus on. Instead, there's the city's response, which is almost always to bring outside people and groups in to help.
>
> *(cited in Ranganatha and Bratman 2019: 130)*

Older activists and leaders east of the river have connections to historic Black power and abolitionist movements and narrate their organizing through abolitionist idioms, while also communicating with fellow residents through these embedded cultural and political registers. Bruce Purnell of the Love More Movement, for instance—a community-based initiative that helps to provide support for seniors and youth—is a fifth-generation descendant of abolitionists who organized the Underground Railroad with Frederick Douglass. Similarly, Black churches, community centers and senior gatherings are centers of gravity for community building and cohesion. But as new waves of real estate development take over, they risk whitening and sanitizing these culturally rooted forms of belonging.

Decolonizing urban greening

Despite an equity-driven progressive agenda embedded in new efforts such as the Bridge Park, Anacostia still risks becoming a new white and upper-class racialized greenscape. Today a partially outsider-driven green virtue seems underpinned by a form of whiteness which risks eventually "activating" formerly no-go zones through diluted memories of the past, but whose unsavory history of racial settlement and legacy (as a deep memory) is left obscured or invisible. In the southeast, it is the profound racialized inequality that allows greening, sustainability and corporate financial capture to advance.

The projection and the imagination deployed around the Bridge Park project risks effacing dissonant green and land experiences. In contrast, developing an abolitionist green practice in Anacostia would mean recognizing the centrality of land recognition, redistribution and control for residents. Put differently, to understand and value the relations of deeply rooted groups in a landscape, one must deal with the history of racialized multi-layer geography. Engaging with that racialized landscape and history would help achieve a different outcome for green and sustainable Anacostia, rather than new racialized landscapes of vulnerability and displaceability. Thus, a final question emerges: what are the best ways to rehumanize a space, to value the soul, memory and history of its residents and create a new sense of (green) ownership and leadership through green spaces?

Keywords

(a) the urban development pattern of the city and neighborhood: race and class segregation; income inequality; recovering city; neighborhood redevelopment
(b) the urban greening of the neighborhood: river restoration; green space upgrading; conversion of grey infrastructure; bridge park; green design
(c) the inequalities at stake: green gentrification; luxury developments; affordable housing crisis; high-end commercial development; displacement

Notes

1 Some of the text in this book chapter has appeared in: Anguelovski, I., A. L. Brand, M. Ranganathan, and D. Hyra. 2021. "Decolonizing the Green City: From environmental privilege to emancipatory green justice" *Environmental Justice.*
2 https://statisticalatlas.com/neighborhood/District-of-Columbia/Washington/Anacostia/Race-and-Ethnicity
3 https://www.bizjournals.com/washington/news/2019/03/01/barry-farm-isdown-to-three-households-a-new.html
4 https://www.sellingdc.com/navy-yard-homes.php
5 https://bbardc.org/wp-content/uploads/2018/10/Equitable-Development-Plan_09.04.18.pdf
6 https://www.bisnow.com/washington-dc/news/economic-development/mayor-bowser-proposes-tif-bond-for-anacostias-reunion-square-development-93498
7 https://www.trulia.com/for_sale/1850_nh/

References

Anguelovski, Isabelle, Anna L. Brand, James J. T. Connolly, Esteve Corbera, Panagiota Kotsila, Justin Steil, Melissa Garcia Lamarca, Margarita Triguero-Mas, Helen Cole, Francesc Baró, Johannes Langemeyer, Carmen. Perez del Pulgar, Galia Shokry, Filka Sekulova, and Lucia Arguelles. 2020. "Expanding the boundaries of justice in urban greening scholarship: Towards an emancipatory, anti-subordination, intersectional, and relational approach." *Annals of the American Association of Geographers* 110(6):1743–1769.

Asch, Chris Myers, and George Derek Musgrove. 2017. *Chocolate City: A History of Race and Democracy in the Nation's Capital*. Chapel Hill, NC: UNC Press Books.

Avni, Nufar. 2018. "Bridging equity? Washington, DC's new elevated park as a test case for just planning." *Urban Geography* 40:1–18.

Cowell, Margaret, and Heike Mayer. 2015. "Anchoring a federal agency in Washington, DC Community: The Department of Homeland Security and St Elizabeths." In *Capital Dilemma: Growth and Inequality in Washington*, edited by Derek Hyra and Sabiyha Prince, 207–224. New York: Routledge.

Hyra, Derek. 2015. "The back-to-the-city movement: Neighbourhood redevelopment and processes of political and cultural displacement." *Urban Studies* 52(10):1753–1773.

Hyra, Derek. 2017. *Race, Class, and Politics in the Cappuccino City*. Chicago, IL: The University of Chicago Press.

Hyra, Derek, Mindy Fullilove, Dominic Moulden, and Katharine Silva. 2020. *Contextualizing Gentrification Chaos: The Rise of the Fifth Wave*. Washington, DC: Working Paper, The Metropolitan Policy Center.

Hyra, Derek, and Sabiyha Prince. 2015. *Capital Dilemma: Growth and Inequality in Washington*. New York: Routledge.

Long, Joshua. 2016. "Constructing the narrative of the sustainability fix: Sustainability, social justice and representation in Austin, TX." *Urban Studies* 53(1):149–172.

Loughran, Kevin. 2014. "Parks for profit: The high line, growth machines, and the uneven development of urban public spaces." *City & Community* 13(1):49–68.

McKittrick, Katherine. 2011. "On plantations, prisons, and a black sense of place." *Social & Cultural Geography* 12(8):947–963.

Nieves, Emanuel, Chuck Collins, and Josh Hoxie. 2017. *The Road to Zero Wealth: How the Racial Wealth Divide is Hollowing Out America's Middle Class*. Washington, DC: Prosperity Now.

Prince, Sabiyha. 2016. *African Americans and Gentrification in Washington, DC: Race, Class and Social Justice in the Nation's Capital*. New York: Routledge.

Rabinowitz, Kate. 2017. *A Decade of Demographic Change in D.C.: Which Neighborhoods Have Changed the Most?* Washington, DC: DC Policy Center.

Ranganathan, Malini, and Eve Bratman. 2019. "From urban resilience to abolitionist climate justice in Washington, DC." *Antipode* 53:115–137.

Richardson, Jason, Bruce Mitchell, and Juan Franco. 2019. *Shifting Neighborhoods: Gentrification and Cultural Displacement in American Cities*. Washington, DC: National Community Reinvestment Coalition (NCRC).

Safransky, Sara. 2014. "Greening the urban frontier: Race, property, and resettlement in Detroit." *Geoforum* 56:237–248.

Summers, Brandi Thompson. 2019. *Black in Place: The Spatial Aesthetics of Race in a Post-Chocolate City*. Chapel Hill, NC: UNC Press Books.

Yiftachel, Oren. 2020. "From displacement to displaceability: A southeastern perspective on the new metropolis." *City* 24(1–2):151–165.

Zhan, Abby. 2019. *Climate of Change: Envisionning Climate Justice from DC's Front Lines*. New York: Parsons School of Design, The New School.

Zippel, Claire. 2016. *DC's Black Residents Increasingly Live East of the Anacostia River*. Washington, DC: DC Fiscal Policy Institute.

13

ADDRESSING GREEN AND CLIMATE GENTRIFICATION IN EAST BOSTON

Galia Shokry and Isabelle Anguelovski

New green assets in the coastal community of East Boston have enhanced its environmental and economic value, while creating new sources of housing vulnerability and social exclusion for historically marginalized groups. These hard-won green investments were intended to provide new community green spaces and mitigate pollution from activities of the adjacent airport. However, as the latest luxury real estate investments multiply along the recently greener, more resilient waterfront, residents are becoming increasingly more vulnerable—this time through green climate gentrification and the creation of "islands of resilience" that increasingly exclude residents of color and privilege new white residents. In this chapter, we investigate how, for those already at risk of increased flooding and flooding-related costs, the transformed coastline presents a new source of displacement due to climate gentrification and the rapid whitening of a Latinx neighborhood.

From environmental injustice to a newly greened and reinvested neighborhood

Home today to a majority Latinx population, East Boston took a critical turn in its environmental trajectory in the mid-1990s when the historically park-deprived and heavily contaminated neighborhood began gaining a series of new large and small green spaces built over the next 20 years.

For decades, East Boston's working-class and minority residents had endured air, water and noise contamination from industrial activities along the Boston Harbor, Chelsea Creek and the nearby Logan Boston Airport (Douglas et al. 2012). It soon became renowned for clipper-style wooden shipbuilding, attracting mostly Irish immigrants who built wharves and rail facilities as well as houses and roads. By the turn of the 20th century, newcomers from Russia, Italy and

DOI: 10.4324/9781003183273-16

Portugal would find work in a growing diversity of industries including coal yards and textile factories. But with the post-war shift to a service economy, many larger industries closed. As Latin American and other immigrants from around the world began to settle in increasing numbers in East Boston, white families began to leave for the suburbs.

The communities of the East Boston peninsula were, for much of the 20th century, upheaved and fragmented by the expanding transportation sector needed to connect it to the mainland and beyond. These included a shipping port, the Sumner Tunnel, a subway connection to the mainland and several major traffic arteries. Perhaps most devastating was the creation of Logan International Airport in the 1920s which inflicted intensive air and noise pollution on neighboring residents, and its subsequent 1960s expansion that displaced residents, tore down trees and removed the beloved Olmsted-designed Wood Island Park enjoyed by locals.[1] Led by Massport, the airport's operator, the construction carried out with state troopers and demolition crews resulted in growing mistrust from residents and catalyzed environmental campaigns in protest.[2]

From 1995 onwards, in addition to initial water clean-up efforts, this organizing work spurred the development of new green spaces such as Piers Park, the East Boston Greenway (Figure 13.1), Bremen Street Park and Bayswater Street Park. Each of these was financed by Massport in an effort to provide 33 acres of green area to compensate residents for the environmental and health impacts of Logan Airport. The parks have been enormously popular with locals, and the Bremen Street Community Garden, which opened in 2007, has provided Latinx and other families the highly popular pastime of growing fresh vegetables, herbs and flowers and the possibility of enjoying a peaceful, green space in a dense and heavily trafficked neighborhood. Eastie Farm is another community garden started in 2015 by residents who turned an overgrown, underused lot into a space for growing fresh food, accessing urban nature, and building new community ties, while offering educational programs on composting, sustainable growing techniques and environmental stewardship and resilience. Much of Eastie Farm's

FIGURE 13.1 East Boston Greenway and Lopresti Park. (Photo by: Isabelle Anguelovski.)

programming is implemented in collaboration with the environmental nonprofit Harborkeepers and the Piers Park Sailing Center.

Most recently, the City of Boston began adding and redeveloping green spaces by partially relying on private developer funding. One such space is Lopresti Park, a remodeled green area with outdoor recreation facilities and playgrounds along the waterfront.

From the mid-2010s onwards, as part of municipal efforts to respond to climate risks, a series of new resilient parks and green infrastructure interventions have also been envisioned (e.g., Piers Park II and III). Such efforts are deemed essential for a neighborhood where estimates warn that half of the land could be flooded during a major storm in the next 50 years (City of Boston 2017). Green resilient infrastructure is indeed included in the 2017 East Boston Climate Ready Plan and the Resilient Boston Harbor plan, two comprehensive municipal initiatives to build elevated berm landscapes, resilient shorelines and flood-mitigating parks along the city's 47 miles of waterfront.

Enhancing the environmental quality and resilience of the East Boston waterfront has also been taken on by developers (e.g., Lendlease with the Clippership Wharf project) in compliance with Massachusetts General Law Chapter 91,[3] the state's "primary tool for the protection and promotion of public use of its tidelands and other waterways." Thanks to this statute, developers have been obliged to deliver new open, public space when they build new housing. In East Boston, these include a living shoreline with the planting of a salt marsh and the creation of a new habitat for marine life and diversity.

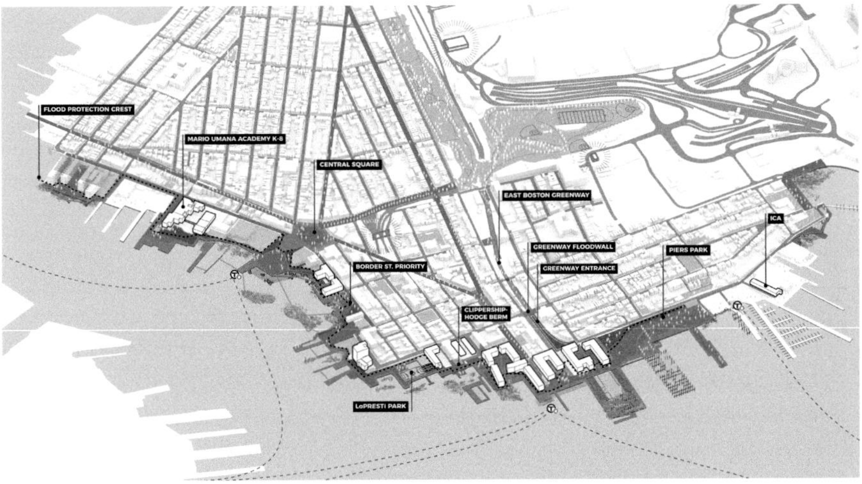

FIGURE 13.2 Planning new green resilient infrastructure for East Boston. (Source: City of Boston.)

New green resilient spaces meet recent gentrification pressures

Despite the potential for generating tremendous social and health benefits for longtime residents, some of these new green projects have become tools for accelerating luxury real estate investment which has garnered concern and criticism from community leaders.

Especially in the context of a cleaned-up and greened waterfront, green gentrification in East Boston is illustrated by the rapid, broad-scale and uncoordinated real estate development projects targeting the neighborhood. Development is indeed anchored by greening and legitimized by new green spaces and parks that are ostensibly for the benefit of the community. A representative from a youth-centered community organization explained,

> No one was willing to take a risk with the land. It was not economically appealing. But Massport had the land turned into a park, and then redeveloped the rest. They built the park as a form of remediation. Then all of a sudden it leased the land after they did the beautiful park and built this monstrosity [a new real estate complex]. Greening served them. Greening serves as motivation; it shows you are a good neighborhood.

Community leaders' interpretation of the role played by land clean-up and greening as a marketing tool, investment strategy and sales pitch for developers is further confirmed by developers themselves:

> The greenway is a green benefit because it connects two aspects of the shoreline. We advertise the greenway as an amenity for residents. A portion of what we opened as public, open, green space is a requirement and we are making it an advantage. We built a story—that we are super resilient, super sustainable. We have an interesting piece with the living shoreline project. Buyers have loved the story of openness, outdoor living and open space.

Some of the long, connective, linear green spaces such as the greenway have paved the way for higher housing prices and new developments. As some developers put it, they "activate" a space and a neighborhood, connect key destinations and help secure the visibility and success of new housing projects while attracting wealthier residents to the neighborhood.

While much of the post-industrial East Boston waterfront land was acquired by developers before the 2008 financial crisis, it has mostly been redeveloped over the last six years. By February 2018, 1,251 units were under construction or planned in East Boston, the large majority of which were market-rate units (Acitelli 2018). On Haynes Street, 2017 city records showed 18 housing developments as either approved, under review or under construction within a 10-block radius (Rios 2017). Among those, Boston East on Border Street is a $71 million residential project built in 2018—branded for its proximity to the green

Harborwalk—with 174 market-rate apartments and 26 affordable units (Acitelli 2018).

Some of the new green developments illustrating rampant gentrification include the Clippership Wharf inaugurated in 2019, a residential project comprised of 478 condo and rental units starting at $2,300 and four acres of open space including Harborwalk access, a beach area and floating docks;[4] and The Eddy,[5] another luxury waterfront residential complex inaugurated in 2016 comprised of 258 rental units starting at $1,800 for studio apartments, a pool deck and a high-end restaurant and bar. The construction of The Eddy included 0.75 acres of outdoor public open space (part of which extends the East Boston Harborwalk).

Facing a multi-faceted green and climate gentrification wave

Today, green gentrification is felt both physically and socio-culturally, as gentrification scholars have argued (Marcuse et al. 2009). Greening is associated with experiences of increased housing pressures and the creation of an exclusive, green resilient neighborhood.

Physical displacement from soaring housing costs, saturated development and recent green infrastructure

Housing affordability has become the first concern for both low-income and middle-class residents in East Boston. Considering that 70% of East Boston residents are renters, current average rental prices of $1,700 are unaffordable for many (Chief of Housing, City of Boston, City Council Committee Hearing Meeting, 18 July 2018). Overall housing prices rose by 52% since 2010, leaving more than 50% of residents currently rent-burdened (M.A. Senator Joe Boncore, City Council Committee Hearing Meeting, 18 July 2018) or unable to purchase a home.

Tenant families are particularly affected by housing vulnerability and eviction, often unable to secure new housing in the area and destabilized by the pressures of moving to another neighborhood. Youth also struggle with adapting to a new school and community of peers. A community organization leader recounts these traumatic experiences:

> There is an incredible amount of stress because of eviction notices and rent doubling. In the past three years we've had a massive increase in the amount of discussion around suicide here and people feeling suicidal, and we've had to get training for our staff to address it.

In a booming development context, East Boston homeowners are also concerned about increased property tax bills and home insurance rates due to flood risks. One resident reported that her flood insurance rate increased from $500 in 2013 to $3,000 in 2018. As an environmental activist summed up, "Residents seem to be afraid of both being wiped out by a storm and by the price of flood insurance." Such housing cost increases are particularly difficult to absorb in East Boston where two-thirds of homeowners are low to moderate income. Put differently, environmental and greening factors play a double role in housing insecurity—both by creating new value and increased direct housing costs and by creating losses and increased housing costs through flooding risks.

As a result, increases in housing prices and costs are the predominant drivers of displacement in the neighborhood. Displacement is illustrated by four trends mentioned by our respondents. These are (1) declining school enrollment and subsequent budget loss; (2) patient loss at local health centers such as the East Boston Community Health Center; (3) increasing East Boston eviction cases in the East Boston Court (which many housing rights organizations and city of Boston officials explain are actually under-reported, especially so for immigrant families facing immigration threats); and (4) the numerous anti-eviction workshops held by housing groups such as City Life/Vida Urbana, Neighbors United for a Better East Boston, or Center for Cooperative Development and Solidarity. One City Life staff member explained:

> Residents are leaving before we call it a displacement because of the pressures that they see; they don't want a housing eviction on their record and they don't want to go through the court process because court can be really intimidating, particularly in an administration that has immigration and customs enforcement inside court houses.

Physical displacement is also accelerated by the number and size of development and construction projects scattered throughout the neighborhood, which residents and local lawmakers perceive as a land and construction oversaturation. During the early and mid-2000s, development at this rate—with little municipal oversight of real estate permits and a lack of strategic neighborhood plans or guidelines—was called "spot zoning" by City Councilor Lydia Edwards (Valencia 2018) or "zoning by variance." Long-term residents experience this as "total and complete exclusion and … lack of control."

Residents from social housing developments feel particularly overwhelmed and spatially compressed by these new developments. For example, the 396-unit mixed-income Boston Housing Authority (B.H.A.) Maverick Landing development sits across from The Mark, a 9-story luxury building currently under development with the promotional tagline: "Welcome to East Boston's new vantage point … The exterior amenities at the Mark are designed as your own perennial sanctuary in the city."[6] The Mark follows a series of large-scale waterfront developments next to Maverick Landing, such as The Eddy to the west, which block existing residents' view of the waterfront and void their sense of connection to the new environmental assets that developers rave about.

Green resilient infrastructure in East Boston also rouses displacement fears among residents and activists, even though the Boston Harbor and East Boston Climate Ready Plans bring special attention to the environmental vulnerability of lower-income residents and pledge to create new wealth opportunities. Residents and experts alike are particularly fearful of the role of private banks and developers as the city seeks funding to execute the plans.[7] When private-sector actors play a strong role in urban projects, the most socially vulnerable residents risk being physically displaced, both by the plan itself and eventually, by the rebranding and selling off of their neighborhood as a new (luxury) green resilient landscape (Anguelovski et al. 2016, Teicher 2018). As new research by Shokry et al. (2020) shows, while wealthier residents benefit from this increased climate security, low-income residents may be displaced to unprotected neighborhoods where they face new climate risks and impacts.

In addition, the heavy, large-scale housing infrastructure of new real estate projects such as Clippership Wharf and The Eddy pose new environmental risks (Figure 13.4). Existing terrains and homes located behind new housing complexes might be exposed to greater flooding and consequently, displacement (Anguelovski et al. 2019). One environmental expert and long-term activist explained the unequal environmental protection created by the new high-end "resilient" development projects.

> We've lost storm buffer space. That means making space where the stormwater can go. If you put concrete everywhere, where will the stormwater go? It turns into flooding. We're experiencing that. And for the people that you create affordable housing for, if the house is flooded, how affordable is it? And what if they have to pay flood insurance?

A climate planner also confirmed the lack of consideration for broader landscape and site needs around the newly built residences, which are referred to as self-protecting "islands of resilience." "It doesn't mean that the area around them, or even the roads that take you to and from those buildings, are safe."

Socio-cultural displacement arising from environmental amenities

Green gentrification in East Boston also manifests in the perceived exclusion of residents from new amenities. New housing units and health-oriented facilities such as bike storage, fitness centers, dog parks and spas seem mostly geared towards wealthy singles and couples rather than families. Residents report feeling increasingly like strangers in their own neighborhood. Other residents experienced a sense of physical and social disconnect with newcomers who seem to ignore them or impose new habits and social practices in parks and along the shoreline. A member of a local nonprofit organization reported,

FIGURE 13.3 Elevated luxury condo development and adjacent green infrastructure in East Boston. (Photo by: Isabelle Anguelovski.)

> There is also a dock for water taxis right in front of the building. When I saw it, my partner and I said: "So they can go directly from the downtown finance district to their house and they never have to actually go through East Boston."

The design of green spaces and their surrounding amenities is also a point of concern. Community organizations have reported that youth feel uneasy in new parks such as Lopresti Park or Harborwalk because of their spatial juxtaposition and contrast with the towering luxury housing located directly across the street from the park. Some youth say they feel unwelcome there, although many remain active users as a vindication of their right to use them.

Furthermore, a sense of environmental disconnection caused by the disrupted interaction with their coastal environment is depriving residents of opportunities to build resilience to climate impacts such as sea level rise and flooding. Many of those spaces were already associated with safety concerns due to residents' reduced familiarity and exposure to them. One environmental organizer explained how they had a

> sked for this coastal "community" to be allowed to be coastal, which means interacting with the water, learning about the environment, accessing any

place ... Being able to walk to The Eddy and Harborwalk and not feel out of place.

These physical disruptions jeopardize place attachment and sense of belonging, increasing risks of social isolation and undermining opportunities to engender meaningful relations.

Feelings of socio-environmental vulnerability in this multi-racial neighborhood also stem from the frequent lack of meaningful inclusion of and communication with residents and environmental stakeholders in the design and construction of green resilient projects, as occurred in the case of the municipal decision to deploy and test a flood wall on the Boston greenway in 2018. As scholars have recently argued, the inclusion of residential testimonies about the racial formation of insecure landscapes and the creation of community-driven resilient green alternatives are essential to avoid "colorblind adaptation planning" (Hardy, Milligan and Heynen 2017).

Today's abundance of recently built shoreline, beachfront, kayak docks and other recreational infrastructure only emphasizes the extent to which minority and immigrant residents have historically been deprived of access to green and blue spaces. Now, despite opportunities through local organizations to develop a greater sense of comfort and safety next to waterways, waterbodies and green spaces (Irwin et al. 2011, Finney 2014), the green and blue gentrification of the shoreline displacing those residents only serves as a reminder of that historical discrimination and exclusion.

Resisting green climate gentrification through anti-displacement organizing and policy-making

In response to green and climate gentrification concerns, organizers in groups such as City Life/Vida Urbana, Grassroots, N.O.A.H. and Harborkeepers have been working together to prevent displacement while building climate resilience.

Housing policies to keep families in their homes

Among the key struggles of tenant organizers is the "just cause" eviction act, also known as the Jim Brooks Community Stabilization Act, which promised to keep good tenants in their homes—in light of skyrocketing rents—while "also putting the brakes on some of the [condo] conversions happening," as one housing organizer explained. The act is meant to restrict landlords' ability to evict tenants where no fault is found, require non-binding mediation for sudden rent hikes, improve data collection on evictions and provide public legal representation for tenants facing eviction. It has faced immense challenges, however, especially from powerful property owners and real estate associations, that have managed to lobby the city council and state legislature into killing the bill. Many saw it as a step toward some form of eventual rent control, which housing advocates see as the most secure way to prevent current rent hikes of up to 200%.

Housing advocates have also turned to inclusionary zoning as a measure to ensure that some percentage—13% in East Boston—is reserved for affordable housing. However, many argue that the percentage should be higher, considering that the remaining 87% of new housing is being built at luxury rates. To make matters worse, in some cases developers are either paying a fine to circumvent the regulation or building off-site. Beyond that, one local organizer said, "First we have to have a conversation about what affordable really means because we have current affordability rates that a lot of residents cannot access."

In June 2018, a citywide ordinance established new guidelines and regulations to prevent the conversion of affordable long-term housing for residents into short-term, more expensive rentals for tourists. But in years prior to its enforcement in January 2019, many families in East Boston had already lost their homes to Airbnb conversions, helping transform the landscape of a working-class neighborhood into a tourist destination. According to Boston City Councilor, Michelle Wu, there were "anywhere from 2,500 to 4,000 units that could be returned to the residential housing stock of Boston" if the law was properly enforced.

The disrepair of East Boston's older housing stock—sometimes due to flooding—has also meant that affordable homes are being torn down and converted into high-rent housing after evicting existing tenants. Housing organizers argue that instead

> the city [should] offer a comprehensive system for landlords to place their tenants in transitional housing while their buildings are torn down and then apply the same benefits given to developers in terms of relief for zoning, to the new resilient buildings that could be built.

On the other hand, one planner also insisted that

> we really need to ensure that places where we see folks most vulnerable are no longer in harm's way and we are not reinforcing putting them in harm's way. And certainly not in flood zones … [that is] to make sure that affordable housing is built in a safe place.

Guaranteeing long-term community ownership and stability

Community land trusts (C.L.T.) and other community stabilizing measures are also advocated as a way to "incentivize use of land rather than speculation". In response to low occupancy rates of new luxury housing and holding practices by developers, activists have suggested that "by essentially tax[ing] land at a higher rate and then maybe decreas[ing] income tax you won't have speculators sitting on land waiting for the market to do this or that, or to turn it over or sell." According to another activist, "If you're just going to have your house sit empty, then you [at least] have to contribute to our free transit system and you're going to contribute to our great public schools!" CLTs are also often advocated as a tool to protect resident-driven greening such as urban farms and community gardens.

These and other policy proposals, such as a minimum income guarantee, and greening efforts by groups like Chelsea GreenRoots which combine greening with cooperative economic development, require steadfast political courage, strong local support and in the case of immigration policy change, coordination at higher levels of government. They also underscore the extent to which climate solutions that protect the most vulnerable must address the myriad ways in which historically marginalized residents grapple with insecurity.

To preserve the greening wins of recent decades, and despite feeling unwelcome in their own neighborhood, locals continue to claim their right to green and blue spaces. Nonprofit organizations such as Harborkeepers are organizing environmental education, safety and stewardship activities to build a new sense of connection and care between long-term residents and their natural environment. "People never thought this could happen in East Boston, kids learning

FIGURE 13.4 Community garden in Bremen Park, East Boston. (Photo by: Isabelle Anguelovski.)

how to make rope, learning how to tie knots, learning about sails, learning about boats," one member said. Yet, as residents are pushed out of their neighborhood, they lose access to these customized programs and face possible displacement to less climate-protected and more underserved areas.

The compounded socio-climate vulnerabilities and injustices of East Boston

New green assets have enhanced the environmental and economic value of East Boston, but they have also created new sources of housing vulnerability and social exclusion for historically marginalized groups while benefiting whiter, more privileged newcomers. At the same time, air and noise pollution from the airport, construction and highways, salt piles and jet fuel tanks along the Chelsea River continue to overwhelm longtime residents as they newly engage in a struggle against plans to install an electrical substation near a flood zone. Such vulnerability and environmental injustice are compounded by the risk of increased flooding and flooding-related costs and projects. While parts of Latinx East Boston have become greener and more resilient thanks to community efforts, it may well be that these environmental elements become a new source of displacement, this time by climate gentrification.

Keywords

(a) the urban development pattern of the city and neighborhood: fast-growing city; tech, financial and health economy
(b) the urban greening of the neighborhood/city: environmental clean-up; new parks and gardens; green climate adaptation
(c) the inequalities at stake: environmental gentrification; environmental injustices; green climate gentrification; remaining contamination

Notes

1 https://www.tpl.org/magazine/parks-lost-and-found%E2%80%94landpeople
2 http://www.openmediaboston.org/content/east-boston-neighborhood-close-destruction-after-40-year-fight
3 https://www.mass.gov/guides/chapter-91-the-massachusetts-public-waterfront-act
4 https://www.elevatedboston.com/buildings/clippership-wharf/
5 https://www.elevatedboston.com/buildings/the-eddy/
6 www.themarkeastboston.com
7 https://www.wbur.org/earthwhile/2019/05/02/moakley-park-east-boston-climate-resiliency

References

2005. "The Widening Racial Scoring Gap on the SAT College Admissions Test." *The Journal of Blacks in Higher Education* 49:88–93.

Acitelli, Tom. 2018. "East Boston's Major New Developments, Mapped." *Curbed Boston.*

Anguelovski, Isabelle, James J. T. Connolly, Hamil Pearsall, Galia Shokry, Melissa Checker, Juliana Maantay, Kenneth Gould, Tammy Lewis, Andrew Maroko, and J. Timmons Roberts. 2019. "Opinion: Why Green "Climate Gentrification" Threatens Poor and Vulnerable Populations." *Proceedings of the National Academy of Sciences* 116(52):26139–26143.

Anguelovski, Isabelle, Linda Shi, Eric Chu, Daniel Gallagher, Kian Goh, Zachary Lamb, Kara Reeve, and Hannah Teicher. 2016. "Equity Impacts of Urban Land Use Planning for Climate Adaptation: Critical Perspectives from the Global North and South." *Journal of Planning Education and Research* 36(3):333–348. doi: 10.1177/0739456x16645166.

Campbell, Scott, and Susan S. Fainstein. 2003. *Readings in Planning Theory.* 2nd ed. Malden, MA: Blackwell Publishers.

City of Boston. 2017. *Climate Ready East Boston.*

Douglas, Ellen M., Paul H. Kirshen, Michael Paolisso, Chris Watson, Jack Wiggin, Ashley Enrici, and Matthias Ruth. 2012. "Coastal Flooding, Climate Change and Environmental Justice: Identifying Obstacles and Incentives for Adaptation in Two Metropolitan Boston Massachusetts Communities." *Mitigation and Adaptation Strategies for Global Change* 17(5):537–562.

Finney, Carolyn. 2014. *Black Faces, White Spaces: Reimagining the Relationship of African Americans to the Great Outdoors.* Chapel Hill, NC: UNC Press Books.

Hardy, R. Dean, Richard A. Milligan, and Nik Heynen. 2017. "Racial Coastal Formation: The Environmental Injustice of Colorblind Adaptation Planning for Sea-Level Rise." *Geoforum* 87:62–72.

Irwin, Carol C., Richard L. Irwin, Timothy D. Ryan, and Joris Drayer. 2011. "The Legacy of Fear: Is Fear Impacting Fatal and Non-fatal Drowning of African American Children?" *Journal of Black Studies* 42(4):561–576.

Marcuse, Peter, James Connolly, Johannes Novy, Ingrid Olivo, Cuz Potter, and Justin Steil. 2009. *Searching for the Just City: Debates in Urban Theory and Practice.* New York: Routledge.

Rios, Simon. 2017. "As Construction and Tides Rise, Group Aims to Make Eastie 'More Resilient' against Climate Change." *WBUR.*

Teicher, Hannah M. 2018. "Practices and Pitfalls of Competitive Resilience: Urban Adaptation as Real Estate Firms Turn Climate Risk to Competitive Advantage." *Urban Climate* 25:9–21.

Valencia, Milton. 2018. "Don't Let Eastie become Southie, Residents Say." *The Boston Globe.*

PART 4

The complex entanglement of greening and multiple other gentrification pressures

Part 4, "The complex entanglement of greening and multiple other gentrification pressures," analyzes how greening is often not a single dimension of neighborhood change and redevelopment but is rather embedded in other dynamics that create additional pressures on the lives of marginalized residents. These might include food gentrification, tourism and commercial gentrification or student gentrification. Especially in circumstances where there are visible associated drivers of redevelopment and injustice, as in Montreal, Dublin, Barcelona and Philadelphia, greening is co-occurring next to other clear drivers of new inequities and gentrification.

In Montreal, the environmental quality of the historically working-class Saint-Henri neighborhood has been improving in recent decades through the "extraordinary greening" of the Lachine Canal and adjacent food gentrification pressures. However, tensions that have arisen between community groups and the local government around the new emblematic, top-down greenspace have led to hopes for a more environmentally restored neighborhood through the "ordinary greening" led by neighborhood residents.

While Dublin hosts the largest park in Europe, the historic working-class neighborhood of The Liberties has the least amount of green space in Ireland and continues to harbor low-quality parks and playgrounds. Yet, as The Liberties embarks on an official Greening Strategy, intersecting environmental inequities and scarcities—in green space and quality, affordable housing—in a context of city-wide uncontrolled real estate growth benefiting hotels and international student housing, represent dichotomous challenges for achieving sustainable revitalization.

DOI: 10.4324/9781003183273-17

Barcelona, even without abundant green spaces in its urban core, is known for its high quality of life, walkable neighborhoods, livability and tourism. As city greening, beautification and traffic-reducing projects strive to provide locals with a better city, those transformations have unexpected and unintended consequences, as they bring in more outside visitors and accelerate gentrification. While the current COVID-19 pandemic is drastically reducing the number of tourists to Barcelona, producing a new experience of green space access for local residents, improved access to green space for all is not a guarantee.

The City of Philadelphia has long engaged large-scale non-profits to implement greening and sustainability planning goals, a strategy that has evolved into an expansive set of programs which today are increasingly targeting climate risks and impacts. In recent years, however, the city and non-profit community have come to recognize that their efforts may not have resulted in the most equitable citywide outcomes, with central and gentrified neighborhoods becoming greener, cooler and more permeable, while more peripheral lower-income neighborhoods of color remain some of the hottest and most flood-prone, therefore posing a dire challenge to residents' health and security.

14

ORDINARY AND EXTRAORDINARY GREENING

Tensions amidst Saint-Henri, Montréal's development boom

Melissa García-Lamarca and Aaron Vansintjan

The neighborhood of Saint-Henri in Montréal's Southwest borough has long been associated with poverty, marginality and squalor. But this is rapidly changing as both extraordinary, large-scale green infrastructures and small-scale, more ordinary forms of greening are expanding across the neighborhood, amidst private luxury housing development and rising rents. Both extraordinary and ordinary greening are also connected to Saint-Henri's transforming foodscape, where new gourmet restaurants and up-scale cafés, a renovated farmers' market and renovated grocery stores are displacing the diners, *dépanneurs* (corner stores) and other food shops long frequented by working-class residents. What happens when, all at once, a community faces food gentrification, small-scale greening projects and large-scale green infrastructure? In this chapter we explore the greening-related tensions and inequities that are unfolding in Saint-Henri, where new multi-scalar greening projects and foodscapes are stitching together a post-industrial landscape to create new—and often exclusionary—forms of urban living. We also show how, at the intersection of these tensions, local community groups have resisted and fought for alternative forms of development on multiple scales.

Saint-Henri in transformation, for whom?

At the heart of Canada's industrial revolution, Saint-Henri is one of Montréal's original working-class neighborhoods. It is seared in the minds of many Canadians by Gabrielle Roy's classic 1945 novel *Bonheur d'occasion* (The Tin Flute), a work that evokes the struggle, hardship and marginality that has long characterized the area. After the Lachine Canal was opened in 1825 and the Montreal & Lachine railroad was inaugurated two decades later, Saint-Henri was established as a key settlement for burgeoning industries and workers who lived

DOI: 10.4324/9781003183273-18

in overcrowded and poor-quality housing. Cotton, iron, steel and metal industries were predominant in Saint-Henri by the end of the 19th century (Lewis 2001) and, after decades of continued growth, the municipality was annexed to the city of Montréal in 1905. Industrialization thrived until the 1920s, generating abysmal working conditions and extensive slum housing as well as a thriving working-class organizing culture.

Saint-Henri's urban fabric was torn apart in the 1960s when the Lachine Canal closed, and the massive expressway and multilevel highway known as the Turcot interchange was built across the neighborhood's northern boundary. This was also a period of heightened activism, with Saint-Henri hosting Montréal's first citizen committees and the growth of many popular institutions across the neighborhood (McGraw 1978, Bélanger and Lévesque 2004, Twigge-Molecey 2014) demanding social housing, community healthcare and neighborhood democracy. No longer used as an industrial transport and energy hub, a local long-term resident explained that "by 1985, the canal was full of stolen cars, stolen bicycles, a few dead bodies, beavers and foxes, deer … you didn't put your foot in the water unless you wanted to watch it dissolve." It was lined by derelict warehouses, abandoned malt silos and empty red-brick hydraulic works—and many of its residents moved to the suburbs if they could afford it.

Fast forward to today, and Saint-Henri's status is rapidly changing. Triggered in part by the decontamination of the Lachine Canal in the 1990s, new luxury housing development has been steadily expanding across the neighborhood, coexisting with the poor-quality housing of long-term working-class residents and significant social and cooperative housing stock. Due to the fact that 77% of a city's financial income in the province of Quebec comes from property taxes, a local borough council underlined to us that it is in a city's "best" financial interest to both encourage luxury housing and increase urban density with new residential buildings. The extensive rezoning and facilitation of real estate development is making the Southwest one of the wealthiest boroughs in the city, although this begs the question of who is benefiting from this rapid wealth generation and the accompanying environmental improvements.

Placing focus on these environmental improvements, this chapter explores key forms of "ordinary" and "extraordinary" greening, concepts used to describe how variegated greening processes overlap, generate tensions between different actors and reshape local residents' lives and their resistance to gentrification. Within the broader context of green and food gentrification—processes referring to how environmental amenities and green developments drive the (re)valorization of a neighborhood and how cultural food spaces may drive, or be affected by, gentrification (Gould and Lewis 2017, Anguelovski 2016)—we unpack how new multi-scalar greening projects and contemporary foodscapes are stitching together a post-industrial landscape to create new, and often exclusionary forms of urban living. We also consider how community groups are resisting and fighting for more equitable forms of development at multiple levels. We show how community activists understand gentrification as part of a broader, multi-scalar ecosystem of development and, in response, orient their similarly ordinary and extraordinary resistance on multiple scales.

Extraordinary greening: Lachine Canal decontamination and the Woonerf

The Lachine Canal and its surrounding landscape began to change dramatically in the 1990s. Over the following decade, more than $100 million in federal and provincial funds were dedicated to extensively decontaminate and regenerate the canal. The 14-kilometer-long waterway now offers one of the best bike paths in Montréal, weaving its way alongside and through many green and blue amenities.[1] Alongside the decontamination process, numerous brownfield sites next to the canal were gradually transformed for residential, recreational and touristic use (Bliek and Gauthier 2007), beginning with the conversion of existing buildings into creative class commercial space and luxury lofts. The canal is an exemplary case of what we refer to as extraordinary greening: a large-scale, emblematic greening project requiring significant funding, largely delivered through a top-down and publicly funded process.

New real estate developments have sparked gentrification concerns amongst local residents, with many feeling that the canal has suffered from a purposeful cycle of public-sector neglect for decades followed by capital infusion from the state, with the private real estate sector—and subsequent private property

FIGURE 14.1 High-end real estate developments along the Lachine Canal. (Photo by: Melissa García-Lamarca.)

owners—capturing the improved environmental value through higher real estate values and prices. A local housing organizer working with Saint-Henri's housing committee, P.O.P.I.R.—the Popular Organization, Information and Uniting Project, a member-based community organization focusing on tenant and housing rights—explained that much of the canal access in Saint-Henri has been either directly or indirectly privatized by these new private condo developments. As he said: "even if the entrance to this project is not privatized, by having no low-income housing around this green project you are kind of privatizing … all the area around it, the access to this park for the people." P.O.P.I.R. grew out of labor and tenant organizing in the 1960s and is one of the oldest housing committees[2] in Montréal, having successfully organized tenants and defended housing rights for decades.

A more recent green addition to Saint-Henri is the St-Pierre Woonerf, a Dutch "living street" concept for pedestrians, cyclists, children, community growing and general green living.[3] In the late 2000s the Southwest borough council pitched the idea to green a wide gray alleyway running parallel to the Lachine Canal from Rue Saint-Rémi to Chemin de la Côte-Saint-Paul that was used for car parking —"a public space that could not be used for anything else" as expressed by a borough councilor—and received funding from the Quebec Public Health Agency to create the Woonerf. This new green infrastructure sought to combat the heat island effect, an issue of concern in the city's increasingly hot summers. Montréal's Urban Ecology Centre led the participatory planning process,[4] and the decision was made to transform the previous back alley "no man's land" into an innovative public space. Lauded with several sustainability awards, the Woonerf is viewed by the borough as a huge improvement from environmental and social points of view.

Local housing organizers and community groups, however, disagree. While the consultation process was largely lauded by the media,[5] the input generated by local youth through an innovative participatory design project carried out in a nearby high school was disregarded (Heffez and Bornstein 2016). Furthermore, there is broad consensus among local groups that the Woonerf, inaugurated in 2013, was very much a top-down project implemented by the borough because it was an innovative, "cute" green project, with no consideration of how it would affect rents. This is of particular concern because this pocket of Saint-Henri was the most poverty-stricken sector of the neighborhood and the last frontier of gentrification. While social housing exists along one small portion of the half-kilometer-long Woonerf, new condo developments have been built with sales pitches boasting their location on the "living street." Furthermore, a local resident and former housing organizer pointed out how the outdoor exercise equipment on a portion of the Woonerf is largely used by white, middle-class condo owners. Once the Woonerf was completed, a lot of criticism emerged from local residents around the cost of the project ($1.5 million), standing in stark contrast to the local schoolyard play space that a community organizer noted "looks like a prison yard." Finally, the name of the space is also alien to local residents, as it is one adopted by the city to emulate

FIGURE 14.2 The Saint-Pierre Woonerf, with exercise equipment in the distance. (Photo by: Melissa García-Lamarca.)

innovative European urban planning trends and market the space as an extraordinary green one.

The Woonerf has generated significant impacts on the surrounding housing stock. Several requests have been made for demolitions near the Woonerf, and many properties along the green street have been converted from multi-unit to single-family condominiums. Both processes are noted by housing organizers as clear signs of gentrification: by increasing the size of properties, the number of available and affordable housing is reduced. An article from 2019 in Canada's *National Observer*[6] further discusses the green gentrification impacts of the Woonerf: for example, through a door-to-door survey, the housing rights group P.O.P.I.R. found that 45% of households that flank the Woonerf have lived there for less than one year.

Ordinary greening: resident-led Green Alleys, housing-environmental conflicts and food gentrification

These extraordinary greening initiatives have occurred alongside more ordinary greening projects and the transformation of long-existing local foodscapes. A number of Green Alleys[7] (*Ruelles Vertes*) have been developed in the past decade across the neighborhood as part of a city of Montréal-wide program that boasts over 700 alleys with the stated goal of expanding urban agriculture in the city. Unlike the Woonerf, this project takes advantage of already-existing alleyways common in most neighborhoods that were previously concretized or underused. Such projects are funded by each Montréal borough, coordinated by local environmental organizations known as *Écoquartiers* and implemented by local residents. It is these residents who must first formally organize and apply for funding to convert their alley into a green one, a process then supported by their local *Écoquartier.* These Green Alleys differ from the Woonerf not only in their bottom-up process but also in the much smaller amount of capital available and scale of transformation.

Tensions have arisen between housing committees and environmental organizations in Saint-Henri over the role local Green Alleys play in processes of gentrification. Both agree that real estate developers use Green Alleys to increase housing prices and market real estate projects. Yet conflict sprouted between environmental and housing rights groups when the latter lumped the Woonerf and Green Alleys together as one and the same. The local *Écoquartier* outlined critical differences between the two in process and outcome. In contrast to what many residents feel was a top-down development in the case of the Woonerf, Green Alleys in Saint-Henri are driven by people living immediately on the alley. Local residents initiate these projects because they are interested in creating safe play spaces for kids or a quiet car-free haven to rest in, with the added bonus that social fabric is built as residents take charge of building the green space. According to the local *Écoquartier,* it is largely housing cooperative members, not private condo owners or upper-middle-class people in the neighborhood, who participate in creating Green Alleys.

Ultimately, local housing committees, community organizations and environmental groups agreed that the core issue is not greening per se, but rather greening with no limits to speculation. In other words, it's fine to support greening projects but action must be taken to keep (green) gentrification from displacing residents and transforming the character of the neighborhood. Indeed, this tension around greening is recognized by many housing committees across Montréal, expressed through a banner at a 2017 gentrification protest reading "your parks are nice but our rents are too high!" As one organizer explained: "it's really cool to have a green space near your house but if you have mold in your apartment you're not really much healthier."

In terms of transformed foodscapes and new drivers of gentrification in Saint-Henri, the most visible example is the arrival of high-end, gourmet restaurants on the neighborhood's main thoroughfare Notre-Dame. Long-term residents complain that many of their favorite diners, clothing stores, second-hand stores and bars have closed since the early 2000s. Many *dépanneurs*, small corner stores that the largely working-class population relied on for food, have also closed or been renovated to appeal to new residents. Today, Notre-Dame Street, lined with brunch restaurants and cocktail bars, looks nothing like it did just one decade ago.

FIGURE 14.3 New brunch restaurants jostling for space amidst the few remaining diners and bars on Rue Notre-Dame. (Photo by: Aaron Vansintjan.)

While many low-income residents admitted that they thought the area was now safer, they did not frequent any of the new restaurants and stores due to their elevated prices. The *Guardian*[8] reported several cases of high-profile vandalism by anonymous anti-gentrification activists, including a break-in of various boutique organic stores with the aim of redistributing food.

Food shopping spaces in Saint-Henri are also upscaling and thereby excluding many long-term local residents. Since the renovation of Atwater Market, an iconic farmers' market on the canal, low-income, long-term residents mentioned that they no longer shop there because it is too expensive and increasingly oriented to newer residents and tourists. Even a large nationwide supermarket chain located in Saint-Henri, I.G.A., has changed the kind of food it offers and its prices. As one community organizer noted, "When the neighborhood began to gentrify, they [IGA] moved closer to the [Atwater] market, they raised their prices enormously. Yes, the IGA looks nicer now, but it's less affordable." On the other hand, western Saint-Henri, which is still predominantly a low-income area, is considered by local community groups to be a food desert. The pressure on housing compounds this issue: residents of a senior housing complex noted that the nearest *dépanneur* was recently replaced by a condominium.

As gentrification transforms the neighborhood, with clear evidence of social, cultural and housing market displacement (Twigge-Molecey 2014), the foodscape of Saint-Henri has also shifted to accommodate newer high-income residents. In this way, the transformation of foodways and the changing housing market have combined to slowly erode community life for low-income residents, who relied on food spaces such as diners, *dépanneurs* and dive bars to meet their friends and stay socially connected. As one longtime resident and community activist noted,

> For me, when I used to walk around in the neighborhood, everyone said hello to each other, but not now. That doesn't happen anymore, there are lots of people who say "I'm moving because I don't really feel at home anymore in St-Henri."

Community resistance and organizing: a cornerstone to ensure (green) equity in Saint-Henri

The borough and broader city government, both formed by the progressive municipal party Projet Montréal,[9] are making attempts to counter gentrification processes that are affecting many neighborhoods in the city, but the limited tools they have at hand are largely ineffective. The inclusion strategy, obliging developers of over 150 housing units to include 15% social housing in their developments, has been dubbed the exclusion strategy by housing activists: not only does it bind social housing development to new speculative construction that raises land values, it also guarantees the displacement of everybody living around it who isn't lucky enough to be selected for the 15% slated for social housing. New policies have been put in place to include affordable housing, but penalties for non-implementation are not enforced, and the "affordable" nature of this housing is questionable. While the city has increased funding for social housing development—from 1 million per year to 3 million per year, with a goal to build 12,000 affordable housing units in 4 years—housing needs remain enormous and funding is insufficient, especially since provincial funding has been frozen for a decade and federal funding for social housing is nonexistent.

The most effective tactics so far to address gentrification in Saint-Henri are community organizing and mobilization, which build on a long and rich history of direct action in the neighborhood (McGraw 1978, Bélanger and Lévesque 2004, Lord 2005). A critical recent victory was halting the conversion of the decades-long abandoned Canada Malting Factory into a mixed-use condo development.[10] The development has so far been stalled thanks to a coalition of dozens of local community groups named *À nous le Malting* (the Malting is ours!) who, among other actions, occupied the offices of different candidates running for local office in the 2013 elections and forced all candidates to sign a moratorium on the redevelopment of the site. *À nous le Malting* continues to campaign for the Canada Malting Factory to become much-needed social housing for the neighborhood .[11]

A coalition of over 20 community-based groups, coordinated through Solidarité Saint-Henri, has also created a popular development plan (P.D.Pop.)[12] for the area. This was developed in contrast to the city's local development plan,

FIGURE 14.4 At a protest, residents sit down to eat around a reimagined depiction of the Canada Malting factory, including a garden, social housing and a cafeteria. The text reads: "my neighborhood is not for sale," "my garden," "a meeting place for everyone." (Photo by: Aaron Vansintjan.)

as a truly neighborhood-based plan that begins with the premise of securing residents' right to stay put. With over 2,000 local residents involved in its creation over years of community planning workshops, the popular development plan proposes (green) infrastructure that local low-income residents need. In other words, rather than prioritizing a "cute" Woonerf, exercise equipment for the white middle class or high-end condo development, they want a functional school playground, swings for kids, social housing and to improve the unhealthy living conditions in existing apartments.

Community resistance has also focused to a large degree on the issue of changing the food system. While housing advocacy organizations and community groups like Solidarité Saint-Henri focus on big initiatives such as the Canada Malting Factory and tenant advocacy, they have also initiated smaller projects focused on improving food access for lower-income residents. These same community groups have started regular collective kitchen events, maintain subsidized fresh vegetable stands in remaining neighborhood *dépanneurs* and organize vegetable stands in social housing. They have also been campaigning for a project to repurpose Saint-Henri's old library, now empty, into a multi-purpose community kitchen and cafeteria. Finally, one of the key victories by community groups was the implementation of a new zoning regulation that limits the number of restaurants on Notre-Dame Street, in an effort to safeguard existing businesses catering to low-income residents and keep commercial rents low.[13]

Meanwhile, long-term community connections and informal, daily community resistance remain strong in Saint-Henri. Many low-income residents hang out on their balconies, sidewalks and in public parks, where they make connections with other residents who share food with them and tell them about deals at the supermarkets. Residents at senior homes have also started pooling their resources, cooking together, sharing taxi rides to the supermarket and organizing weekly cafeteria events in shared spaces.

Aside from the Green Alleys initiatives, low-income residents organize their own urban agriculture projects as a way to meet neighbors and use available empty space. In one case, an elderly resident began a guerilla community garden behind her apartment, which then led to connections and neighborhood meetings in the garden, eventually catalyzing organized resistance in the early 2000s against the expansion of the Turcot interchange in western Saint-Henri—including an occupation of the Ministry of Transport in Quebec by residents.

FIGURE 14.5 An abandoned guerilla garden on an empty lot. The resident who started and maintained it has since been evicted and had to leave the neighborhood. As he noted bitterly, "I've seen new people who told me they decided to live on [my street] simply because they saw the garden … I wanted to make the neighborhood a more pleasant place, I did, and people decided to take it over." (Photo by: Aaron Vansintjan.)

Ordinary and extraordinary greening, ordinary and extraordinary community resistance

This chapter highlighted how the ordinary and extraordinary greening, intimately connected to (food and green) gentrification processes in Saint-Henri, is transforming the neighborhood's built environment, foodscape and everyday life. We saw how the different forms of greening and new foodscapes overlap and solidify into a lived reality in long-term local residents' lives as they experience not only what might be termed "traditional" forms of gentrification—increased rent and physical displacement—but also through the erosion of their community spaces and the spectacular transformation of their built environment. Regarding the erosion of community spaces, our findings on the neighborhood's changing foodscapes parallel Twigge-Molecey's (2014) evidence of social and cultural displacement in the neighborhood. Long-term ties and networks have been ruptured as diners, *dépanneurs* and dive bars close and markets become inaccessible—and thus exclusionary—to low-income residents. Similarly, while the spectacular (extraordinary) green transformation of the Lachine Canal and the Woonerf improve the environmental quality of the neighborhood, they either directly or indirectly privatize access (in the case of the canal) or trigger new luxury developments, building demolitions and/or housing conversions. In both cases, green interventions do not meet long-term residents' needs for things like improved play spaces in local public schoolyards or healthy homes.

We also illustrated how green gentrification in Saint-Henri is not driven by a single activity or intervention but is constituted by interlocked and overlapping dynamics and corresponding strategies of resistance. Indeed, our chapter shows how many long-term residents and activists respond to and resist gentrification in both "extraordinary" and "ordinary" ways. For example, recognizing the way (green) gentrification affects housing, long-term residents and activists mobilize for tenant rights *and* make larger demands like that of turning the Canada Malting into a multi-purpose space and limiting the number of new up-scale restaurants in the neighborhood. Equally, they focus on "ordinary" interventions, such as ensuring more access to appropriate food spaces and using those to build social spaces for low-income residents, which had been eroded through gentrification. Residents themselves also intervene in the environment through informal urban gardening projects, occupying public space, and collectively addressing issues of food access and the need for social space. This shows how even as green gentrification transforms the built environment at different scales, many activists and long-term residents are responding holistically, taking into account the multiple effects it has on community, housing and lived experience.

More change is undoubtedly coming to Saint-Henri. Another extraordinary greening project plans to remove the multilevel highway Turcot interchange and develop a large green space.[14] If local grassroots plans and mobilizations are not taken to heart, and if the city does not take greater steps to ensure equity in the neighborhood and limit new, large-scale real estate developments, this new greening project may only reinforce growing urban (green) inequities and compound existing processes of (green) gentrification.

Keywords

(a) the urban development pattern of the city and neighborhood: recent fast-growing; real estate development boom; de-industrialization
(b) the urban greening of the neighborhood: canal decontamination and regeneration; new green infrastructure; green space
(c) the inequalities at stake: insufficient affordable housing; income inequalities; displacement; green gentrification; food gentrification

Notes

1 https://www.pc.gc.ca/en/lhn-nhs/qc/canallachine/activ/sentiers-trails
2 https://metropolitiques.eu/Organizing-Against-Projet-Gentrification-Housing-activism-in-a-white.html
3 https://ville.montreal.qc.ca/portal/page?_pageid=7757,100311674&_dad=portal&_schema=PORTAL
4 https://www.ecologieurbaine.net/fr/detail-projet/woonerf-saint-pierre
5 https://www.journaldemontreal.com/2013/09/08/woonerf-saint-pierre-un-poumon-vert-pour-saint-henri; https://www.pressreader.com/canada/montreal-gazette/20130909/281505043895806; https://www.ledevoir.com/opinion/blogues/le-blogue-urbain/381278/comment-transformer-une-ruelle-sordide-en-un-magnifique-parc
6 https://www.nationalobserver.com/2019/01/23/features/did-green-development-project-drive-rent-montreal-neighbourhood
7 https://wwf.ca/biopolis-projects/green-alleys-of-montreal/
8 theguardian.com/world/2016/jun/01/montreal-gentrification-st-henri-looting-vandalism
9 http://en.projetmontreal.org/
10 https://www.renwickdev.com/project-page---canada-malting
11 https://montrealgazette.com/news/local-news/st-henri-group-works-to-put-social-housing-on-old-canada-malting-site
12 https://www.solidarite-sh.org/pdpop/
13 https://www.theguardian.com/cities/2016/nov/16/one-in-six-rule-can-montreal-canada-fight-gentrification-banning-restaurants
14 https://montreal.ctvnews.ca/turcot-interchange-project-to-get-two-new-green-additions-1.3983916

References

Anguelovski, Isabelle. 2016. "Healthy Food Stores, Greenlining and Food Gentrification: Contesting New Forms of Privilege, Displacement and Locally Unwanted Land Uses in Racially Mixed Neighborhoods." *International Journal of Urban and Regional Research* 39, no. 6: 1209–1230.

Bélanger, Paul R., and Lévesque, Benôit. 2004. "Le mouvement populaire et communautaire: de la revendication au partenariat (1963–1992)." In *Le Québec en jeu. Comprendre les grands défis*, edited by Gérard Daigle et Guy Rocher, 713–747. Montréal: Les Presses de l'Université de Montréal.

Bliek, Desmond, and Gauthier, Pierre. 2007. "Mobilising Urban Heritage to Counter the Commodification of Brownfield Landscapes: Lessons from Montréal's Lachine Canal." *Canadian Journal of Urban Research*, 16, no. 1: 39–58.

Gould, Kenneth, and Lewis, Tammy. 2017. *Green Gentrification: Urban Sustainability and the Struggle for Environmental Justice*. Abingdon and New York: Routledge.

Heffez, Alanah, and Bornstein, Lisa. 2016. "Youth Fusion's Urban Environment Project: Increasing Youth Participation in Urban Planning through Place-Based Environmental Education." *Children, Youth and Environments*, 26, no. 2: 110–127.

Lewis, Robert D. 2001. "A City Transformed: Manufacturing Districts and Suburban Growth in Montreal, 1850–1929." *Journal of Historical Geography*, 27, no. 1: 20–35.

Lord, Kathleen. 2005. "Permeable Boundaries: Negotiation, Resistance, and Transgression of Street Space in Saint-Henri, Quebec, 1875–1905." *Urban History Review*, 33, no. 2: 17–29.

McGraw, Donald. 1978. *Le développement des groupes populaires à Montréal, 1963–1973*. Montréal: Editions coopératives A. Saint-Martin.

Twigge-Molecey, Amy. 2014. "Exploring Resident Experiences of Indirect Displacement in a Neighbourhood Undergoing Gentrification: The Case of Saint-Henri in Montréal." *Canadian Journal of Urban Research*, 23, no. 1: 1–22.

15

ENVIRONMENTAL INEQUITIES IN FAST-GROWING DUBLIN

Combined scarcity of green space and affordable housing for The Liberties

Isabelle Anguelovski, Panagiota Kotsila, Dave Moore, and Mick Lennon

The centrality of the historic and culturally rich neighborhood of The Liberties has captured the attention of developers and investors during the last decade, as land and housing have become increasingly scarce and valuable in the city of Dublin. The uptake and implementation of a Greening Strategy (2015) for the area has been an important step in acknowledging the historic lack of quality public green space that the deprived and underserved communities populating The Liberties have endured. However, it has not been able to counterbalance the mass gentrification[1] trends that the neighborhood has been experiencing as a result of elite real estate developments and visitor-oriented economies, jeopardizing the long-term ability of residents to benefit from a newly greened neighborhood.

Balancing two crises: greening and housing access

Dublin hosts the largest city park in Europe (Phoenix Park, 707ha), yet its historically working-class neighborhood of The Liberties has had the least amount of green space per capita in the whole of Ireland, compounded by its low quality in terms of parks and playgrounds. Such discrepancy dates back to the beginning of the 20th century, when Dublin's industrialization was accompanied by dense housing construction in response to rural migration and urban social needs (Connolly et al. 2018).

The Liberties, in the south inner city of Dublin, was a neighborhood of manufacturing and industry—particularly weaving, brewing and distilling—up until the 1960s. The remnants and/or continuation of this legacy today constitute core tourist attractions. This history was accompanied by horrendous living conditions for workers, with notorious slums widespread in the 19th century, later followed by worker housing developments built by the rich families of Guinness (brewers) and Power (distillers) (Prunty 1998). During the 1970s and 1980s, most inner-city

DOI: 10.4324/9781003183273-19

neighborhoods experienced disinvestment, as the prevailing policy had the effect of pushing jobs and population to the suburbs of Dublin (Kelly 2007), leaving thousands unemployed or forcing them to move. This fractured the strong community bonds that characterized The Liberties and partially contributed to the drug crisis of the early 1990s. During the economic boom that followed, and arguably until today (interrupted by a period of economic crisis 2008–2011), a radical revalorization of inner-city land was enabled through developer-friendly policies and area-based plans (ibid.). But new office developments, retail spaces and large-scale private residential apartment complexes were not accompanied by a much-needed increase or improvement of public green spaces in areas like The Liberties.

Over the years, housing priorities relegated the construction of green and open spaces in the inner city, allocating new park investments for the outskirts, and thus creating persistent conditions for the unequal distribution of green space in the city. Consequently, today, many of Dublin's inner-city neighborhoods fall far below the WHO recommendation of nine square meters of green space per resident, while for The Liberties, this statistic drops to lower than one square meter per resident.[2] Coupled with historically low investments in green space, The Liberties has been plagued by high amounts of post-industrial empty or derelict space and, in the last decade, by a housing crisis affecting long-term and/or socially vulnerable residents. As was recently reported, just between June and

FIGURE 15.1 Social housing in The Liberties lacking green space (Photo by: Isabelle Anguelovski)

September 2018, 415 Dublin families (including 893 children) became homeless. Between March 2017 and March 2018, rents rose by 12% throughout the city.[3]

The 2015 Liberties Greening Strategy: from ambitious visioning to timid implementation

In 2015, to respond to the green space crisis in The Liberties, the Dublin City Council published the Liberties Greening Strategy (Dublin City Council 2015), a strategic planning document proposing a widely lauded improvement strategy to address residential needs through new large and green spaces and parks, bike lanes and infrastructure and green streets. "With population densities as high as 18,000 per km^2 in places, the inner city has a social, economic, indeed moral claim for substantially greater 'green' investment," according to planner and architect Paul Kearns and Motti Ruimy (2014).

However, to date, the Greening Strategy acts solely as a visioning guide and not a legally binding document; it harbors no political power. Nevertheless, the Department of Parks and Landscape Services has since balanced several greening initiatives at once: those of greatest scale built by the City Council, such as Weaver Park and Bridgefoot Street Park, which have since grown into successful actualizations, and five community-driven gardens, some of which are contested.

Weaver Square Park on Cork Street

The debut of Weaver Park in October 2017 stands out as the first successful park space to arrive at full fruition in The Liberties; an achievement won as a result of a long and arduous battle by local activists and strong-willed city councilors. Weaver Park is a 0.5-ha attractive, multi-purpose park which features a large playground complex, ample green space and skate amenities partially designed by local skaters. Cleanly paved paths separate these different spaces that serve varied users and aspirations, fostering an open, permeable space with pedestrian access to each section. Unlike previous trends of fenced parks in Dublin, Weaver Square Park maintains visual and physical connectedness with the street and neighborhood, making it safer and more welcoming.

However, an unsettled conflict concerns the allotment and community gardens growing adjacent to the new park space, which face displacement by a city-backed, rapid-build, low-cost housing project. In an area with limited green space, the gardens in Weaver Square had served as a lush space for social and community gatherings and for addressing the pressing social needs of more than 100 at-risk youth over the course of 4 years (Lennon and Moore 2019). As a spokesperson for the Weaver Garden group shared:

> The square was a really ugly sight—the worst kind of inner-city eyesore. The allotments changed the tone of the place, it was seen as a real progressive thing that the City Council had supported, and now they're going to

FIGURE 15.2 Weaver Park in the Liberties (Photo by: Isabelle Anguelovski)

> get rid of that ... Everyone on the allotments agrees that there is a housing crisis, but to lose the gardens for [these] units, a lot of which are one-bedroom apartments, it seems like a lot to lose.

Here, beyond regretting the loss of their gardening spaces, garden activists contend that the number of social housing units to be built—55 in total—is insufficient to address deeper social and housing needs and that volumetric construction, where parts for each home are manufactured in standardized form at an offsite facility,[4] will eventually create poor-quality, stigmatized housing.

In that conflict, the Dublin City Council is caught in a political dilemma. While they agree that the community gardens contribute much social and cultural value to a rapidly urbanizing inner-city environment, they argue, too, that it is "difficult to get the right balance between the urgent need for housing and the demand to retain open spaces such as this one."[5] The densely populated area of Dublin 8, which includes The Liberties, hosts the second highest number of people in homeless emergency accommodation (785 people in 2019),[6] and ultimately, the Council opted to reclaim the land next to Weaver Park to work towards addressing the housing crisis.

Gardeners were required to vacate the site at the end of 2018 and might have expected that council housing construction adjacent to Weaver Park would commence quickly. Yet the site remained idle and overgrown for several months, a situation that caused acute discomfort for local councilors who had reluctantly acceded to the plan.[7] Meanwhile the gardening community was split between two other sites that were insufficient to accommodate all previous allotments and the community gardening project.[8] As the battle for green space bows to that for social housing, community gardeners worry that local authorities will fail to find them a new suitable site while pushing for more low-quality social housing throughout The Liberties.

Bridgefoot Street Park and community gardens

On the northern side of The Liberties, the Bridgefoot Street space began hosting active community gardens in 2007 but faced an uncertain fate due to several evictions and reoccupations over an extended period, contributing to mounting tensions between the municipality, local residents and social justice advocates. Residents regularly tended to the gardens, which occupied around a quarter of the derelict site and served as a close-knit social community space—albeit not without internal tensions. When the gardeners were officially told in 2017 that they would face displacement in lieu of construction of a new park, the prospect of a larger green space, designed inclusive of community input, was offered as mitigation for their loss. Nevertheless, activists still saw the need to campaign against a proposed rezoning of the lot as "Z5" (which would have allowed for both housing and park space) and succeeded in preserving the recent zoning designation of the space as "Z9" which stands for amenity or open space.

Offering more than twice the ground area as the Weaver Square site, Bridgefoot Street's current plan promises a park with "lawns, flowers, trees, furniture, lighting, allotments, the displaced community gardens, a terrace, play space and play equipment and more."[9] Under the new park plan, the community growing space will be geographically shifted and broken up into "organic" shapes for a "playful" landscape, according to statements from the designers' planning document. Dermot Foley Landscape Architects[10] maintain that their intention is "that this atmosphere is cherished and preserved," but final plans dashed gardeners' hopes for a design that would keep their community-maintained portion of the park intact. Material-testing and processes of consultation with the local community had commenced in 2016 but, after gardening at the site was curbed, progression of the park was much delayed, even more so due to the COVID-19 pandemic. Construction ultimately began in June 2020, with the park expected to open to the public in 2021.

Competing green and housing interests in The Liberties

The 2015 Greening Strategy has served as a motivating framework for the development of two key public parks and the preservation of some community gardens. However, it may have unduly also catalyzed the rapid rise of new and large-scale high-end construction, and thus the ensuing gentrification in The Liberties—compromising the creation of new public green amenities while accelerating the displacement of residents.

Indeed, when investors heard of the Greening Strategy and of new development incentives for The Liberties, many realized the new value of this centrally located, well-connected historic neighborhood. Global firms jumped on the opportunity to transform large vacant sites into lucrative real estate operations. As Dublin City Councilor Tina MacVeigh expresses:

> The Greening Strategy was to make it look like this is a good area for you to invest in: "Build your student accommodation here, because next door you're going to get a park. You're going to get a nice tourist trail. We're going to clean up the mess." Nothing for the residents.

Since its adoption, the Greening Strategy has become a legitimizing and marketing tool for investment and new developments, much of it illustrated by the construction of high-end short-term housing complexes, which MacVeigh also links to a "gentrification that is killing the city"; a city "that is gentrified for a tourist infrastructure, for luxury student accommodation, apart-hotels, hotels and Airbnb."

High-end short-term housing developments

Across Dublin, luxury towers of student accommodation—where prices can reach €1,640 per month for a one-bedroom[11]—are being constructed at alarming rates. The Liberties bears the brunt of this rapid development, with 25% of all student housing being built across the city having taken place in Dublin 8, which includes The Liberties. These single-resident layouts, complete with their own drainage and temperature control, not only inhibit the possibility for future up-zoning but perpetuate inefficiencies for system servicing, electricity use and water utilities. This rapid-build influx of luxury student housing excludes lower-income and most Irish students who cannot afford high rents. A recent study conducted by Dublin City Council of new student accommodations found that 80% of them were rented exclusively to international students. These massive luxury complexes threaten to supersaturate the student housing market, even when there are not enough wealthy international students to rent them out to, and local students cannot afford them.

The housing issue expands beyond luxury accommodation designed for students. Looking up Dublin's online property listings yields approximately only 1,200 residential units for rent, while Airbnb boasts over 9,500 available accommodations, as reported in June 2020.[12] Airbnb contributes further to gentrification and residential displacement, as recent regulations on its activities in Dublin acknowledge (July 2019). The Liberties also hosts several new large-scale luxury hotels, including the €50 million Hyatt Hotel and the €40 million Aloft Marriot Hotel, and large redevelopment sites such as Newmarket Square. As our interviews and observations have revealed, none of those new facilities create new social connections or benefits within the neighborhood and with local businesses, and small business owners report that wealthy visitors do not spend time or disposable income in the neighborhood.

Furthermore, recent housing investments are explained by the attractiveness of The Liberties due to lower land prices, proximity to the city center, and renovated touristic sites such as the Guinness Factory and St Patrick's Cathedral. Overall, The Liberties has become a hot spot for development and quick returns. As a local developer reported in 2019:

> What happened was once we got the student accommodation up and running, money then started to attract itself to this area. Now I'm talking a fluctuation of values of up to 20 times increases from 2008 … Bigger money is now beginning to attract itself in, and we then went for planning permission for a big [redevelopment] scheme across here.

Investors in Dublin benefit from 2017 national planning regulations which allow large housing developments of more than 100 units to bypass local planners, politicians and communities by going directly for planning assessment to the Irish planning board (An Bord Pleanála).[13] This can considerably cut costs incurred by developers, such as the interest on bank loans, via time savings consequent on short-circuiting the local planning system (Lennon and Waldron 2019). Additionally, since December 2018, new planning guidelines allow for taller buildings in the city center, due to what were deemed by some as "overly restrictive maximum heights."[14]

Finally, although Dublin City Council mandates the inclusion of 10% social housing for every 100 residential units added according to the Part V requirements under the Planning Act, student accommodation and hotels are not required to be zoned as residential. This further complicates the issue that, after these buildings are complete and quite likely hosting a mix of high-end temporary residents and expensive (vacant) units, they cannot be redeveloped easily to accommodate families or local residents. According to interviewees, they run the risk of being abandoned after profit returns have been reached by the investors, turning into the area's new high-end ghettos.

Today, the speed and rate of development of tourism and other short-term accommodation in The Liberties have spurred acute gentrification and displacement.[15] As residents report, international tourists and visitors are living in housing that should be meant for residents. An additional shortcoming of this development model is that with the current COVID-19 crisis hotels and student flats will remain empty, as these are housing units that cannot easily be converted into long-term residential homes.

Internationally led gentrification and displacement

The heavy proliferation of luxury developments marketed towards wealthy international students and other transient populations from the creative sector plays a major role in the displacement of current residents in The Liberties. First, they are built on land which could be used for long-term housing developments, through a mix of market-price, affordable and public units. Second, the value of the land adjacent to these sites has risen exponentially and compromises long-term residents' abilities to afford increased prices, and, in turn, benefit from new greening.

The proliferation of such global investment groups mirrors international tech investments through Dublin, where companies such as Google, LinkedIn and Facebook have built their tax-free headquarters in the Dublin Docklands area for a new international class of workers. Those high-paid international workers either sleep in the new short-term luxury rentals throughout the city or on private apartment campuses built by those same tech companies. Interestingly, smaller apartments are also those most profitable for developers to build, and real estate companies to rent, as they maximize the underlying value of the land. In contrast, Dublin residents have to endure poor social or public housing quality and a decline of public housing construction, while 3,489 people live in homeless emergency accommodation in the city (as of 2017).[16] As of 2020, international investors are looking for even more new spaces to build, including in Dublin 8 and The Liberties.

Dilemmas over privately financed and exclusive green spaces

While those short-term accommodations are only affordable to the international wealthy classes, they also tend to exclude residents from enjoying the green amenities cleaned up and/or built on site. In The Liberties, several green spaces zoned as public land have been revitalized and are now maintained by private developers. Additionally, zoning ordinances require that developers include a certain percent of green space for every high-rise erected. However, many of those spaces are manicured green amenities that are rather unwelcoming to playing children and free community usage, manifesting as socially exclusive, gated or purpose-restricted spaces. One controversy emerged in fall 2018 when the students' complex New Mill gated off its green area despite the permitting conditions imposed on the site by law.[17]

Even in those cases where green spaces are legally open to the public, lower-income or minority residents are more likely to feel out of place and culturally excluded when surrounded by wealthy and foreign visitors, as members of the Dublin City Council acknowledge. New green spaces can be zoned as public, but still feel private and socially exclusive. Here, the Department of Parks and Recreation is presented with a dilemma: either slowly implement The Liberties Greening Strategy, constantly in competition with the Housing or Public

Infrastructure Department within the City Council, waiting for bits of funding here and there and competing with other types of investments, or let private developers finance and maintain new green spaces. International investors thus represent new financial assets for crafting and maintaining much-needed green spaces.

Rapid social costs for The Liberties

Across The Liberties, as the growth of public green spaces fails to match the rapid high-end constructions, the health of local residents (Dublin City Council 2011)—especially youth disaffection, obesity and mental health—is raising concerns.[18] Without fields or facilities for sports or well-kept, populated and publicly available open green spaces and playgrounds, children have only a few places in The Liberties to call their own.

Rising gang presence places youth in an ill-equipped position to deal with the mounting economic depression taking hold over these increasingly gentrified residential communities. The emotional and physical tolls of derelict streets, the threat of eviction as reported by working-class residents and city officials, social exclusion from city infrastructure, mounting gray space and high rates of homelessness only widen the gap between low-income residents and incoming wealthy transient populations who stand to benefit from these transformations of city infrastructure.

What are the possible solutions to this dilemma? To adjust for this, Dublin City Council could proliferate usage of Development Contributions included under Section 48 of the Planning and Development Act, 2000 (as amended), which allows local planning authorities to require contributions to community improvement by developers when granting planning permissions under Section 34 of the same act. Taking advantage of Section 48 would not only slow the alarmingly rapid construction of high-density luxury complexes but also serve as a policy instrument for significant community enhancement through the implementation of sustainable infrastructure, improved green spaces or rezoned mixed-use lots apportioned and reserved for social housing. However, the use of these provisions is quite limited, especially so in lower-income areas, since local authorities are reluctant to become liable for the maintenance of newly enhanced green spaces and fear that local residents might engage in poor "uncontrolled" social activities that will compromise those spaces (Lennon and Moore 2019).

To date, local community activism and resistance strategies for urban justice have been limited, partially as a result of the numerous social issues and political campaigns similarly vying for city residents' attention. Even after the economic crash, Dublin saw a fast recovery of the real estate sector combined with deepening financialization of urban space and austerity policies. However, the issue of mortgage debt was never politicized as in other countries (e.g., Spain), and instead family homelessness was at the core of resistance narratives during these years. As housing activists from Dublin assert, referencing one of the Irish State's founding documents:

> Certainly the affirmation in the [1919] Democratic Program [drafted by the Labor Party] that all right to private property must be subordinated to the public right and welfare is in direct and absolute conflict with the reality of life in this city and in this country today where more than 10,000 people are homeless, over 3000 of whom are children, and 500,000 people are in housing distress enduring extortionate rents, living in arrears, are unable to secure affordable homes, and where public lands are used to benefit private commercial interests rather than for decent public housing.
> *(Justice activist, Dublin, 1 May 2019)*

FIGURE 15.3 A call for mobilization for affordable housing and against evictions in Central Dublin (Photo by: Isabelle Anguelovski)

In 2014, the platform Housing Action Now sought to incorporate a broader vision for housing struggles and was in conversation with the P.A.H. movement in Spain, while also being influenced and strengthened by a parallel movement against the privatization and pricing of water in the country (Right2Water). This resulted in the formation of the Irish Housing Network (2015) which engaged in direct actions of occupations and detected increasing levels of homelessness by exposing and emphasizing the number of vacant buildings (Hearne et al. 2018). Some vulnerable neighborhoods have developed strong representative structures and have managed to influence public housing redevelopment strategies. In Rialto, bordering The Liberties, for example, a cooperative is managing affordable and market-price units redeveloped around new green and sports facilities. Yet other areas, including The Liberties, remain weak in terms of organization, representation and influence (Norris and Hearne 2016). Recent community support for increasing affordable housing units has taken shape in the form of public lobbying and campaigns by regional nonprofit coalitions, but makes little impact as political narratives sink deeper into compliance with neoliberal growth policies. Hopeful initiatives include the recent activism of the National Homeless and Housing Coalition[19] and one of its most vocal leaders, Dublin City Councilor Tina MacVeigh.

Towards a dichotomous green-ish neighborhood?

In a neighborhood rich in strong local culture and historic value, underserved populations are likely to watch their schools and other public amenities shut down as displacement pushes local families out. Greening efforts have managed to insert some rays of hope and change with regards to the creation of open, participatory and welcoming green spaces in the neighborhood. At the same time, informal green spaces of allotments and lush community gardens fall under constant threat of development. As community spaces and opportunities for affordable housing make way for glowing high-rises, and local family businesses transition to catering almost exclusively to tourists or simply shut down, the residents of The Liberties face mass gentrification, threat of eviction and loss of cultural value: a life struggling to get by day to day, watching themselves slowly disappear as actors in the story of a green-ish Liberties.

Keywords

(a) the urban development pattern of the city and neighborhood: fast-growing; post-industrial redevelopment, post-crisis regeneration; tech-driven
(b) the urban greening of the city and/or neighborhood: new neighborhood parks; greening strategy; redevelopment of abandoned or underused land
(c) the inequalities at stake: deep inequities in access to green space; green gentrification; student-centered gentrification; housing crisis; pseudo-public green spaces

Notes

1 https://planetgeogblog.wordpress.com/2020/04/30/the-changing-face-of-the-liberties-gentrification-of-dublin-8/
2 https://www.dublininquirer.com/2016/08/10/bridgefoot-street-should-there-be-housing-a-park-or-both
3 https://www.theguardian.com/cities/2018/nov/29/empty-dublin-housing-crisis-airbnb-homelessness-landlords
4 https://www.dublininquirer.com/2019/06/26/council-briefs-inchicore-library-revamp-chapelizod-housing-and-more
5 https://www.thejournal.ie/weaver-garden-dublin-city-council-housing-4429500-Jan2019/
6 https://www.rte.ie/news/dublin/2019/0808/1067575-third-of-irelands-homeless-accommodated-in-dublin/
7 https://www.dublininquirer.com/2019/05/22/council-briefs-weaver-square-allotments-plans-for-cherry-orchard-and-traffic-in-rathmines
8 https://politicalecologytcd.wordpress.com/2019/10/23/is-green-the-new-black-the-longevity-of-community-gardens-in-dublins-recovering-economy/
9 See: https://naturvation.eu/nbs/dublin/bridgefoot-street-park
10 http://www.dublincity.ie/sites/default/files/content/RecreationandCulture/DublinCityParks/Documents/Design%20Rationale%20Rev%20C%20PART%208.pdf
11 https://www.studyinternational.com/news/ireland-student-accommodation-luxury-afford/
12 https://www.hostyapp.com/airbnb-statistics-laws/dublin/
13 https://rebuildingireland.ie/news/new-regulations-to-fast-track-large-scale-housing-developments/
14 https://www.housing.gov.ie/sites/default/files/publications/files/urban_development_and_building_height_guidelines_for_planning_authorities_december_2018_0.pdf
15 https://www.irishtimes.com/life-and-style/travel/ireland/all-these-hotels-go-up-and-what-do-we-get-back-nothing-1.4162924
16 https://www.homelessdublin.ie/info/figures?type=accommodation-usage&year=2018
17 https://dublininquirer.com/2019/03/27/council-briefs-the-future-of-kilmainham-mills-blocked-off-green-space-and-more
18 See: https://www.youth.ie/programmes/youth-health/youth-mental-health-signposting-tool/
19 https://www.theguardian.com/cities/2018/nov/29/empty-dublin-housing-crisis-airbnb-homelessness-landlords

References

Connolly, James J. T., Tatjana Trebic, Isabelle Anguelovski, and Erin Wood, eds. 2018. *Green Trajectories. Municipal Policy Trends and Strategies for Greening in Europe, Canada and United States (1990–2016)*. Barcelona: Barcelona Lab for Urban Environmental Justice and Sustainability & ICLEI-Local Governments for Sustainability.

Dublin City Council. 2011. "Dublin City Socio-Economic Profile Indicator Catalogue." Dublin. https://consultation.dublincity.ie/admin/lecp/supporting_documents/Dublin City Socio Economic Profile.pdf.

———. 2015. "The Liberties Greening Strategy."

Hearne, Rory, Cian O'Callaghan, Cesare Di Feliciantonio, and Rob Kitchin. 2018. "The Relational Articulation of Housing Crisis and Activism in Post-Crash Dublin, Ireland." In *Rent and Its Discontents: A Century of Housing Struggle*, 153–67. Lanham, MD: Rowman & Littlefield International.

Kearns, Paul, and Motti Ruimy. 2014. *Beyond Pebbledash:... And the Puzzle of Dublin*. Cork: Gandon Editions.

Kelly, Sinéad. 2007. "The Liberties Transformed: The Emergence of New Commercial and Residential Spaces in Inner Dublin." *Journal of Irish Urban Studies* 6: 89–118.

Lennon, Mick, and Dave Moore. 2019. "Planning, 'Politics' and the Production of Space: The Formulation and Application of a Framework for Examining the Micropolitics of Community Place-Making." *Journal of Environmental Policy & Planning* 21(2): 117–33. https://doi.org/10.1080/1523908X.2018.1508336.

Lennon, Mick, and Richard Waldron. 2019. "De-Democratising the Irish Planning System." *European Planning Studies* 27(8): 1607–25.

Norris, Michelle, and Rory Hearne. 2016. "Privatizing Public Housing Redevelopment: Grassroots Resistance, Co-Operation and Devastation in Three Dublin Neighbourhoods." *Cities* 57: 40–46.

Prunty, Jacinta. 1998. *Dublin Slums, 1800–1925: A Study in Urban Geography*. Newbridge: Irish Academic Press.

16

BARCELONA'S GREENING PARADOX AS AN EMERGING GLOBAL CITY AND TOURISM DESTINATION

Panagiota Kotsila, Emilia Oscilowicz, Filka Sekulova, Margarita Triguero-Mas, Jordi Honey-Rosés, and Isabelle Anguelovski

Following an ambitious urban renewal program that began in the late 1980s, Barcelona has become a major tourism destination in the Mediterranean. As an emerging global city, planners in Barcelona today are greening the city through park improvements, street redesigns and street traffic-calming initiatives. While striving to provide locals with green amenities, these beautification projects have also produced unexpected consequences, as they bring in more visitors and accelerate gentrification. Currently, few city greening projects can avoid discussions of gentrification, and neighborhood groups have begun to organize in opposition to the touristification that has made their neighborhoods unaffordable and/or exclusive. While the current COVID-19 pandemic has drastically reduced the number of visitors to Barcelona—producing a new experience for local residents—the impacts on local street life, public space, housing and gentrification trends remain uncertain.

Green livability as an attraction point: Barcelona's green appropriation by visitors and global capital

Barcelona is known for its high quality of life, vibrant street life and walkable neighborhoods, a reputation that has benefited its inhabitants but also attracted an increasing number of visitors since the early 2000s. In 2018 alone, Barcelona received over 9 million overnight visitors,[1] while between 2015 and 2019 the number of residents living in the city center declined by 11%.[2] As visitors fill the city streets, residents are leaving. This shift is largely attributed to a rental market that pushes out long-term residents and homeowners by converting housing into highly profitable holiday apartments. Long-term residents are also fleeing from overcrowded central neighborhoods such as *Ciutat Vella* (Old City) which are arguably no longer safe or healthy due to tourist-driven gentrification

DOI: 10.4324/9781003183273-20

(Anguelovski et al. 2020). The attractiveness of Barcelona has thus turned into a double-edged sword, often tightly connected to neighborhood regeneration and green interventions, such as park improvements, street redesigns and street traffic-calming initiatives that are combined with new greening.

After Francisco Franco's lengthy dictatorship (1939–1975), Metropolitan Authorities in Barcelona embarked on a series of city greening efforts to create new public space and improve neighborhood livability, especially in degraded areas (Connolly et al. 2018). Much of their vision was inspired by the 19th-century Catalan engineer and planner, Idelfons Cerdà, who designed a city extension that aspired to improve living conditions, create healthier urban environments and address deep social hierarchies and inequalities. In a post-Franco era, creating a more livable Barcelona for all was gradually accomplished through improved infrastructure, social services, local equipment and public spaces, all while promoting tourism and reducing crime throughout neighborhoods (Anguelovski 2013). Some examples of large-scale greening during the 1980s include Parc Joan Miró on the site of an old slaughterhouse, Parc de la Creueta del Coll on the site of a former quarry and a number of other parks developed in post-industrial sites (Connolly et al. 2018). But most green interventions in the 1980s were small-scale and born out of a dialogue between the municipality and local social movements and neighborhood associations (Jonas and While 2007).

This approach changed towards the end of the 1990s and early 2000s when green space provision became more intertwined with the redevelopment of large post-industrial areas in or close to the city center (i.e., Poblenou) and, notably, anchored around new real estate developments. In 1996, and after much civic pressure and support, Barcelona adopted the Agenda 21 program, a roadmap for sustainable development that drew from the 1992 Rio de Janeiro Earth Summit. The 2000s were marked by large-scale traffic reconfiguration and planning of new green space and infrastructure that also addressed water permeability, noise reduction, and micro-climate regulation needs (Barcelona City Council 2013). Arguably, however, the sustainable city agenda and its greening rhetoric has been used to promote economic transformations throughout the city (Jonas and While 2007). As our cases studies—the Old City and the Eixample neighborhoods—show, whether intentionally or not, much of this new greening and beautification has enhanced the process of touristification and the exclusive features of particular neighborhoods in the city.

A rebranding of Barcelona was instrumental in allowing this gradual transformation from the 1980s to the early 2000s. Both through longstanding campaigns (*Barcelona Posa't Guapa* 1980s–2009) and more recent smart city narratives and initiatives (Anguelovski 2014; March and Ribera-Fumaz 2016), the city has been calling out to visitors, investors and the creative sector. In terms of governance, the so-called "Barcelona model" of urban remodeling and rebranding took shape in the aftermath of the Olympic Games (1992) and was characterized by public–private and inter-governmental arrangements, the promotion of major events to attract international visitors and the association of public interest with private money. In other words, Barcelona experienced an entrepreneurial neoliberal turn in governance (Martí-Costa and Tomàs 2017), while also advancing its sustainability agenda through, inter alia, various types of greening projects. Both these processes, filled with tensions and contradictions, laid the groundwork for the evolution of Barcelona as a touristic attraction. The impacts of an increasingly non-regulated touristified city are still today vividly experienced in centrally located neighborhoods, where foreigners (both visitors and mostly Western European and North American immigrants) are often better positioned in housing market dynamics, causing locals to feel increasingly excluded and displaced by inflated prices (Cocola-Grant and Lopez-Gay 2020).

These trends have not gone without resistance from a number of resident-based initiatives that have contested the ongoing housing speculation, insecurity and "theft" of public space caused by the massive influx of tourists, as well as the related quality of life and health issues. Indeed, these concerns heavily shaped the critical stance towards tourism of the leftist coalition that won the municipal elections in 2015 (Barcelona en Comú). In the aftermath of the *Indignados* movement (2011–2012), a number of alternative community gathering and organizing spaces (re-)emerged, including urban gardens (e.g., *Hort Indignat* and *La Vanguàrdia* gardens in Poblenou; *Forat de la Vergonya* in Casc Antic; *Jardins del Emma* self-managed park in the Eixample), that follow a long tradition of community management in Barcelona dating back to the anarchism and cooperativism of the early 1930s. Residents facing home evictions also found the support of housing justice groups such as the Platform of Mortgage-Affected People (P.A.H.) (García-Lamarca forthcoming), from which the current mayor Ada Colau emerged. The most acute manifestation of this has been an alarming number of evictions executed throughout the city in the 2010s (despite a slight decrease from 3,286 in 2013 to 2,519 in 2017),[3] 75% of which are due to rent defaulting and much of it due to the increased financialization of housing in Barcelona monopolized by investment funds such as Blackstone.[4]

The following two case studies are emblematic examples of a city-led and a citizen-led greening initiative in the city center. Their trajectories amidst urban development and change in Barcelona serve to shed light on equity issues, particularly how these interventions have been used and who they have benefited throughout the years.

Neighborhood activism for green space and the impacts of tourist-driven gentrification in *Ciutat Vella* (Old City)

The bottom-up case of Pou de la Figuera in *Ciutat Vella* illustrates how community activists successfully fought for and co-produced new open spaces. During the 1990s, a contracting company continually abandoned rubbish and construction materials at a particular spot in the neighborhood, which led residents to call the place the *Forat de la Vergonya* or "Hole of Shame" (Anguelovski 2014). Over time, the community occupied and reclaimed the space by planting vegetation, adding benches and creating a community garden. The community fought hard to transform the hardscaped plaza with underground parking—as originally intended—into an open green space. The final design was a collaborative exercise between city planners and community members, including the design and construction of a new social center co-managed by neighbors and city employees. The redevelopment project also included additional social housing units for the surrounding neighborhood. In 2007, the site was renamed *Pou de la Figuera* and officially recognized as a public green space (Anguelovski 2013).

Yet the inspiring activism at Pou de la Figuera has not been immune to the pressures of tourism in the area. Over the last two decades, streets and commercial spaces throughout the Old City have been transformed to accommodate souvenir shops, luxury fashion retailers or trendy take-out food establishments. Many long-term residents and neighborhood activists have been facing the negative impacts of touristification on everyday life (A.B.D.T. 2020). Small local grocery stores are closing or struggling to stay in business, whereas trendy tapas restaurants and boutique fashion shops proliferate. Families with children are leaving the neighborhood as census data shows a 13% drop in the number of households with children from 2007 to 2016 (Barcelona Metropolitan Census 2007; 2016). Additionally, the proliferation of short-term rentals is shrinking the housing stock for locals and contributing to rising rents, demonstrated in the 14%

FIGURE 16.1 Green space and children's playground in Pou de la Figuera. (Photo by: Isabelle Anguelovski.)

increase of Airbnb units in the neighborhood from 2015 to 2019. In interviews from 2019 with local residents, one mother describes how her neighborhood has drastically and negatively changed as a result of tourism:

> The small, local grocery shops and artisans from the neighborhood are closing and have left. It's changing to tourism. This changes the prices, because the prices are for the tourists, not the locals ... [Tourists] come in late at night, they yell, and they don't think about the neighbors. This really affects our health. Little by little, the neighbors are being pushed out of their neighborhoods.

This type of tourism gentrification is driven by capital flows, market changes and land speculation. Local residents are unable to compete in the housing market, and many are displaced and forced to move to less expensive peripheral neighborhoods. As a result of this exodus, the sense of community has also significantly eroded (Oscilowicz et al. 2020).

Residents also link tourism to an increasing lack of safety, as delinquents tend to target gawking visitors and linger in spaces where tourists are gathered. Further interviews with local residents suggest that residents have stopped using several green spaces out of fear of crime:

FIGURE 16.2 A children's playground in Plaza Allada Vermell, adjacent to Pou de la Figuera green space and playground. (Photo by: Emilia Oscilowicz.)

> You don't feel safe in these spaces. A woman walking alone with her children feels unsafe ... It has changed so much. Recently with my children, we passed by some kids who were fighting violently and there was blood. I know that this affects [my daughter], she asked me later what happened. She asked if they are going to kill people, and I have to tell her it will be okay. She has nightmares.

In late December 2019, violence in the neighborhood escalated when a young man was stabbed to death in the park, and resurged again in June 2020 following a period of strict confinement. This tragedy, coupled with the negative consequences of tourism and real estate pressures, has significantly eroded the sense of community in the neighborhood and made residents feel like outsiders despite the success of the grassroots green intervention Pou de la Figuera that has served the community for years.

Commercial gentrification through greening around Passeig de Sant Joan

The redevelopment of the Passeig de Sant Joan promenade (2010–2015) in the Eixample Dret neighborhood near the Old City aimed to increase ecological and social connectivity within a broader greening strategy for the city (City of Barcelona 2013). Though the Passeig was conceived in Cerdà's 1859 Plan to connect the city to the waterfront, the construction of the fenced Ciutadella Park obstructed its access and the Passeig became a transit road that connected the Old City with the districts of Eixample and Gràcia (Hoyos Núñez 2012).[5] The new green corridor widened sidewalks from 12 m to 17 m, reduced vehicle lanes, introduced rest areas and playgrounds, planted native tree species and shrubs of various heights and installed semi-green permeable pavement for improved stormwater management and the cooling effects of evapotranspiration.

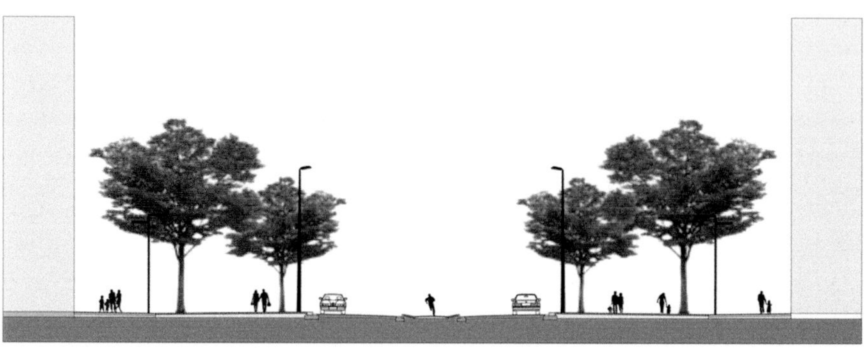

FIGURE 16.3 Outline of the boulevard section of Passeig Sant Joan in Eixample. (Image courtesy of Hoyos/Santacruz 2012.)

The redevelopment of the Passeig was a pending issue for the neighborhood well before it turned into a street greening project. Two moments are key in demonstrating how decisions around this renovation were highly driven by concerns over local economy dynamics.

First was the decision to start from the lower section of the corridor that was flagged for redevelopment (Plaça Tetuan-Arc de Triomf), due to demands by neighbors and business associations. Yet despite this section of the Passeig being comprised mostly of wholesale stores with Chinese-origin owners—many referred to it as the "Chinese neighborhood"—none of these vocal groups included any members of the Chinese community. Since 2010, Chinese wholesale shops have slowly disappeared, and in 2017, 47 out of a total of 176 businesses along this section were hip restaurants and cafés. This transformation has been made easier by stricter loading and unloading regulations, allowing more sidewalk space to be used for bar terraces, and by the increasing price of real estate. Nearly 90% of the restaurants and bars that exist there today (37 of 42) have outdoor seating; only 8 of those have been operating for more than 10 years (Giraldo Malca 2017).

The second was the decision on how to distribute space for pedestrians, bicycles, cars and consumers. The selected design introduced extended green areas for resting, playing and socializing on the widened sidewalks on either side of the street, replacing the idea of a central "rambla" flanked by car lanes. The decision was made by the district government without substantial consultation with residents, with the promise to boost the local economy (Kotsila et al. 2020). As one of our interviewees described:

> The participatory process began with a pre-determined outcome. The project had already been presented. There were direct discussions between local state-managers ... A decision was made (for a boulevard design) and the participatory process was based on this basic idea. There was no debate over rambla vs. boulevard.

Although the redevelopment project coincided with a process of commercial gentrification in the area, the effect that tourism and property speculation have had on real estate in Barcelona during and after the renovation (2009–today) does not allow for a clear picture of the extent to which green infrastructure has contributed to this type of gentrification. The case of Passeig Sant Joan does, however, illustrate that well-intentioned green interventions can be linked to changes in the types of businesses, land use or commercial ownership in a neighborhood. The benefits of these improvements (e.g., access to play areas, public green space, improved temperature micro-climate) can be capitalized by powerful businesses and enjoyed by those who can afford to live or spend time around these new infrastructures. Local real estate advertising frequently highlights the enhanced quality of life provided by green amenities. The Passeig promotes a model of "consuming while enjoying" that does not necessarily coincide with

FIGURE 16.4 Passeig Sant Joan. (Photo by: Francesc Baró.)

the needs, habits, and realities of long-term residents. As a local government representative explained in an interview:

> The activity that you can generate at the sidewalks, playgrounds might not be as green as the Passeig above [older section], but it has more synergy with "ground floor" business activity. This produces a greater flow of people, in an area that, to my liking, was quite depressed in terms of economic activity and movement of people. (In person interview, 2016)

Moreover, the case of Passeig Sant Joan illustrates how an attractive green neighborhood may hide underlying dynamics of racial exclusion, which in this case was evidenced through the indirect removal of Chinese-owned businesses.

Public responses to speculation and exclusion

In 2014, the Catalan philosopher Marina Garcès described how tourism expropriates citizens from their own city, while filling urban spaces with individuals who are oblivious to their surroundings and their impacts. The City of Barcelona today emphasizes equity through greening and climate change adaptation and mitigation measures, attempting to attend to the needs of vulnerable social groups while also including citizens in designing and implementing interventions.[6] But Barcelona's greening paradox, in the context of an emerging global city and tourism destination, highlights the importance of accompanying policies that can safeguard housing rights and regulate the use of touristic establishments and attractions in order to guarantee truly equitable and just urban greening and urban planning.

After 2015, the new city government of progressive mayor Ada Colau filed a provisional moratorium on tourist accommodation licenses that was later replaced by the Special Plan (P.E.U.A.T.) in 2017 and which led to the closure of more than 4,500 illegal tourist flats. During the same year, the citizen-led Neighborhood Assembly for Sustainable Tourism (A.B.T.S.) was founded, demanding a decrease in tourism and the redistribution of benefits generated by the sector. These demands were, albeit partly, reflected in the development of the Strategic Tourism 2020 Plan for Barcelona (Blanco-Romero et al. 2018). These measures were also complemented by anti-displacement policies (rental mediation services, the establishment of the Unit Against Residential Exclusion (U.C.E.R.) to enforce housing rights in 2016) and new social or public housing provision, and alternative models such as cohousing transfer-of-use from the municipality to housing cooperatives paying special attention to the projects. As of 2019, any new housing development in Barcelona must designate 30% of units for accessible social housing.[7] The decision responded to substantial civic pressure from housing rights groups such as the P.A.H., the city's Neighbors Association (F.A.V.B.), and the D.E.S.C. social and cultural observatory. The year before, in 2018, the General Metropolitan Plan (P.G.M.) was already modified to allow for more social housing on consolidated urban land, with the proposal to implement officially protected housing at a homogeneous, appraised price in the city.

In the midst of the current COVID-19 pandemic, the city has also announced in July 2020 that it will give €1,200 per month to tourist apartment landlords who agree to house vulnerable families. It also plans to expropriate 194 empty apartments (possibly up to 426 in the near future) collectively owned by 14 corporate landlords (including the B.B.V.A. bank; the U.N.-denounced private equity Blackstone-subsidiary Budmac; and the Sareb government-owned "bad bank" and asset manager) unless they are designated as low-income housing units within the next month.[8]

Despite these efforts, Barcelona remains exposed to the risk of urban greening projects accelerating touristification and displacing local interests, as other cases in neighborhoods further away from the city center illustrate. Even the emblematic Superblocks traffic pacification project (*Superilles*, in Catalan), which is conceived to include adjacent social housing units, is not exempt from surrounding real estate speculation and family displacement (Oscillowicz et al. 2020).

Reviving Barcelona's early vision of a livable and green city for all?

Barcelona teaches us that city greening and beautification projects do not necessarily make communities more resilient or robust. The case of Pou de la Figuera in the Old City shows how community mobilizations are capable of building new equitable green spaces, but gentrification and touristification trends threaten the longevity of such grassroots efforts. While neighbors in the area managed to halt a large development project and demanded a revised planning process that contributed significantly to the redesign of the site, decades later community ties have deteriorated as a result of intense touristification and gentrification pressures that have displaced residents and made others feel like they don't belong. The redevelopment of Passeig de Sant Joan demonstrates that green street projects can have negative socio-economic consequences for residents and business owners, especially for racialized and vulnerable groups. Finally, the current COVID-19 pandemic is expected to drastically reduce the number of visitors to Barcelona, producing a new experience for local residents and uncertain impacts on street life, use of public space, the housing market and the economy. With such a reliance on tourism, Barcelona is likely to be hit hard by the pandemic, with lasting yet unknown impacts (Honey-Rosés et al. 2020).

Acknowledgments

The authors are grateful for feedback from Zaida Muxí on an earlier draft.

Keywords

(a) the urban development pattern of the city and neighborhood: long-term growth, post-crisis fast recovery; urban renewal; tourism-driven development; tech and design economy
(b) the urban greening of the city and/or neighborhood: green resilient infrastructure; green corridors; grassroots green spaces; new parks and park improvements; green beautification
(c) the inequalities at stake: green appropriation; green gentrification and displacement

Notes

1 Statistic Yearbook of the City of Barcelona 2019, Tourists and Overnight stays realized by the Statistical Department of the Barcelona City Council: https://www.bcn.cat/estadistica/catala/dades/anuari/cap13/C1302010.htm
2 https://www.euronews.com/2019/07/23/barcelona-s-tourist-headache-as-residential-population-declines-11-since-2015
3 (Observatori DESC)—https://observatoridesc.org/es/node/4320
4 (Observatori Metropolità de l'Habitatge de Barcelona)—https://www.ohb.cat/wp-content/uploads/2019/07/Habitatge_metropolis_2018.pdf

5 Passeig de Sant Joan marked the edge of what was called the "Golden Quadrant" of Cerdà's grid because it is where the most luxurious homes of the elite were built in the late 19th century.
6 Barcelona City Council. (2018). Pla clima 2018–2030. Available online: https://www.barcelona.cat/barcelona-pel-clima/sites/default/files/documents/pla_clima_cat_maig_ok.pdf
Barcelona City Council. (2013). Barcelona Green Infrastructure and Biodiversity Plan 2020 (original title: Plan del verd i de la biodiversidad de Barcelona 2020). Available online: https://ajuntament.barcelona.cat/ecologiaurbana/sites/default/files/PlanVerde_2020.pdf
7 https://www.barcelona.cat/infobarcelona/es/tema/vivienda/acuerdo-entre-entidades-y-grupos-politicos-para-que-el-30-de-las-nuevas-viviendas-sean-protegidas-2-2_744446.html
8 https://www.lavanguardia.com/local/barcelona/20200714/482307954875/barcelona-amenaza-expropiar-pisos-vacios.html

References

Anguelovski, Isabelle. 2013. "From Environmental Trauma to Safe Haven: Place Attachment and Place Remaking in Three Marginalized Neighborhoods of Barcelona, Boston, and Havana." *City & Community* 12 (3): 211–37.

———. 2014. *Neighborhood as Refuge: Community Reconstruction, Place Remaking, and Environmental Justice in the City*. MIT Press.

Anguelovski, Isabelle, Margarita Triguero-Mas, James J. T. Connolly, Panagiota Kotsila, Galia Shokry, Carmen Pérez Del Pulgar, Melissa Garcia-Lamarca, Lucia Argüelles, Julia Mangione, and Kaitlyn Dietz. 2020. "Gentrification and Health in Two Global Cities: A Call to Identify Impacts for Socially-Vulnerable Residents." *Cities & Health* 4 (1): 40–49.

Assemblea de Barris pel Decreixement Turístic (ABDT). 2020. "Assemblea de Barris Pel Decreixement Turístic (ABDT)." 2020. https://assembleabarris.wordpress.com/ (accessed 04/05/2020).

Barcelona City Council. 2013. "Barcelona Green Infrastructure and Biodiversity Plan 2020 (Original title: Plan del verd i de la biodiversidad de Barcelona 2020)." Medi Ambient i Serveis Urbans - Hàbitat Urbà. Available online: https://ajuntament.barcelona.cat/ecologiaurbana/sites/default/files/PlanVerde_2020.pdf

Blanco-Romero, Asunción, Macià Blázquez-Salom, and Gemma Cànoves. 2018. "Barcelona, Housing Rent Bubble in a Tourist City. Social Responses and Local Policies." *Sustainability* 10 (6): 2043.

Cocola-Gant, Agustin, and Antonio Lopez-Gay. 2020. "Transnational Gentrification, Tourism and the Formation of 'Foreign Only' Enclaves in Barcelona." *Urban Studies* 57 (15): 3025–43.

Connolly, James J. T., Tatjana Trebic, Isabelle Anguelovski, and Erin Wood, eds. 2018. *Green Trajectories. Municipal Policy Trends and Strategies for Greening in Europe, Canada and United States (1990–2016)*. Barcelona: Barcelona Lab for Urban Environmental Justice and Sustainability & ICLEI-Local Governments for Sustainability.

García-Lamarca, M., and M. Kaika. 2016. "'Mortgaged Lives': The Biopolitics of Debt and Housing Financialisation." *Transactions of the Institute of British Geographers* 41 (3): 313–27.

Giraldo Malca, Ximena. 2017. *The Importance of Public Participation in the Planning of Urban Green Infrastructure for the Distribution of Ecosystem Services: The Case of Passeig de Sant Joan in Barcelona*. Universitat Autonoma de Barcelona (UAB).

Honey-Rosés, Jordi, Isabelle Anguelovski, Vincent K Chireh, Carolyn Daher, Cecil Konijnendijk van den Bosch, Jill S Litt, Vrushti Mawani, Michael K McCall, Arturo Orellana, and Emilia Oscilowicz. 2020. "The Impact of COVID-19 on Public Space: An Early Review of the Emerging Questions–Design, Perceptions and Inequities." *Cities & Health*, 1–17.

Hoyos Núñez, Felipe Alberto. 2012. *El Modelo Barcelona de Espacio Público y de Diseño Urbano: Passeig de Sant Joan (1979 - 2011)*. Universitat de Barcelona.

Jonas, Andrew E. G., and Aidan While. 2007. "Greening the Entrepreneurial City: Looking for Spaces of Sustainability Politics in the Competitive City." In *The Sustainable Development Paradox: Urban Political Economy in the United States and Europe*, edited by Krueger, R., and Gibbs, D. C., 123–59. New York: Guilford.

Kotsila, Panagiota, Isabelle Anguelovski, Francesc Baró, Johannes Langemeyer, Filka Sekulova, and James J.T. Connolly. 2020. "Nature-Based Solutions as Discursive Tools and Contested Practices in Urban Nature's Neoliberalisation Processes." *Environment and Planning E: Nature and Space*, January, 251484862090143. https://doi.org/10.1177/2514848620901437.

March, Hug, and Ramon Ribera-Fumaz. 2016. "Smart Contradictions: The Politics of Making Barcelona a Self-Sufficient City." *European Urban and Regional Studies* 23 (4): 816–30.

Martí-Costa, Marc, and Mariona Tomàs. 2017. "Urban Governance in Spain: From Democratic Transition to Austerity Policies." *Urban Studies* 54 (9): 2107–22.

Oscilowicz, Emilia, Jordi Honey-Rosés, Isabelle Anguelovski, Margarita Triguero-Mas, and Helen Cole. 2020. "Young Families and Children in Gentrifying Neighbourhoods: How Gentrification Reshapes Use and Perception of Green Play Spaces." *Local Environment* 25 (10): 765–86.

17

COMPETING RISKSCAPES OF CLIMATE CHANGE, GENTRIFICATION AND ADAPTATION IN PHILADELPHIA'S HUNTING PARK NEIGHBORHOOD

Galia Shokry

While Philadelphia has been a frontrunner in deploying green infrastructure to reach sustainability goals that increasingly target climate risks and impacts, in recent years there has been growing recognition that these efforts have lacked equitable outcomes. Central gentrifying neighborhoods have become greener, making them cooler in summertime and more permeable during storms, while peripheral, lower-income neighborhoods of color remain some of the hottest and most flood-prone, posing significant risks to the health and security of residents. Hunting Park is a culturally and organizationally vibrant neighborhood that faces shifting and combined social and environmental riskscapes, including decades-long disinvestment, encroaching gentrification and widespread climate vulnerability. This chapter unpacks and explores Hunting Park's struggles and successes in protecting vulnerable residents from growing displacement pressures just as it is incorporated into a new and potentially "risky" green resilience planning paradigm.

From an industrial city to urban transformations and accelerated gentrification in Philadelphia

In Philadelphia's central city neighborhoods, tree-lined streets, historic two- to three-story row houses and quaint shop façades convey a sense of vibrant community, though like many rust belt cities, Philadelphia still grapples with the loss of its once-powerful industries, related environmental degradation and uneven development (Adams et al. 2008; Brownlow 2006). Amidst these struggles, climate inequity is a new and rising concern. It has become increasingly apparent that Philadelphia residents are differentially protected, and that this protection depends on starkly divergent neighborhood histories, as well as enduring racially and ethnically targeted policies and practices of (dis)investment (Shokry et al.

DOI: 10.4324/9781003183273-21

2021; Shokry, Connolly and Anguelovski 2020). This is one part of the shifting and combined riskscapes—that is, the multiple, interacting socio-spatial risks, as well as practices and perceptions vis à vis risks (Neisser and Müller-Mahn 2018)—those communities are facing. Meanwhile, installing green resilient infrastructure has become a key municipal strategy in climate adaptation to urban heat island effect, flooding, stormwater runoff and water pollution while investing in neighborhood beautification (Uittenbroek, Janssen-Jansen and Runhaar 2016).

Hunting Park is in many ways emblematic of these competing social and environmental riskscapes that marginalize vulnerable groups as they strive to build vibrant communities under changing climate conditions. It is a young and racially and culturally diverse Upper North Philadelphia neighborhood, several streets above El Centro de Oro (*the Golden Block*), the commercial corridor that is the beating heart of the city's Latinx community. In 2018, nearly 60% of Hunting Park's population identified with the Latinx/Hispanic ethnicity, and just under 40% identified as Black non-Hispanic.[1] While the city's overall trends reflect comparatively lower demographic percentages (e.g., approximately 15% of all Philadelphians identified as Hispanic in 2018), median household income citywide is nearly $23,000 higher than that of Hunting Park.[2]

Hunting Park's current mix of residents follows from the legacies of deindustrialization, redlining and gentrification of the city center. Since the late 19th century, Hunting Park was already a largely immigrant neighborhood, comprised mainly of residents of Italian and Polish descent who came to work in the manufacturing hub that made Philadelphia known as the "Workshop of the World." The area boasted industries from chemical, steel and paper-making to packaged pastries, sleds, clothing and thermometers. It was also home to critical public services and Proctor Electric, a large manufacturer of household appliances (Wilson 2008). In the early 1950s, Black and Latinx residents would also begin to migrate there in search of economic opportunities.

But as these industries closed or left North Philadelphia for the suburbs and overseas, so did capital investments in the area. All over Philadelphia, redlining practices had made it especially difficult for African Americans to obtain affordable housing. Many along with Latinx residents were forced to settle in these increasingly disinvested areas. As their numbers increased in the 1970s and 1980s, white residents would move out to Philadelphia's peripheries and suburbs. Likewise, speculators' growing interest in Center City and its surrounding neighborhoods had set off the process of displacing longtime Black and Latinx residents (Wilson 2008). In the Spring Garden neighborhood, this widespread practice spelled disaster for the majority-Latinx community (Beauregard 1990), who were displaced to areas such as Hunting Park. According to a local organizer, this northward shift has been ongoing since the 1980s and 1990s:

> We've seen that shift from our Hispanic population moving upwards more towards our area ... a lot of [Puerto Rican] families were pushed out of that [center and art museum] area with the development and forced up

north. Our concern is that our Hispanic and minority families, not only Hispanic families, are being almost pushed out of Philadelphia.

All in all, gentrification in Philly has accelerated since the 1990s, making it today the fifth most gentrifying city in the US according to the National Community Reinvestment Coalition (N.C.R.C.) (Richardson, Mitchell and Franco 2019). The N.C.R.C. study concurrently found that more than 12,000 Black residents moved out of gentrifying neighborhoods and that Philadelphia was among the top 12 cities with the highest rates of displacement for Latinx residents. The resulting widening racial wealth gap in Philadelphia is illustrated by Latinx residents having the highest poverty rate at 37.9% (PEW Charitable Trusts 2016) and Black residents having a median income two-thirds that of white residents with twice the rate of unemployment.

An enduring legacy of environmental losses and threats for vulnerable residents

Hunting Park has also undergone dramatic environmental transformations, some of them accelerated by these social and economic shifts. Once a forested landscape sustaining Native American settlements, in the late 19th century, industrialization led farmland from a colonial-era agricultural estate to be replaced by factories, residences and shops, evolving into today's asphalted landscape. According to the Hunting Park Revitalization Plan, in the early 20th century, the large park after which the neighborhood was named underwent expansion, eventually incorporating a small lake, a swimming pool, tennis courts, a carousel and other amenities.

By the late 1960s, economic restructuring was transforming the neighborhood in a new way. Care for the park was abandoned, its most notable facilities left to deteriorate or even outright demolished. Historical research conducted for the Hunting Park Neighborhood Strategic Plan reveals that funding for affordable housing, once attained thanks to the efforts of Hunting Park activists, was stolen by Mayor Rizzo's administration to purchase vacated industrial facilities and sell them off to private interests. This coupled process of deindustrialization and racism left behind a legacy of vacant land, with over 400,000 such lots across Philadelphia alone.

Some local industries—from chemical manufacturing to auto body shops—remain in Hunting Park, which is recognized as an Environmental Justice Area by the Pennsylvania Department of Environmental Protection (D.E.P.). In 2009, local activists fought and won against the expansion of the Mt. Cayuga waste site and the operations of Richard S. Burns & Company, a construction and demolition waste recycling and transfer station. Furthermore, an unzoned, unlicensed junkyard is the site of another community struggle to reappropriate land along an expired railroad right of way to create a community land trust and a more connective green fabric. Residents call this the *Campo Verde* or *Campus Green*, which is also the name of the citizen action group.

In the late 2010s, Hunting Park still lacked sufficient, safe and quality green spaces and remained saddled with numerous abandoned lots that attracted litter, pests and related public health issues. According to a 2018 local study by the Philadelphia Office of Sustainability (O.O.S.), Hunting Park residents today rate their health lower than the average Philadelphian. The area has higher risk rates of child obesity, diabetes, high blood pressure and high cholesterol in adults. Asthma, a respiratory condition linked with environmental exposures including airborne particulate matter, is also prevalent according to a study conducted by the Chemical Heritage Foundation. These issues, stemming in part from disinvestment, pollution and lack of green space, overlap significantly with climate issues in the neighborhood and are worsened by them.

Addressing socio-environmental vulnerability and historic inequities through green infrastructure

Despite racialized climate and adaptation inequities, Philadelphia has become a model for wide-scale urban green stormwater infrastructure. Starting in the early 2000s, the Philadelphia Water Department (P.W.D.) began to tackle flooding, stormwater runoff and drinking water pollution with green interventions, making it an early adapter in green watershed planning in the United States. Efforts to reduce storm and wastewater overflows by 85% by 2036 were part of a movement to use softer, lower-impact methods to renature the city while promoting multifunctional and cost-effective infrastructure with economic development benefits. P.W.D. also aimed to reduce urban heat island effects, flooding and air pollution by incorporating green resilient interventions in private yards, vacant lots, parks, schools and university campuses (Uittenbroek, Janssen-Jansen and Runhaar 2016). In 2016, a citywide climate adaptation framework, Growing Stronger: Toward a Climate-Ready Philadelphia, highlighted green stormwater infrastructure as a key tool for reducing climate vulnerability.

Today city-driven stormwater management interventions are accompanied by environmental stewardship initiatives sponsored by local nonprofits or by the City of Philadelphia to encourage homeowners to green their streets and homes. For example, as part of the Philadelphia Rain Check program, run by the nonprofit Pennsylvania Horticultural Society, homeowners can cost-share in reducing urban environmental risks by purchasing rain garden kits and downspout planters to install in their private outdoor spaces. Other programs such as Tree Philly and Tree Tenders provide street and yard trees for free to Philadelphia residents in exchange for their commitment to maintaining them.

However, inequitable outcomes were perhaps inscribed into those very first projects and the subsequent plans adopted by the city which tended to highlight green branding and economic attractiveness over inequities and social vulnerabilities. Due to their high visibility and low implementation risks, the P.W.D. selected the most centrally located neighborhoods for their demonstration projects, making it easier to gain political backing (Edwards and Bulkeley 2017). In

2009, the Greenworks sustainability plan conceived to boost the city's revival by branding Philadelphia as the greenest city in the U.S. (Garcia-Lamarca et al. 2019) and increasing its economic competitiveness. In 2011, the adoption of the Green City, Clear Waters (G.C.C.W.) plan set in motion a 25-year land and water strategy for improved urban permeability through green stormwater management and improved public access to water corridors. Stormwater management regulations for new developments and major retrofits, as well as parcel-based stormwater fees and grants, were put in place to incentivize private residential and nonresidential properties to install green stormwater infrastructure and reduce impervious surface areas.

To respond to growing equity concerns and address overlapping and competing social and environmental riskscapes, in 2017, the Office of Sustainability introduced the idea of a *Greenworks Equity Index* as evidence of its commitment to equity but has yet to publish the index as of 2021. At the end of 2019, the Philadelphia Department of Parks and Recreation launched its *Future of the Urban Forest* planning process for a ten-year equitable tree planting strategy aiming to understand, through community engagement, how the city's tree canopy could be expanded more equitably. In 2018, the Office of Sustainability also launched the *Beat the Heat* initiative, wherein the city will partner with lower-income and minoritized communities in the hottest neighborhoods to expand tree-planting and green stormwater infrastructure. Citywide trends indicate that neighborhoods with low socio-economic status are hotter than their wealthier counterparts and could benefit from greening to mitigate urban heat inequities (Pearsall 2017). The pilot project has been located in Hunting Park, one of the hottest neighborhoods in Philly, with temperatures up to 22°F higher than others. A preliminary study suggests that this was mostly due to more roads and paved surfaces: 75% of the neighborhood land cover, compared to 52% in Philadelphia overall. In contrast tree canopy is only 9%, compared to 19% in Philadelphia and 48% in neighborhoods like Chestnut Hill. "If you were walking down any of these streets [15 years ago] and you were leaving school and the sun was out, it was hot, there was no shade," according to a local community development coordinator (see Figure 17.2).[3]

In search of solutions, local players have been partnering with nonprofit and municipal programs to gain access to trees at low or no cost and encouraging residents to join the initiative. The Esperanza NeighborCare program, a mini grant that provides up to $1,000 in materials for residents to green, clean and beautify their blocks, has benefited 112 blocks so far and aims to achieve the goal of 30% canopy cover by 2050 established by *Greenworks*. Their efforts are especially impressive given "the neglect that took place over the past 50 years making people so resistant to trees when we needed them the most," said one city employee. According to a community-based activist,

> Some blocks used to have trees back in the day, but some died, and others were cut down. So, with all the problems trees caused, we needed to

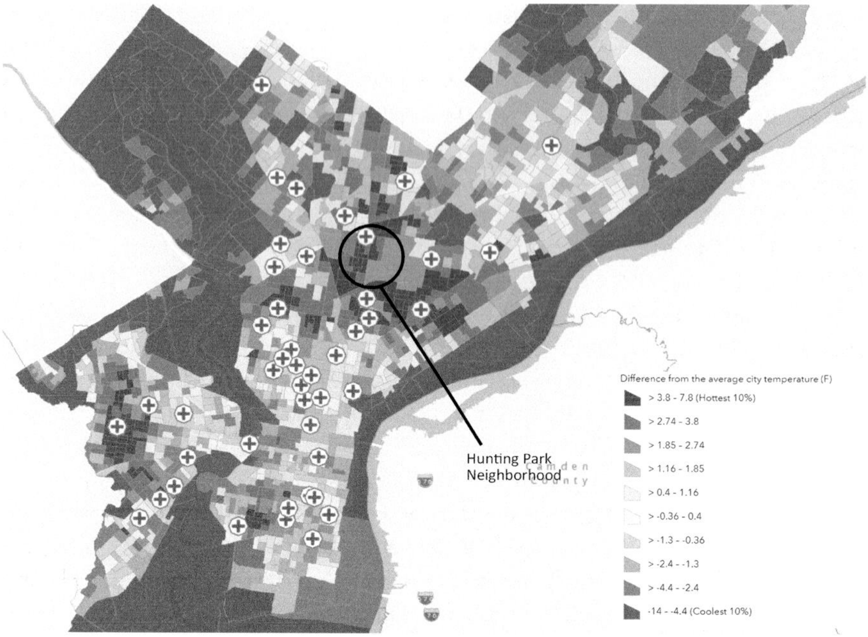

FIGURE 17.1 Heat exposure as measured and mapped by the City of Philadelphia's online HeatVulnerability Index. Hunting Park is among the hottest neighborhoods (indication added by author).

FIGURE 17.2 Two streets in Hunting Park: (left) no greening; (right) new planters and street trees provided by the City's TreePhilly program, installed and cared for by residents. (Photo by: Galia Shokry.)

> convince people to get trees planted—to get them to let go of that trauma of trees breaking sidewalks and all that.

Local activists have spearheaded other impressive and transformative projects that have improved permeability through green stormwater elements, providing shade for residents on hot summer days and safe community spaces. Hunting Park United is a group of volunteers that dedicated themselves to cleaning up and transforming the park, so that today it hosts community events and provides sports and other healthy activities to approximately 400 youth per year. Before their efforts took off, it was "basically a no-go zone," according to the group's co-founder, with dilapidated facilities, overgrown trees, scarce lighting and no viable playground in the entire 87 acres of green space. He had only positive memories of the park until his teens when the crack epidemic of the 1980s led to degraded conditions. "When we were playing on the field, you had to go down before practice, walk and remove drug paraphernalia and objects off the field just so the kids could play."

The group was approached in 2008 by the Fairmount Park Conservancy, a historically high-powered Philadelphia-wide parks nonprofit, to revitalize the park. Through a community-driven focus, they created an intercultural and intergenerational green space. "We were successful in working along with the conservancy [and city of Philadelphia and the multiple agencies and donors] and invested $5.5 million in the park in the last ten years," according to Hunting Park United's co-founder. The Philadelphia Water department further installed eight new rain gardens for enhanced stormwater management and an Olympic-sized pool where kids can stay cool in the summertime. "We decided we wanted to take back [the park] and give people a safe place, maybe to practice or be a part of something special and it worked." Today the space is a welcoming multi-purpose park with climate-resilient infrastructure.

The greening efforts by Hunting Park neighborhood groups are also aimed at strengthening social cohesion in the community and changing the relationship that residents have with their public places. Another local organizer imagines a Hunting Park with trees on every block leading to smaller interspersed green spaces. He became involved more recently in cleaning up abandoned lots which he says are about beautification, but also about making spaces safer. "We don't just want to plant the trees for you, we want to plant the trees *with* you, and a relationship builds when you're trying to plant the trees." Despite many achievements, there is dismay that with the challenges of locating lot owners so that the community can legally adopt these spaces, the creation of more actively communal spaces has been limited. After recent gun violence on a treeless street in Hunting Park, the local organizer urged that things have "to change after an event like that … They need something on that block for neighbors to come together." In that regard, greening has the potential to contribute to restoring wellbeing, healing trauma and rebuilding community and relationships (Anguelovski 2013).

FIGURE 17.3 Resident activists voluntarily clean up vacant lots and unmaintained properties across Hunting Park. (Photo by: Jose Ferran.)

The unresolved battle over gentrification

All in all, without clear and measurable equity goals, Philadelphia's resilience-enhancing policies and programs, which heavily depend on private investment and development, have become increasingly greening-oriented, protecting central gentrifying neighborhoods from climate risks and impacts and triggering private greening by developers. This is especially true since the G.C.C.W. plan was passed. According to an advocate for expanded green stormwater infrastructure:

> I would say the water department has a pretty robust engagement, a community outreach department actually … and then I think it's also being driven by private development. So, we're seeing [G.S.I.] in areas where there is development happening and there [in the city center] it's happening because they have to do it.

Gentrification also intersects with climate vulnerability throughout the city. A study by Shokry et al. (2020) reveals that the percentages of Black and Latinx residents have significantly risen in some of the most impervious and least climate-protected areas of Philadelphia, while the percentages of whiter and wealthier residents have increased in cooler, more permeable areas that have also been gentrifying. From a look at the study's maps, in addition to being one of the most climate-vulnerable areas in Philadelphia, Hunting Park also appears to have seen a moderate increase in residents of color while facing new gentrification pressures.

Some local organizations feel the looming threat of gentrification and displacement from recent and ongoing greening in Hunting Park that could make it more attractive and increasingly less affordable. Hunting Park is also bordered by an active Opportunity Zone, where new tax incentives meant to boost private development may not necessarily generate benefits for vulnerable residents, as recent research argues (Looney 2018). The city's longer-standing tax abatement program for new development and major renovations is seen by many as having fueled rapid and widespread gentrification and displacement while defunding public schools and other programs. "I think that tax abatement definitely has been a benefit to those who are better off and a burden to families that do not have enough sometimes to live," said one local organizer. In Hunting Park where homeownership has dropped from 70% in the 1990s to 45% (according to recent estimates by local organizers), most renters are cost-burdened, public schools are ailing and many households are underbanked, a fairer collection and distribution of taxes would go a long way. Curbing unfettered growth while stabilizing neighborhoods through strong social protection policies would support efforts by the Office of Sustainability to better protect socially and climate-vulnerable neighborhoods. "It's just hard politically because … it's hard on development, and the city wants development. [And too much development] pretty much goes completely against everything our office does."

In fighting displacement, the Hunting Park Community Collaborative coalesces multiple grassroots organizations with diverse missions to ensure that neighborhood change happens on residents' terms. One local organizer believes that they can prevent undesired development by working together and staying vigilant,

> that way it doesn't come as a surprise and are affected like some other communities have been [by unwanted development]. We have a great coalition of organizations that all work together, attending each other's meetings and sharing resources. It changes the idea of somebody just coming in and taking over.

In this way they can prepare ahead for new development activities.

Esperanza, a Hispanic, faith-based nonprofit serving the community for over 30 years, is planning to secure more land for community land trusts to help stabilize renters by coalescing multiple nationwide investors. In recent years, they rehabilitated an abandoned factory into 38 units of affordable rental housing and 5,000 square feet of new commercial space, helped small entrepreneurs access small grants and rebates for storefront improvements to sustain the neighborhood commercial corridor and purchased a vacant lot to create a community garden, among other actions. As local homeowners increasingly face costly home repairs and a lengthy waitlist for assistance from the City home repair program, they may be forced to move. A partnership with Habitat for Humanity has resulted in funding to rehabilitate 20 homes. Addressing climate vulnerability through weatherization and greening is a key part of these efforts, so that by saving energy, residents also save money and can stay in place. According to the organization, these efforts must not be in vain: "So, what happens if our community shifts? Who are we serving, you know? We want to protect not only the cultural identity of the neighborhood but all of those services that come with it."

The long-term benefits are an important objective for the Office of Sustainability's *Beat the Heat* initiative, which has turned Hunting Park into a model for strengthening ground-up climate resilience through public support in low-income, minority neighborhoods facing climate risks. These efforts are supported by research proposing more interactive methods for incorporating community perspectives on the benefits of green infrastructure to make G.I. planning more equitable and adapted to community needs (Heckert and Rosan 2018). According to an O.O.S. employee, "The problem is that [equity is] usually an afterthought. I want us to be talking about this: What are the ways that we're going to affect displacement?" Achieving equity on the ground through community-led and publicly supported efforts is a slow process, however. "It was community-based, community-driven, community-informed," says one local activist. "I wouldn't say it was actually fully implemented though." Presumably, due to a lack of public funding to implement plans and maintain support for the equity initiative, the City "stopped being present." This leaves the neighborhood vulnerable to unmitigated climate risks and reliant on attracting future private investment and development in order to become greener and more climate resilient. Both options carry with them heightened displacement risks for

FIGURE 17.4 The 5th Street Community Garden, *La Cosecha Latina*, received private and nonprofit sponsorship to purchase and transform a vacant lot. (Photo by: Galia Shokry.)

vulnerable residents, thereby combining with and shifting existing social and climate riskscapes.

Moving from green resilience to a more just socio-environmental resilience

Hunting Park, Philadelphia's new model neighborhood for equitable climate resilience planning, faces shifting and combined social and environmental riskscapes. This chapter explored how green climate resilience planning may be perceived as a "risky practice" in a context of encroaching gentrification and longtime disinvestment. It has shown how the kind of infrastructural investment seen in central neighborhoods has arrived slowly and unevenly in this mostly Latinx and Black neighborhood, requiring sustained civic action to achieve important but insufficient gains. Grassroots organizations have managed to create new schools and affordable housing units, improve public safety, upgrade and bring quality programs to the neighborhood's eponymous park and expand social services for health and childcare. However, this research also uncovers that an encroaching Opportunity Zone and unjust urban policies, such as extended tax abatements for developers, may mean that new greening, resilience and beautification efforts will only contribute to exacerbating existing riskscapes and engendering new displacement challenges. Green resilience programming must take intentional action to challenge unjust urban policies while centering social justice and anti-displacement measures in climate adaptation efforts, or longtime residents will lose out on the many gains they have been fighting for.

Keywords

- the urban development pattern of the city and neighborhood: post-industrial redevelopment; recovering city; recent gentrification
- the urban greening of the neighborhood: green resilient infrastructure; green stormwater infrastructure; revitalized parks and gardens; vacant lot greening
- the inequalities at stake: deep inequities in access to green; climate vulnerability; climate gentrification; green gentrification; insufficient affordable housing; poor-quality housing; vacant lots

Notes

1 Percentages derived from 2018 A.C.S. five-year estimates by recreating neighborhood boundaries using census tracts.
2 See Philadelphia Inquirer: https://www.inquirer.com/news/poverty-median-household-income-philadelphia-temple-university-graduate-hospital-20191219.html (accessed 22 January 2021).
3 See City of Philadelphia Heat Vulnerability Index on E.S.R.I., https://phl.maps.arcgis.com/apps/webappviewer/index.html?id=9ef74cdc0c83455c9df031c868083efd (accessed 2 February 2021).

References

Adams, Carolyn, David Bartelt, David Elesh, Ira Goldstein, Joshua Freely, and Michelle Schmitt. 2008. *Restructuring the Philadelphia Region: Metropolitan Divisions and Inequality*. Temple University Press. https://www.jstor.org/stable/j.ctt14btd2h.

Anguelovski, Isabelle. 2013. "From Environmental Trauma to Safe Haven: Place Attachment and Place Remaking in Three Marginalized Neighborhoods of Barcelona, Boston, and Havana." *City & Community* 12 (3): 211–37. https://doi.org/10.1111/cico.12026.

Beauregard, Robert A. 1990. "Trajectories of Neighborhood Change: The Case of Gentrification." *Environment and Planning A* 22 (7): 855–74. https://doi.org/10.1068/a220855.

Brownlow, Alec. 2006. "An Archaeology of Fear and Environmental Change in Philadelphia." *Geoforum* 37 (2): 227–45. https://doi.org/10.1016/j.geoforum.2005.02.009.

Edwards, Gareth A. S., and Harriet Bulkeley. 2017. "Urban Political Ecologies of Housing and Climate Change: The 'Coolest Block' Contest in Philadelphia." *Urban Studies* 54 (5): 1126–41. https://doi.org/10.1177/0042098015617907.

Garcia-Lamarca, Melissa, Isabelle Anguelovski, Helen Cole, James J. T. Connolly, Lucía Argüelles, Francesc Baró, Stephanie Loveless, Carmen Pérez del Pulgar Frowein, and Galia Shokry. 2019. "Urban Green Boosterism and City Affordability: For Whom Is the 'Branded' Green City?" *Urban Studies*, December, 004209801988533. https://doi.org/10.1177/0042098019885330.

Heckert, Megan, and Christina D. Rosan. 2018. "Creating GIS-Based Planning Tools to Promote Equity Through Green Infrastructure." *Frontiers in Built Environment* 4 (May). https://doi.org/10.3389/fbuil.2018.00027.

Looney, Adam. 2018. "Will Opportunity Zones Help Distressed Residents or Be a Tax Cut for Gentrification?" *Brookings* (blog). February 26, 2018. https://www.brookings.edu/blog/up-front/2018/02/26/will-opportunity-zones-help-distressed-residents-or-be-a-tax-cut-for-gentrification/.

Neisser, Florian, and Detlef Müller-Mahn. 2018. "Urban Riskscapes—Social and Spatial Dimensions of Risk in Urban Infrastructure Settings." In *Urban Disaster Resilience and Security: Addressing Risks in Societies*, edited by Alexander Fekete and Frank Fiedrich, 347–59. The Urban Book Series. Cham: Springer International Publishing. https://doi.org/10.1007/978-3-319-68606-6_20.

Pearsall, Hamil. 2017. "Staying Cool in the Compact City: Vacant Land and Urban Heating in Philadelphia, Pennsylvania." *Applied Geography* 79 (February): 84–92. https://doi.org/10.1016/j.apgeog.2016.12.010.

PEW Charitable Trusts. 2016. "Philadelphia's Changing Neighborhoods: Gentrification and Other Shifts since 2000." http://www.pewtrusts.org/-/media/assets/2016/05/philadelphias_changing_neighborhoods.pdf.

Richardson, Jason, Bruce Mitchell, and Juan Franco. 2019. "Shifting Neighborhoods: Gentrification and Cultural Displacement in American Cities » NCRC." https://ncrc.org/gentrification/.

Shokry, Galia, Isabelle Anguelovski, James J T Connolly, Andrew Maroko, and Hamil Pearsall. 2021. "'They Didn't See It Coming': Green Resilience Planning and Vulnerability to Future Climate Gentrification." *Housing Policy Debate*.

Shokry, Galia, James JT Connolly, and Isabelle Anguelovski. 2020. "Understanding Climate Gentrification and Shifting Landscapes of Protection and Vulnerability in

Green Resilient Philadelphia." *Urban Climate* 31 (March): 100539. https://doi.org/10.1016/j.uclim.2019.100539.

Uittenbroek, Caroline J., Leonie B. Janssen-Jansen, and Hens A. C. Runhaar. 2016. "Stimuli for Climate Adaptation in Cities: Insights from Philadelphia – An Early Adapter." *International Journal of Climate Change Strategies and Management* 8 (1): 38–56. https://doi.org/10.1108/IJCCSM-06-2014-0069.

Wilson, Catherine E. 2008. *The Politics of Latino Faith: Religion, Identity, and Urban Community*. NYU Press.

PART 5

(Fragile) green justice victories and gray zones in the just green city

In our final part, "(Fragile) green justice victories and gray zones in the just green city," we offer alternative scenarios and lenses on more equal and just green futures. Each chapter articulates what greening can mean for equity when it goes hand in hand with social policies that seek to guarantee and/or protect a right to place for fragile residents. The cases of Portland, Copenhagen, Nantes and Vienna show that such successes can promise more just green futures while not being exempt from contradictions.

Portland, one of the whitest cities in the U.S., is often considered a pioneer of sustainability planning. However, its traumatic history of racial exclusion and discrimination, together with the inequitable distribution of environmental amenities and advanced gentrification, raises concerns about procedural environmental justice in the city. In response, Latinx communities are mobilizing to defend community-controlled greening, even as they face increasing pressures to include the African American community.

In Copenhagen, the world's self-proclaimed greenest and healthiest city, the green redevelopment of working-class districts such as Nørrebro represents a leap forward in ecological terms with some noteworthy equity initiatives. However, those also contain increasing contradictions as municipal efforts to increase the city's attraction value, combined with the privatization of public housing and large investments from international capital funds, are causing affordability issues and increasing exclusion for racialized minorities.

Thanks to 30 years of urban re-naturing ("Nature en ville"), social equity policies and high-quality public transit systems, Nantes has become one of the greenest yet still relatively affordable cities in Western Europe. Yes, as gentrification concerns are looming, Nantes' tale raises the question of what types of re-naturing and social policies are now needed to keep Nantes a green, equitable and inclusive city for all—rather than a museum of green, as it has branded itself.

DOI: 10.4324/9781003183273-22

Lastly, while in other cities urban transitions to environmental sustainability are often either motivated by neoliberal strategies of urban development or end up triggering gentrification, Vienna shows a model characterized by the radical notion (today) that housing is not a commodity, but a universal right, even in a city praised for its greening and livability. Yet, as exceptional as the Viennese case is, it is not alien to the challenges posed by a pressing neoliberal political environment that is slowly recommodifying housing and shifting toward new international real estate investment.

18

A GREEN, LIVABLE COPENHAGEN IN THE SHADOW OF RACIALIZING, NEOLIBERALIZING POLITICS

Rebecca L. Rutt

Copenhagen is often portrayed as a sustainable and livable city, echoing Denmark's longstanding reputation as a leader in environmental and social welfare protections, including affordable housing. Yet, neoliberal and racializing trends at the national level are undermining social and environmental justice across the country. In Copenhagen, these intertwining trends manifest in the dominance of a green growth rhetoric in the city's urban sustainability and livability agendas alongside the dismantling of housing protections, leading to skyrocketing prices and the uprooting of more vulnerable residents. Focusing on the Nørrebro neighborhood, this chapter sheds light on local manifestations of national trends through several recent struggles over the right to the neighborhood. These struggles illustrate how processes of ecological gentrification are shaped by broader trends of neoliberal and racializing politics that manifest in urban governance, but that the tireless efforts of resident-activists may stem the tide and avoid the co-option of urban green spaces.

Copenhagen: the livable green city?

Over the past 15 years, Copenhagen has come to epitomize Denmark's longstanding reputation as a global leader in tackling environmental issues. The city was anointed as the European Green Capital in 2014 and aspires to become Europe's first carbon-neutral capital. Copenhagen now prides and brands itself as a livable green city. As a renowned welfare state, affordable housing, particularly in urban areas, has also long featured as a backbone of true livability in the country.

However, major national shifts, especially since the early 2000s, threaten these accomplishments. Ushered in by an economically liberal, socially conservative administration headed by Anders Fogh Rasmussen, the first shift was

DOI: 10.4324/9781003183273-23

the foregrounding of the logics of neoliberalism, which amongst other impacts has deeply undermined accessible housing. Another is the rise of racist rhetoric, already visible in the spread of a cultural anxiety in the country since the 1990s, but made explicit in Rasmussen's first speech as prime minister, in which he cited a growing "cultural war of values" (Rytter and Pedersen 2014). Both shifts led some scholars to describe Denmark as a post-welfare state that no longer prioritizes the needs of the most vulnerable (Simonsen 2015). Both shifts also profoundly influence municipal politics, practices and landscapes and raise questions about who is served by the current sustainability and livability agendas.

Copenhagen's contemporary livable green city and sustainability agendas, for example, feature a neoliberal discourse rich in innovation, smart technology and green growth. A 2009 strategy introduced a "vision of being the world's eco-metropolis," noting

> More green and blue areas ... invite people to take part in urban life. A varied urban life gives us an advantage [over] other large cities. It attracts tourists, creative people and firms. It creates economic growth as well as painting a positive picture of both the city and business and cultural life.
>
> *(K.K. 2009, 5)*

The celebrated Copenhagen 2025 Climate Plan also proclaims, "We must show that it is possible to combine growth, development and increased quality of life with the reduction of CO_2 emissions." Alongside clean energy and gray infrastructure projects, urban greening is thus receiving increased attention as Copenhagen brands itself the "cool green city" and aspires toward continued growth. A network of blue and green infrastructure including parks and waterfronts extends through the city, underpinned by goals outlined in the Urban Nature Strategy 2025.

Copenhagen's work toward becoming a sustainable city has, however, occurred alongside a drastic rise in real estate prices, reaching record levels in 2020.[1] This trend is cause for celebration by growth-oriented officials, and the authors of the Urban Nature Strategy follow suit, noting the "positive influence on property values" of green space proximity and highlighting that "the value of a home increases by up to 10% on average with each additional 10 hectares of park or urban nature area within a 500-meter distance." But such positive interpretations of rising housing prices fail to acknowledge the social justice implications of ecological gentrification that such strategies are likely to produce (Anguelovski 2015). The context for ecological gentrification is ripe in Copenhagen, where housing justice has been aggressively undermined following the uptake of neoliberal "new public management" and rollout policy interventions since the 2000s. Such interventions are rapidly commodifying Danish affordable housing models, namely nonprofit and private housing associations.

Housing politics have also become embroiled in racist and racializing politics, which intertwine with neoliberal impacts including gentrification to particularly burden residents of color. A marked rise in low-income, unemployed, immigrant households occupying Danish nonprofit housing triggered the Rasmussen administration to expand Denmark's unfortunately titled "ghetto" policy, which has been developed by subsequent administrations. The policy targets areas that fulfill highly controversial criteria including a minimum of 50% of residents with "non-Western" nationality or heritage, and additional criteria related to employment, education, income levels and criminal records. It employs several strategies toward "social mixing" (Lees 2008), expelling some to make space for more affluent residents. Under the latest 2018 "ghetto" plan,[2] residents of areas on the list can be subjected to stricter punishments for certain crimes and a compulsion to send small children to daycare. The strategy is widely criticized for stigmatizing more than supporting these areas.

Nørrebro: a legacy of solidarity threatened by gentrification

Nørrebro emerged as a working-class neighborhood in the mid-19th century. A rich community culture flourished; the neighborhood was an early hub for the national workers' movement and, by the 1960s, a stronghold for housing justice activists. The burgeoning population turned to backyards and building corridors to construct more living space, and the increasing spatial and socioeconomic challenges to the mid-20th century led Nørrebro to become one of Copenhagen's first areas to undergo contemporary urban renewal. Renewal manifested largely as a top-down demolition and reconstruction process—an undertaking that alienated residents and stimulated unrest across the neighborhood, especially in the 1970s and early 1980s. The municipality did not prioritize open, green spaces for this low-income neighborhood and even small parks had to be fought for by resident-activists. Despite the renewal, Nørrebro remains dense. With over 80,000 residents in just 4.1 km^2, it is the most populated district in the city—over twice as dense as Copenhagen's second most populated district.[3] Unsurprisingly, Nørrebro has only 7 m^2 of public green space per person, notable in comparison to the city average of 39 m^2 per person.[4]

Since the 1980s, neighborhood demographics evolved with an influx of people from beyond European borders, a trend facilitated by the availability of lower-income housing and tradition of social solidarity. Nørrebro hosts many of the city's shelters, welcoming Danish, European and increasingly non-European houseless migrants. Home to 2 of Denmark's 15 "ghettos," Nørrebro is also categorized by Copenhagen's municipality as 1 of 6 vulnerable areas in the city, according to indicators including high unemployment, low income and education and—most troubling due to its racializing implications—the share of residents labeled as non-Western. This label is applicable even to those born and raised in Denmark. Facing unemployment and social exclusion as a result of such political narratives, some neighborhood youth turn to gangs for income and a sense of belonging (Pedersen and Lindstad 2011). This has perpetuated police presence and recurring attempts by the municipality to introduce security, often through physical renovation (Rutt and Loveless 2018).

Among the most ethnically and culturally diverse areas of central Copenhagen, Nørrebro is labeled by the official tourism website as "colorful" and "multicultural."[5] A recent explosion of shops, bars and cafes has led international press like *Vogue* magazine to describe it as home to "The Coolest Spots in Copenhagen."[6] These changes have occurred in part through a strategic commodification of diversity that enrolls ethnically diverse neighborhoods into growth and tourism development agendas (Schmidt 2017). Alongside the assault on housing protections, it is unsurprising that Nørrebro has experienced sharply rising housing prices, with the cost of owner-occupied housing doubling over the past decade.[7]

Green(ing) Nørrebro?

Nørrebro's development-oriented attention of late articulates in diverse ways with its current green spaces and the greening rhetoric of the city's sustainability and livability agendas. Several examples below of recent struggles over the right to the neighborhood demonstrate the localized impacts of neoliberal and racializing politics that inform these agendas and that ecological gentrification emerges both through the presence of valued green spaces as well as their governance.

The People's Park: multiple environmental injustices

Nørrebro's People's Park, a small green space with a rich but troubled past, has been long embroiled in struggles for environmental justice and municipal aspirations for urban development. The struggles have evolved from distributive dilemmas to procedural conflicts, all within the broader politics of recognition that reflects national trends.

As mentioned, municipal authorities cleared parts of Nørrebro including what would become the People's Park in the early 1970s to accommodate new housing. Yet residents quickly came to value the newly open areas and acted to secure them, mobilizing local materials and labor to cultivate beloved public spaces. In a classic struggle for distributive justice, major conflicts erupted between municipal authorities and resident-activists frustrated with the top-down approach to renewal. Conflict fatigue on the part of urban authorities and the successful reclaiming of several areas for construction likely explains the survival of People's Park, though securing it as a public green space demanded sustained efforts by residents in the years to follow (Rutt and Loveless 2018).

The People's Park received official recognition in the local zoning plan in the early 2000s. Shortly thereafter, municipal officials determined the park needed a makeover to bolster their aspirations of urban security and inclusivity. Lessons learned from earlier conflicts made resident engagement, rather than top-down action, an obvious imperative. Renewals over the following years exhibited well-meaning attempts at procedural justice, with some failures in the mid-2000s, and subsequent successes, particularly in the latest renewal of 2014 (ibid.).

When the 2014 renewal initiated, dubious resident-activists pointed out that too often, the "space for all" described by landscape architects and urban planners usually meant "space for the white middle-class" (Balfelt 2016). This process, led by the artists of Kenneth Balfelt Team, remedied past mistakes by allocating sufficient time for resident engagement and meeting the diversity of user groups in ways that worked for them. By the end, some of the new features even explicitly served vulnerable park users, including non-European houseless migrants. Amenities included track lighting to leave some parts in shadow (challenging typical assumptions of lighting as essential to security) and armless benches. This renovation was celebrated by residents and national media alike.

Yet even well-intentioned urban managers do not act in isolation, and efforts at meaningful inclusion in urban green development can be inhibited or co-opted by wider political agendas. In the years since, increasing resident complaints to City Hall against houseless park users have driven stricter police enforcement of anti-camping ordinances. The ongoing demographic shifts seem to be ushering in residents with less connection to the neighborhood's identity and different understandings of who belongs in renewed public spaces.

The history of the People's Park shows how urban development is an inherently political and contested practice and that the work of resident-activists is critical to the normalization and defense of open, green spaces in traditionally working-class neighborhoods (Rutt and Loveless 2018). Yet despite its inclusive array of amenities, the park's renewed version is nonetheless being co-opted, due

FIGURE 18.1 The People's Park. (Photo by: Simone Grytter/Kenneth Balfelt Team.)

to and in the service of gentrification, resulting in the expulsion of more vulnerable park users.

Transforming Hans Tavsens Park into a city-branding green amenity

In 2016, a consortium of Danish landscape architects and engineers won the prestigious design competition Nordic Built Cities Challenge with a renovation proposal for Nørrebro's Hans Tavsens Park. The proposal was inspired by a severe flooding event in 2011 that caused millions of Danish kroner in damages. The "innovative green design" was widely praised not only for its climate resilience aspects but also for offering "a deep connection to nature" alongside local ownership, co-creativity and a sense of community (Tubridy 2020).

The proposal entails a capital- and infrastructure-intensive renovation that soon raised local concern. Many residents expressed love for the existing park, and while acknowledging the need for better storm water management, worried over the extent of change depicted in the digital artistic rendering (Tubridy 2020). Certain aspects of the proposal were later assessed as unrealistic in terms of sustainability claims, as they would require additional infrastructure to be feasible (ibid.). Claims of co-creation and participation have also been challenged: research identified more affluent residents engaged in participatory processes that

FIGURE 18.2 Digital rendering of a renovated Hans Tavsens Park. (Photo by: The Soul of Nørrebro by S.L.A. and Rambøll. Visualization by: Beauty and The Bit.)

in fact offered little scope for influence (ibid.). Attempts to discuss the project's potential for gentrification were also stifled.

The project is currently on hold largely due to the municipality's uncertainty over the project's value against other priorities and financial constraints. Residents are not complaining. Tubridy (2020) describes this renovation as aiming foremost at establishing a "high-profile green amenity with branding power." This produced tensions between entrepreneurial green design and claimed socio-ecological priorities, undermining the latter. Within the context of the gentrifying neighborhood, the pursuit of this park renovation in the future seems likely to gentrify both through process and outcome.

Superkilen: "extreme" or tokenistic participation

Nørrebro's 750-meter long Superkilen linear park has been widely praised by prestigious institutions including the American Institute of Architects and Aga Khan Trust for Culture. Commissioned in 2008 by the private philanthropic organization Realdania and Copenhagen municipality, the famed architectural firm B.I.G.-Bjarke Ingels Group took the lead on the $10 million project to promote intercultural tolerance and social cohesion in the neighborhood, citing the unrest and divisions following Denmark's 2008 Mohammed cartoon controversy as motivation.

The design process ensured from 2009 to 2010. Described by B.I.G. as "an exercise in extreme public participation," it aimed to garner representations of the neighborhood's diversity to adorn the park's open spaces (K.K. 2011). Via newspapers, radio, the internet and e-mail, the architects asked inhabitants to suggest symbols of their culture, resulting in 108 objects representing 60 different nationalities dispersed across the park, including palm trees from China, basketball hoops from Mogadishu and a Moroccan fountain. The municipality described the outcome as a "surrealist collection of global urban diversity that in fact reflects the true nature of the local neighborhood, rather than perpetuating a petrified image of homogenous Denmark" (ibid.). Superflex, the design firm that led the participatory process, described the benefits: "Before, people would gather behind the buildings. Now Superkilen is the place where people go. There are also lots of tourists. Now anytime during the day you see more people using it and it's just more safe" (GreenL.U.L.U.s E.R.C. project[8]).

Yet Superkilen has also been criticized for failing to achieve the intended social cohesion and for increasing private housing prices. Given the national trends discussed in this chapter, social cohesion may be too much to expect from any renewal initiative. Daly (2019) suggests that such failure emerged from "a divorcing of the design concept from the everyday life of Nørrebro's multi-ethnic community ... Superkilen focuses too much on spatial representation and too little on spatial practice." GreenL.U.L.U.s research similarly found that the participatory process emphasized choices of objects and colors over the broader use of space and relations between diverse users. Within the overarching

FIGURE 18.3 Superkilen park. (Photo by: Jens Friis Lund.)

context of housing politics, it is no wonder that housing prices around the park since completion in 2012 have risen steeply, and as this chapter makes clear, this burden falls hardest precisely on those ostensibly represented in the new park.

Nørrebro's urban farming initiatives: from formalization to gentrification

Urban community gardening is relatively new in Denmark, garnering the attention of municipal officials mostly over the past decade (Rutt 2020). Estimates suggest there are around 15 volunteer-led community gardens in Copenhagen—many of these in Nørrebro—although many have also come and gone. In Copenhagen, various manifestations of neoliberalism, from formalization processes to austerity pressures, articulate with convivial urban farming initiatives in contradictory and recursive relationships that both nurture and endanger them (ibid.).

For example, volunteer associations are expected to adopt a hierarchical governance structure. This formalization is critical to engendering legitimacy vis-à-vis municipal authorities and funders, without which most gardens cannot survive. But formalization alienates those who crave alternative governance forms or lack the skills to engage in technical applications for permissions and grants, leading to the homogenization of demographics in some gardens that favors formally educated and typically higher-income residents (Rutt 2020).

Volunteer farming initiatives are discursively celebrated by Copenhagen officials in municipal promotional materials and simultaneously undermined by austerity measures such as the recent cutbacks on municipal support to volunteer associations to navigate bureaucracy. Rutt (2020) also documents the persistence of reluctant disengagement of well-meaning volunteers, which can be understood through the broader context of the neoliberal city. Rising housing costs, for instance, push volunteers to prioritize wage labor over volunteer work despite their good intentions, resulting in the collapse of some initiatives. Yet the neighborhood's surviving gardens also contribute to its gentrification as desirable green spaces in the context of a lowly regulated housing market.

FIGURE 18.4 Nørrebro community roof garden. (Photo by: Jens Friis Lund.)

Rutt (2020) asserts that community gardening in Nørrebro is a beacon of grassroots greening with anti-capitalist values at heart. Yet these projects face tremendous and often unacknowledged threats from the larger context of neoliberal governance norms, austerity and housing politics, which intertwine to undermine the very life of garden projects.

NordPolen: a green space under threat

Nørrebro's NordPolen (North Pole) park, home to beloved trees including a 200+-year-old hawthorn, has been at the center of a struggle to preserve the little green space left against municipal ambitions to capture it for new construction. Years of tug-of-war in a "long, convoluted and unsightly" planning process embroiled numerous residents and local organizations who lobbied to protect it.[9] A resident explained, "the population density in Nørrebro—and the tensions it gives rise to—make it urgently important to safeguard the few recreational areas that bring together residents from all walks of life." She added that fewer open spaces "increase frustration and discouragement—a consequence that will in every way be reflected in residents' desire to take care of the district."[10] Residents-activists' "hard work and raging behind the scenes" eventually resulted in a change of heart in 2019 by City Hall to leave the green area untouched.[11]

Then in 2020, a 2013 municipal agreement to utilize NordPolen for road expansion unexpectedly came to light, threatening what residents had just achieved. A local environmental organization representative pointed out, "[This] underlines the widespread perception among citizens that it's completely impossible to cooperate with the municipality."[12] Locals mobilized anew, and after several months, politicians committed to dismiss the 2013 proposal and maintain NordPolen as a green space.

Nuuks Square: a questionable housing development

Recent developments at Nørrebro's Nuuks Square also illustrate how resident activism is critical to reshaping urban development trajectories toward local justice. Despite its density and minimal green space, Nørrebro is being pressured to host new housing infrastructure following Copenhagen's growth aspirations. In 2018, City Hall broadly supported a proposed 22-story, 75-meter-high residential tower for the square. The proposal was then released for public consultation. Ninety-six percent of the 463 responses, however, were opposed. Despite the proposal's digital render featuring a revamped, green public space, opponents cited the destruction of the horizon and the long shadow the tower would cast. Opposition also manifested in lively online debate, poster campaigns and plans for protest. The politicians capitulated and local headlines celebrated: "Massive local resistance has toppled the tall residential tower right in the middle of Nørrebro."[13]

FIGURE 18.5 Tower at Nuuks Plads digital rendering. (Photo by: Over Byen Arkitekter.)

Following the tower's rejection, the state real estate company sold the square to private property developers who proposed a new construction with a maximum height of 24 meters, on par with surrounding buildings. By occupying more surface area, the new proposal features less open public space than the tower proposal. Still, both proposals as digitally rendered feature green elements and refer to social and environmental goals. Will urban entrepreneurial green design aspirations for this square supersede the claimed socio-ecological aspirations, echoing the case of Hans Tavsens Park (see Tubridy 2020)? The new proposal's use of existing buildings over demolition seems appropriate from a sustainability perspective, as does the maximum height from a social perspective. A developer representative clarified their approach, which seems to recognize the procedural justice imperative: "We take the task very humbly and hope that there will be support."[14] Critical aspects also include the price of new housing and the effects of the renewed square on the housing prices of the surrounding area.

Resisting gentrification in Nørrebro

Although recent developments threaten the quality of life of current residents, Nørrebro's identity as a beacon of social solidarity remains strong. Reflecting the tradition of housing justice activism since the 1960s, resident-activists in Nørrebro co-developed Almen Modstand (Common Resistance) in 2018, a nationwide, nonprofit-housing resident coalition demanding an end to the sale and speculation of nonprofit housing, greater influence in policymaking and the abolishment of neoliberal, racializing policy mechanisms.

Activists explain that such policy mechanisms "are used to force public housing associations to demolish and sell public housing, as well as to privatize public property."[15] They note the hypocrisy of demolishing even newly renovated nonprofit housing by the current self-described "green" Social Democratic government. Common Resistance asserts, "public housing is for everyone—rich and poor, young and old—and therefore public housing *is* the mixed city where everyone can live." The network recently succeeded in attaining the required 50,000 signatures to bring their demands to the floor of the Danish Parliament, and all eyes will be on the coming debate.[16]

Continuing struggles for justice

The stories above reveal how ecological gentrification must be understood not only in terms of housing prices but also through urban green space governance. These green spaces appear, for now, to have at least the discursive if not material support of municipal officials, often thanks to the work of resident-activists. The persistence of struggles also makes clear that such wins must be continually fought for.

Yet, none of Nørrebro's green spaces are exempt from co-option into broader political agendas. The examples here illustrate how the effects of urban greening cannot be understood outside the manifestations of larger neoliberal and racializing politics, which are decisive in producing a context ripe for ecological gentrification. In Copenhagen, these politics mobilize sustainability and livability agendas while undermining housing protections, collectively driving out lower-income and often racialized residents. Housing justice activists are explicitly tackling this context through collective political action. Their success would be an important contribution to mitigating the broader conditions spurring ecological gentrification in Copenhagen and across the country.

Keywords

- the urban development pattern of the city and neighborhood: sustainability planning; livable city planning; green growth; urban neoliberalization; racialization
- the urban greening of the neighborhood/city: climate resilience; smart technology; urban farming; entrepreneurial greening; green design
- the inequalities at stake: green gentrification; housing exclusion; environmental injustice; racializing politics

Notes

1 TV 2 Lorry. 2020. "Rising house prices have become first-time buyers' nightmare." Accessed 20 December 2020. https://www.tv2lorry.dk/koebenhavn/stigende-boligpriser-er-blevet-foerstegangskoeberes-mareridt-faa-boligoekonomens-bedste
2 The plan, titled "One Denmark without a parallel society – No ghettos in 2030" is available here: https://www.regeringen.dk/nyheder/2018/ghettoudspil/
3 Mit Nørrebro. 2019. "Nørrebro i tal." Accessed 15 December 2020. https://mitnorrebro.dk/noerrebro-i-tal/
4 Ibid.
5 Visit Copenhagen. 2017. "Nørrebro: colourful, casual and young at heart." Accessed 15 December 2020. www.visitcopenhagen.com/copenhagen/culture/multicultural-norrebro
6 Lewis, Victoria. 2016. "All of the Coolest Spots in Copenhagen Are on This Street." *Vogue*. Accessed 10 December 2020. https://www.vogue.com/article/jaegersborggade-copenhagen-guide-coolest-street
7 Politiken. 2018. "Få overblikket bydel for bydel." Accessed 15 December 2020. https://politiken.dk/forbrugogliv/boligogdesign/art6622615/Prisfesten-p%C3%A5-ejerlejligheder-er-rykket-til-de-billige-bydele-i-K%C3%B8benhavn
8 See the Horizon 2020 and European Research Council funded GreenL.U.L.U.s project website at: http://www.bcnuej.org/projects/greenlulus/
9 Trudsø, Mette. 2020. "Grønne kvadratmeter i De Gamles By vinder over trafikprojekt." Mit Nørrebro. Accessed 10 December 2020. https://mitnorrebro.dk/groenne-kvadratmeter-i-de-gamles-by-vinder-over-trafikprojekt/
10 Trudsø, Mette. 2019. "Helhedsplan for De Gamles By får tæsk af beboerne på Nørrebro." Mit Nørrebro. Accessed 10 December 2020. https://mitnorrebro.dk/helhedsplan-for-de-gamles-by-faar-taesk-af-beboerne-paa-noerrebro/
11 Ibid.
12 Ibid.
13 Habermann, Anders. 2018. "Nuuks Plads prøver igen: Nyt forslag til lokalplan klar." Accessed 10 December 2020. https://mitnorrebro.dk/nuuks-plads-proever-igen-nyt-forslag-til-lokalplan-klar/#:~:text=Et%20forslag%20om%20et%2075,nyt%20forslag%20til%20pladsens%20fremtid.
14 Vemb, Christian. 2020. "Nyt byggeri på Nuuks Plads bliver lavt og indrammet i stedet for højt." Mit Nørrebro. Accessed 10 December 2020. https://mitnorrebro.dk/nyt-byggeri-paa-nuuks-plads-bliver-lavt-og-indrammet-i-stedet-for-hoejt/
15 Almen Modstand 2020. *Citizen Proposal*. Available at: http://almenmodstand.dk/borgerforslag/ Accessed 06 January 2021.
16 See more at: https://politikenbyrum.dk/Nyheder/art8026712/Folketinget-skal-behandle-borgerforslag-om-afskaffelse-af-ghettoloven

References

Almen Modstand. 2020. "Citizen proposal." http://almenmodstand.dk/borgerforslag/ (accessed January 06, 2021).

Anguelovski, Isabelle. 2015. "From toxic sites to parks as (green) LULUs? New challenges of inequity, privilege, gentrification, and exclusion for urban environmental justice." *Journal of Planning Literature* 31(1): 23–36.

Balfelt, Kenneth. 2016. "Udvikling af Folkets Park af Kenneth Balfelt Team og Spektrum Arkitekter." Video, 31:33. https://vimeo.com/207077598

Daly, Jonathan. 2019. "Superkilen: Exploring the human–nonhuman relations of intercultural encounter." *Journal of Urban Design* 25(1): 65–85.

Habermann, Anders. 2018. "Nuuks Plads prøver igen: Nyt forslag til lokalplan klar." *Mit Nørrebro*. https://mitnorrebro.dk/nuuks-plads-proever-igen-nyt-forslag-til-lokalplan

-klar/#:~:text=Et%20forslag%20om%20et%2075,nyt%20forslag%20til%20pladsens%20fremtid (accessed December 10, 2020).

KK. 2009. "A metropolis for people. Copenhagen Municipality." Københavns Kommune, (KK). https://kk.sites.itera.dk/apps/kk_pub2/pdf/646_mIr0dQ6Wdu.pdf (accessed July 25, 2021).

KK. 2011. "Superkilen information sheet." https://s3.amazonaws.com/media.archnet.org/system/publications/contents/10680/original/DTP103065.pdf?1475509472 (Accessed 060121).

Lees, Loretta. 2008. "Gentrification and social mixing: Towards an inclusive urban renaissance?" *Urban Studies* 45(12): 2449–2470.

Lewis, Victoria. 2016. "All of the coolest spots in Copenhagen are on this street." *Vogue Magazine*. https://www.vogue.com/article/jaegersborggade-copenhagen-guide-coolest-street (accessed December 10, 2020).

Mit Nørrebro. 2019. "Nørrebro i tal." https://mitnorrebro.dk/noerrebro-i-tal/ (accessed December 15, 2020).

Pedersen, Maria Libak and Lindstad Jonas Markus. 2011. "Første Led I Fødekæden? En undersøgelse af børn og unge i kriminelle grupper." Rapport. Denmark Justitsministeriet.

Politiken. 2018. "Få overblikket bydel for bydel." https://politiken.dk/forbrugogliv/boligogdesign/art6622615/Prisfesten-p%C3%A5-ejerlejligheder-er-rykket-til-de-billige-bydele-i-K%C3%B8benhavn (accessed December 15, 2020).

Rutt, Rebecca Leigh. 2020. "Cultivating urban conviviality: Urban farming in the shadows of Copenhagen's neoliberalisms." *Journal of Political Ecology* 27(1): 612–634.

Rutt, Rebecca Leigh and Loveless, Stephanie. 2018. "Whose park? The forty-year fight for Folkets Park under Copenhagen's evolving urban managerialism." *People, Place and Policy* 12(2): 99–117.

Rytter, Mikkel and Pedersen, Marianne Holm. 2014. "A decade of suspicion: Islam and Muslims in Denmark after 9/11." *Ethnic and Racial Studies* 37(13): 2303–2321.

Schmidt, Garbi. 2017. "Going beyond methodological presentism: Examples from a Copenhagen neighbourhood 1885–2010." *Immigrants and Minorities* 35(1): 40–58.

Simonsen, Kirsten. 2015. "Encountering racism in the (post-)welfare state: Danish experiences." *Geografiska Annaler: Series B, Human Geography* 97(3): 213–222.

Trudsø, Mette. 2019. "Helhedsplan for De Gamles By får tæsk af beboerne på Nørrebro." *Mit Nørrebro*. https://mitnorrebro.dk/helhedsplan-for-de-gamles-by-faar-taesk-af-beboerne-paa-noerrebro/ (accessed December 10, 2020).

Trudsø, Mette. 2020. "Grønne kvadratmeter i De Gamles By vinder over trafikprojekt." *Mit Nørrebro*. https://mitnorrebro.dk/groenne-kvadratmeter-i-de-gamles-by-vinder-over-trafikprojekt/ (accessed December 10, 2020).

Tubridy, Daniel. 2020. "Green climate change adaptation and the politics of designing ecological infrastructures." *Geoforum* 113: 133–145.

TV 2 Lorry. 2020. "Rising house prices have become first-time buyers' nightmare." https://www.tv2lorry.dk/koebenhavn/stigende-boligpriser-er-blevet-foerstegangskoeberes-mareridt-faa-boligoekonomens-bedste (accessed December 20, 2020).

Vemb, Christian. 2020. "Nyt byggeri på Nuuks Plads bliver lavt og indrammet i stedet for højt." *Mit Nørrebro*. https://mitnorrebro.dk/nyt-byggeri-paa-nuuks-plads-bliver-lavt-og-indrammet-i-stedet-for-hoejt/ (accessed December 10, 2020).

Visit Copenhagen. 2017. "Nørrebro: colourful, casual and young at heart." www.visitcopenhagen.com/copenhagen/culture/multicultural-norrebro (accessed December 15, 2020).

19

WILL "EXTRAORDINARY GARDENS" AND SOCIAL HOUSING ENSURE NANTES IS GREEN AND AFFORDABLE FOR ALL?

Francesc Baró and Isabelle Anguelovski

Situated in western coastal France, at the confluence of the Loire and the Erdre rivers and with a population of approximately 300,000 residents, Nantes has over the last 30 years successfully transformed its urban development trajectory from investments in shipyard industries to new technologies, culture and art and high-quality public transport systems, resulting in the second-highest employment growth in France. Thanks to long-term policies on urban renaturing (*nature en ville* in French) and social equity, Nantes has also become one of the greenest yet still relatively affordable cities in Western Europe (Anguelovski et al. 2018). However, as the city accelerates its ambitious green agenda, real estate prices and gentrification have drawn increasing concern and civic mobilization. As we examine these tensions in the latter part of our analysis, we raise the question of what types of urban greening and social policies are needed to keep Nantes a green, equitable and inclusive city.

From an industrial shipyard city to a European Green Capital

Since the 17th century, after the establishment of the French colonial empire, Nantes has stood out as one of the most important ports in France thanks to its strategic location in the Loire estuary. Some of its darkest history was marked by its leading role for nearly half of the 18th century in the French Atlantic slave trade. During most of the 20th century, Nantes was still an industrial city mostly dedicated to shipbuilding. But the 1970s global recession brought a large wave of deindustrialization to France that led to the closure of many of its factories. As a result, in 1987 after a period of economic stagnation, Nantes decided to close its shipyard industry.

The 1990s marked the reorientation of Nantes' development through the conversion of former industrial lands into cultural centers, biotechnology firms,

DOI: 10.4324/9781003183273-24

creative industries and a focus on greening, sustainable development, social cohesion and solidarity and livability (S.A.M.O.A. 2009). In this regard, over the past 30 years the municipality has created attractive, high-quality landscaped neighborhoods, illustrated by the substantial part of the city budget dedicated to green spaces: 6% of the total city budget, more than twice what it spends on security, for instance (Ville de Nantes 2017). From 1984 to 2015, green spaces doubled, reaching 57 m^2 of green space per capita for a total of 100 municipal parks. The city's green space strategy, which combines small neighborhood parks, green corridors and large city-wide historic parks, prioritizes an equality-driven approach, which ensures universal access to green space within 300 m of any resident's home.

Currently, Nantes has 38 urban projects related to green space involving 22 landscape agencies (Paysage Actualités – Le Moniteur 2019) and has set an ambitious plan (l'Étoile Verte et Bleu[1]) to connect most green and blue spaces through 5 major landscape corridors. In the mind of Jacques Soignon, the city's former Director of the City's Green Space and Environment Service (S.E.V.E. in French), the role of his department, and of landscape architects in general, has become increasingly relevant in the design and implementation of urban renewal projects in Nantes and in projecting the image of a green and livable city both on a national and international level. In his view, Nantes is unique in its greening approach:

> Most European cities have attractive [public] space, but it is a very traditional type of green space. It's just green spaces (parks) or shopping markets. Each city has attractive spaces, but the approach is so traditional. I feel that attractive spaces in Nantes are very new and more interesting.

Beyond greening, Nantes has articulated a wider sustainability strategy around a healthy, welcoming and socially equitable eco-city, where car use is minimized thanks to an interconnected network of electric tramways, buses and bike lanes (see Wilding et al. 2017). Throughout the broader region, the public investment of circa 30 million euros per year in renaturing projects has also created a vertically integrated and inclusive green economy with the involvement of professions across social classes and training skills—from gardeners and urban farmers, to landscape architects, to horticulture and permaculture specialists. This strategy culminated with the European Green Capital award in 2013.

Long-term commitment to building a green and equitable city

Nantes' ability over the last 30 years to become a green and equitable city with reasonable affordability rests on a long-term municipal commitment to place green space restoration and expansion at the center of all urban projects (i.e., brownfield, post-industrial, underused or empty sites) and to accompany all green renewal interventions with diverse types of social housing projects and municipal or metropolitan regulations. Affordability and equity are also ensured through a municipal land use conversion and rezoning scheme that uses a special planning status called the Zone d'Aménagement Concertée (Z.A.C.) (Integrated Development Zone), which enables municipal planners to control all project developments and locations rather than leaving private developers as free agents to petition for construction licenses without following comprehensive city-driven mapping and planning.[2]

One of the first and largest Z.A.C. regeneration projects has been deployed in Île-de-Nantes, a 460-hectare fluvial island located on the Loire, in front of the city's Old Town.[3] Since the late 1990s, this former industrial shipyard district and brownfield site has undergone major transformations into a multi-purpose, high-tech, creative, cultural and environmentally centered neighborhood. One particular aspect of this urban transformation is that the overall management of the project was assigned to the Société d'Aménagement de la Métropole Ouest Atlantique (S.A.M.O.A.), a local corporation mainly comprised by public investors. The regeneration of the Île-de-Nantes follows three objectives: (1) to build a new metropolitan center for the city; (2) to restore the Loire as a central place in the life of its residents and renew the network of public and green spaces; and (3) to build an inclusive, mixed city for diverse populations and uses.

Some of the projects already implemented in Île-de-Nantes include Les Berges du Faubourg, a 7-hectare longitudinal green space built along the Loire's riverbanks hosting environmental, social and gardening amenities, and the Prairie-au-Duc eco-district. Inaugurated in 2009 and conceived as a "neighborhood for all in a contemporary park," the area also includes the urban park Parc des Chantiers, a major touristic attraction due to its integration with the internationally renowned Les Machines de l'Île cultural and artistic site.[4] Overall, this intervention created new high-tech and creative industry centers, social housing (1,800 housing units, constituting 23% of all housing built), affordable housing (1,650 housing units, making up 22% of all housing built), universities and other green spaces and eco-districts. Yet some of the major projects planned on the Île-de-Nantes are still to be developed, mostly in the southwestern part of the island, including the new neighborhood of République, the new University Hospital Centre (C.H.U.) and major new green areas such as the Parc de Loire and the Parc de la Prairie-Au-Duc. According to an urban planner from S.A.M.O.A., the implementation of these new parks is posing challenges especially in terms of their future maintenance costs:

> Our big challenge is that today we are just starting to work on this one, the Prairie au Duc park project, and will carry it out in 2023–2024. But our difficulty with the SEVE is that we have these ambitions that are strong.

Through the Île-de-Nantes and other large regeneration projects, the creation of eco-districts (*éco-quartiers*) has been instituted as a holistic, long-term

FIGURE 19.1 New greenways along the Loire River in Île de Nantes. (Photo by: Francesc Baró.)

socio-environmental approach used by the Municipality of Nantes to combine open and green space creation, population density management and ambitious affordable and social housing objectives (Ville de Nantes and Nantes Métropole 2012). For instance, Prairie-au-Duc eco-district's housing prices were slated to be 20–25% lower than the Nantes average and 10–15% less than downtown Nantes, an objective which contrasts with many ecological enclaves that cater to higher-income residents. To ensure affordability in such green interventions, the project also planned for housing maintenance costs and fees paid by residents to be 40% of the cost of those in market-based housing (García-Lamarca et al. 2019).

Diverse housing schemes and financing mechanisms have ensured the ability of those eco-districts to fulfill social objectives alongside environmental goals and their accessibility to both working-class and lower middle-class residents. In the Bottière-Chénaie Z.A.C. eco-district in the northern side of the city, new gardens and a 5-hectare linear park were built in tandem with 2,400 housing units comprised of 30% social housing, 30% affordable housing and 40% free market housing (*accession libre*). In other eco-districts throughout the city, affordable housing schemes also include the Social Rental Mortgage (Prêt Locatif Social – P.L.S., a mortgage for the construction or rehabilitation of a unit to be rented as social housing) and social housing purchase (*accession sociale*).

In addition to urban greening and neighborhood redevelopment, Nantes has long placed housing at the center of urban planning and policy. The city has an established history of developing social housing and integrating such housing projects with urban greening. Neighborhoods such as Malakoff, Dervallières or Bellevue, which were built in the 1950s and 1960s as large districts of social housing (*grands quartiers d'habitat social*) in response to acute housing shortage, have also benefited from new parks such as the Parc des Dervallières or Parc de Procé. Nantes' current social integration strategy is to scatter social housing across the urban fabric in an attempt to integrate all residents within the urban space and avoid the stigmas and discrimination associated with these concentrated forms of social housing. However, despite Nantes' efforts to integrate such neighborhoods into the broader urban and social fabric of the city and use new tramway and bus lines to increase connectivity for residents, problems of insecurity and territorial stigmas still persist. Since 2012, the city classifies Bellevue and Malakoff as "priority safety zones" to justify the increased presence of and investment in police forces.

Underpinned by national social and affordable housing laws, such as the 2000 *Loi S.R.U.* and the 2014 Loi Dufflot, Nantes has put in place an ambitious legal

FIGURE 19.2 Allotment gardens in the Bottière-Chénaie eco-district, Nantes. (Photo by: Francesc Baró.)

requirement of (at least) 33% of social housing and 24% of affordable housing for new real estate projects, higher than most French cities, through the new Programme Local de l'Habitat 2019–2025 (P.L.H., Local Housing Program; Nantes Métropole 2018). The P.L.H. includes a diversity of housing schemes that seek to offer affordable accommodation to all inhabitants, considering their needs and means in terms of income, age and other vulnerability aspects. Nantes' social housing stock is owned and managed by 15 different local landlord authorities (*bailleurs sociaux*). The oldest and largest landlord (with circa 30,000 social housing units) is Nantes Métropole Habitat (N.M.H.). N.M.H. is also a relevant actor in greening policies, as the agency owns 64 hectares of publicly accessible green space in the Nantes metropolis. In N.M.H.'s view, green spaces play a key role for the quality of life and health of their tenants, as these promote physical activity, social cohesion and adaptation to climate change. N.M.H.'s C.E.O., Marc Patay, highlighted this role:

> We have about twenty gardens that are community gardens, that are really built with the residents and that are made available to them, to our tenants, because we know that gardening helps to create social bonding. It helps to get along better with your neighbors. It's a collective project. We were talking about the elderly earlier. It also helps to get people together.

Lastly, social inclusion in Nantes is also associated with civic participation in urban greening. The municipality perceives green space as an opportunity to co-design, co-produce and co-manage urban parks, gardens, esplanades and greenways together with residents. Green space planners are committed to encouraging citizens' cultural, social and artistic ownership of different green spaces through collaborative arts exhibits or family gardens, such as the Jardin des Plantes (Botanical Garden) or the recent Parc des Oblates.

New challenges in Nantes: gentrification, contested projects and the beautification of green spaces

In recent years, Nantes is facing new challenges that jeopardize its status as a green and affordable city. First, large internal migrations from other parts of France are creating social concerns. According to Agence d'Urbanisme de la Région Nantaise (A.U.R.A.N.), the demographic growth between 2011 and 2016 was estimated at 8,983 inhabitants per year at the metropolitan level and 3,770 at the city level (A.U.R.A.N. 2019). An important percentage of these arrivals are young professionals from the Paris region, Île de France, seeking larger homes and a better quality of life, as reported in the media (e.g., *Le Monde* 2019). This trend is creating significant pressure on the real estate market, with prices skyrocketing in neighborhoods such as Chantenay and residents reporting that they are being gentrified and displaced not "just" to other neighborhoods in Nantes but further away to the periphery. Many also accuse the city of accelerating these migrations by marketing Nantes as a green city and a creative-class oriented economy. To respond to this persistent population growth, the P.L.H. plans have included the construction of at least 6,000 new housing units per year through the metropolitan area during the period 2019–2025. As this ambitious goal has naturally triggered the proliferation of urban development projects across the city, some of them have become contested by grassroots groups due to their social and environmental impacts and their encroachment on peripheral, agricultural or informal green land.

One of these contested projects is located in the sector of Bois-Hardy, in the postindustrial neighborhood of Bas-Chantenay. The entire area is affected by a Z.A.C., with the sector of Bois-Hardy poised to receive a new residential development of about 400 housing units. In response, a group of 150 neighbors have created the Collectif du Bois-Hardy[5] in defense of a semi-natural space that residents use as an open community garden area of about 1.5 hectares which the project threatens to remove. In this collective, activists are not organizing against the development project per se, but argue that the new housing should only be built on already urbanized areas in order to preserve community gardens that play such an important role in creating social cohesion. They also consider that the project is of disproportionate scale, both in terms of the number of housing units and the height of the new buildings, given that it is a low-density neighborhood mainly composed of historic, small working-class houses:

> The purpose of the Bois Hardy collective is precisely to ensure that the gardens are a place, a medium of encounters and social cohesion between this mix of population that is historic and those that will arrive. We need a large enough area so that newcomers who will live here can have access to garden plots in the same way as the old ones. But for that, there have to be enough gardens. Otherwise there will be a conflict of use. We are going to generate a future conflict and that is absolutely not what we want.

Another contested project similar to Bois-Hardy is the Z.A.C. Doulon-Gohards project.[6] The Z.A.C., located next to the old neighborhood of Vieux-Doulon in the eastern outskirts of the city, includes the development of 2,700 housing units, several public facilities (e.g., a school), commercial areas and 4 urban farms into 180 hectares of former agricultural land during the period 2017–2034. Currently, circa 1.5 hectares of this land is occupied by a collective of 20 people called the *Jardin des Ronces*[7] who have built several community gardens, beehives and even a bread oven. Their opposition to the Z.A.C. is mainly driven by a perception of a project centered on economic profit and on bringing high-income residents attracted by the idea of living in an experimental urban farming area. In their view, the projected urban farms are a "showcase" of Nantes' green image detached from the farming reality and encouraging the encroachment of semi-rural land for urbanization purposes:

FIGURE 19.3 Current community gardens in Bois-Hardy, Nantes. (Photo by: Francesc Baró.)

> We, for example, are not supposed to stay [in the new project]. This (zoning) map will evolve and over the next five years more buildings will be built. So probably in 15 years there will be 2,700 new housing units, and several new [experimental] urban farms, but once the ZAC is finished, will the farms remain? We don't know. What's really clever of them is that they are using these urban farms to make purchasing a home more attractive.

In addition to green housing development projects, civic concerns are mounting against what residents see as a "monumental," "manicured" or "aseptic" renaturing strategy. Their concerns are somewhat reflected in the words of the city's green space department which articulates that greening is part of Nantes' "green show." A clear example of such a strategy is the new urban park Jardin Extraordinaire,[8] located in the former quarry site of Carrière Miséry and also part of the Z.A.C. Bas-Chantenay together with the Bois-Hardy sector. Inspired by the fantastic stories of Jules Verne of an "astonishing nature," as stated in the touristic brochure, and with an artificial 25-meter waterfall, it is expected to host the Arbre aux Hérons[9] (Herons' tree) in 2022, a large artificial mechanical tree with suspended gardens linked to the above-mentioned Les Machines de l'Île cultural site.

FIGURE 19.4 Community gardens of Jardin des Ronces, Nantes. (Photo by: Francesc Baró.)

In response, activists from the Commune de Chantenay[10] civic group have started to resist the project, viewing it as a tourist attraction—a green museum—that will trigger gentrification in the neighborhood. They argue that the Carrière Miséry site was already used as an informal green space by some residents, with nature taking over the space, and that these existing social and ecological values were ignored by the city during the consultation process. A member of the Commune de Chantenay expressed their concerns about what they perceived as the real economic motivation of the new park:

> The garden is really just a pretext. They took this small part of the city to make a garden—more of an art piece than an actual garden—to promote the Herons' Tree project, since it is not even a real tree [and rather a tourist attraction]. So, at first, they [City Council] only talked about the Herons' Tree and at some point they thought, "maybe it will look out of place … so let's make a garden nearby." It was complicated for us because it seems wrong to fight against a garden. But it's important to look at what kind of garden it is and who it is meant for.

A final relevant civic mobilization is the one led by the grassroots group Du Rififi dans la Bergerie,[11] established in 2017 against the redevelopment of the

FIGURE 19.5 The new urban park of Jardin Extraordinaire and the yet-to-be-developed former quarry Carrière Miséry, in Bas-Chantenay, Nantes. (Photos by: Francesc Baró.)

Park of Moutonnerie and the gentrification of the Dalby neighborhood, just east of the Nantes train station. The project initially involved the transformation of the park (currently conceived as an open neighborhood public space), the creation of 80–100 housing units and the relocation of the Maison d'Accueil de Jour (M.A.J.), a social facility for the most vulnerable (e.g., homeless people) located next to the park. The group organized several peaceful actions during 2017 and 2018 to raise the attention of the media and the City Council, who subsequently decided to keep the M.A.J. in its current location. Currently, the group remains vigilant about the redevelopment of the park and its surroundings.

In general, many of these groups also contest the planning, design, development and management of urban greening projects in Nantes. In their view, the city is adopting a very centralized greening strategy to become a live green museum, in which civic groups and residents are merely involved in some programming activities and only if they are part of official, recognized city networks. They feel that the *concertation* processes, the tradition of civic and social dialogue, organized by the City of Nantes are contradicted by a reality in which local planners do not grant support to more informal, spontaneous greening practices.

The future of Nantes as a green capital

Nantes has developed an ambitious greening agenda since the 1990s articulated through dedicated housing strategies and funding to allow working-class residents and moderate-income residents to benefit from a more livable and greener city. It seeks to make greening inclusive through a wide variety, quantity and density of green spaces, and by making greening a driver of inclusive territorial and housing development. But the growing urbanization of Nantes and its vision to become a green capital that builds on the value of urban nature to attract sustainably minded middle-class residents, as well as new investors and development, jeopardize social achievements and drive urban sprawl in peripheral areas of the city. Questions also remain about the ability of Nantes' socio-environmental projects to more consistently value its ethnic minorities (8% of residents) and build on their identities and uses of the city to further the city's vision for a green city. Ultimately, Nantes faces losing its identity of a green and socially equitable city to one that is green, privileged and broadly gentrified.

Keywords

(a) the urban development pattern of the city and neighborhood: former shipyard center; post-industrial development; long-term stable growth; metropolitan growth
(b) the urban greening of the city and/or neighborhood: land clean-up and greening; new connected green and blue spaces, monumental greening; city-wide policy of eco-districts; former European Green Capital
(c) the inequalities at stake: increased real estate prices; accelerating gentrification; urban expansion to peri-urban fringes

Notes

1 See more information at: https://jardins.nantes.fr/N/Paysage/Etoile-Verte/Nantes-L-Etoile-Verte.asp
2 For more information on Z.A.C.s see: https://www.est-ensemble.fr/cest-quoi-une-zac
3 See more information here: https://www.iledenantes.com/
4 See more information at: https://www.lesmachines-nantes.fr/
5 See their website at: http://lcboishardy.free.fr/
6 See more information at: http://www.nantes-amenagement.fr/projet/doulon-gohards/
7 See their website at: https://lesronces.noblogs.org/
8 See more information here: https://www.nantes-tourisme.com/fr/parc-et-jardin/jardin-extraordinaire
9 See more information here: https://www.nantes.fr/arbreauxherons
10 See their website at: https://fr-fr.facebook.com/lacommunedechantenay/
11 See their website at: https://quartierdalby.noblogs.org/

References

Anguelovski, Isabelle, Argüelles, Lucía, Baró, Francesc, Cole, Helen, Connolly, James, García-Lamarca, Melissa, Loveless, Stephanie, Pérez del Pulgar, Carmen, Shokry, Galia, Trebic, Tatjana, and Wood, Erin, 2018. *Green Trajectories: Municipal Policy Trends and Strategies for Greening in Europe, Canada and United States (1990–2016)*. BCNUEJ book (260 pp.). Available at: http://www.bcnuej.org/wp-content/uploads/2018/04/GreenTrajectories.pdf

AURAN, 2019. *Aire urbaine de Nantes: une forte croissance démographique durablement installée! - « Les Synthèses de l'Auran » # 39*. Available at: https://www.auran.org/publications/aire-urbaine-de-nantes-une-forte-croissance-demographique-durablement-installee-les

Garcia-Lamarca, Melissa, Anguelovski, Isabelle, Cole, Helen, Connolly, James, Argüelles, Lucía, Baró, Francesc, Loveless, Stephanie, Peréz del Pulgar, Carmen, and Shokry, Galia, 2021. Urban green boosterism and city affordability: For whom is the 'branded' green city? *Urban Studies* 58(1): 90–112. doi:10.1177/0042098019885330

Le Monde, 2019. *Nantes, c'est le nouveau Montreuil »: paroles de Parisiens qui ont fui « la capitale de l'invivable* (press article). Available at: https://www.lemonde.fr/campus/article/2019/02/28/jeunes-cadres-les-nantisiens-ces-ex-parisiens-en-quete-d-une-vie-meilleure-a-nantes_5429252_4401467.html

Nantes Métropole, 2018. *Programme Local de l'Habitat 2019–2025*. Available at: https://metropole.nantes.fr/files/pdf/logement/prog-local-habitat-2025.pdf

Ouest France, 2018. *La carte des projets urbains contestés à Nantes par les citoyens* (press article). Available at: https://www.ouest-france.fr/pays-de-la-loire/nantes-44000/la-carte-des-projets-urbains-contestes-nantes-par-les-citoyens-5588697

Paysage Actualités – Le Moniteur, 2019. *Nantes Métropole Nature* (special issue). Available at: https://www.lemoniteur.fr/article/a-nantes-la-planification-et-l-action-fusionnent-dans-le-paysage.2042085

SAMOA, 2009. *Île de Nantes: Éco-quartier de la Prairie au Duc. Un quartier dans un parc*. Report. Île de Nantes: Société d'aménagement de la métropole Ouest Atlantique (SAMOA).

Ville de Nantes, 2017. *Budget primitif, rapport de présentation*. Available at: https://www.nantes.fr/files/live/sites/nantesfr/files/PDF/ville-denantes/institution/budget/rapport_presentationBP2017Vdef.pdf

Ville de Nantes and Métropole Nantes, 2012. *Rapport de développement durable de Nantes Métropole 2011–2012*. Report. Nantes: Ville de Nantes and Nantes Métropole.

Wilding, Helen, Gould, Ronald, Taylor, Julia, Sabouraud, Annette, Saraux-Salaün, Partricia, Papathanasopoulou, Dionysia, de Blasio, Antonio, Nagy, Zsuzsanna, and Simos, Jean, 2017. Healthy cities in Europe: Structured, unique, and thoughtful. In de Leeuw Evelyne, Simos Jean (eds.), *Healthy Cities: The Theory, Policy, and Practice of Value-Based Urban Planning*. Springer, New York.

20

PRIORITIZING GREEN AND SOCIAL GOALS

The progressive Vienna model in jeopardy

Carmen Pérez del Pulgar

While in other cities urban transitions to environmental sustainability are often motivated by neoliberal strategies of urban development that may end up triggering gentrification, in Vienna we observe that a strong commitment to greening the city is not displacing longtime residents. The case of Vienna is representative of a model inspired by the radical notion that housing is not a commodity but a universal right. This is the result of a complex social and administrative architecture of policies rooted in the "Red Vienna" period (1918–1934) that have endured until today. In this chapter we review the main urban policies and funding mechanisms that have prevented gentrification in Vienna even in the aftermath of important climate and environmental redevelopment projects. In this sense, Vienna is an exceptional case that other cities can learn from, though it is not exempt from the challenges posed by a pressing national and regional neoliberal environment, budget cuts, private-development-driven economy and international real estate investment.

An early pro working-class urban development anchored in municipal socialism

In an antithetical manner to the country's political isolation and shrinking economic development after WWI, a unique municipal socialism developed in Vienna during the interwar period. Although interrupted during WWII, the municipal welfare consensus continued in the post-war period following the welfare state expansion trends of other Western European states. Initially triggered by the urgent need to respond to the severe housing scarcity and the deplorable living conditions for the working class inherited from the Hapsburg Empire, the local authority during the period of Red Vienna (1919–1934) initiated a radical transformation of the city in terms of education, sanitation and

DOI: 10.4324/9781003183273-25

democratization. Unlike early 19th-century Marxism, which advocated for the dissolution of the city, Red Vienna's early, 20th-century Austro-Marxism envisioned the non-commodified modern city as an emancipating project (Schwarz, Spitaler and Wikidal 2019). Tax policy, land use policy and the provision and design of housing and public space were meticulously and profoundly altered. From 1923 to 1933, the implementation of progressive taxation and rent regulations to restrict private housing culminated in the construction of 64,000 city-owned social rental dwellings of a very high architectural quality.

The ideological, fiscal and bureaucratic legacy of that experimental period produced a solid consensus and complex path-dependent system of policies securing a large stock of decommodified housing and public spaces that has lasted, at least partially, until today (Kadi 2015). At present, few cities in Europe, and even fewer in North America, have as comprehensive decommodification strategies as Vienna, which has numerous measures in place to keep housing decommodified and affordable while projecting new urban areas and carrying out extensive urban renewal interventions such as the restoration of obsolete infrastructure, urban greening, waterfront development, new eco-districts and the ecological retrofitting of old buildings (Cucca 2019).

Despite having an excellent baseline social infrastructure, Vienna is facing new challenges. The municipal corporation is juggling between resisting a pressingly neoliberal national and European political environment and fostering its

FIGURE 20.1 Council housing post-WWII. (Photo by: Carmen Pérez del Pulgar.)

rapid repositioning into the heart of the European Union and the European urban nexus since the fall of the Iron Curtain and Austria's entry into the European Union in 1995. Vienna is increasingly attracting headquarters of international companies, supranational organizations and international universities, which is intensifying the demand for high-end housing and international housing investments as a lucrative investment product (Cucca 2019). The city is also experiencing abundant immigration and a large increase in population. As a result, new trade-offs between social justice, rapidly rising land prices, urban renewals and environmental innovations are emerging (Kadi and Verlic 2019).

The green path of the most livable city in the world

Vienna's significant greenspaces are in large part physical remnants of its imperial and monarchical past. Stately public spaces such as the Viennese Forest, the Prater Park, Schönbrunn Palace gardens and the Belvedere Palace gardens are important legacy greenspaces. Originally constructed at the city boundaries, these spaces currently lie almost entirely within central historic Vienna. Other greenspaces are the result of Red Vienna policies which bolstered the development of cooperative greenspace. In the aftermath of World War II, the reconstruction of greenspaces once again yielded new public spaces.

The more contemporary municipal efforts toward building a green city began in the 1960s with a four-decade period of mega-project development. The Donaupark, Kurpark Oberlaa and Wienerberg projects all transformed brownfield sites into large public parks during this time, while the Donauinsel project addressed flood protection with the creation of a new channel and recreational island.[1] The redevelopment of brownfield sites continued through the 1990s, 2000s and even the current decade, converting former metal warehouses, cable factories, slaughterhouses and other aging industrial and commercial sites into public parks. In the 2000s, parking facilities were also converted into parks, including the old car park next to Simmering Prison, which was turned into the Hans-Paulas-Park "piazza," playground and sports lawn in 2003. In the most notable recent example of Viennese brownfield redevelopment, an old railway station was replaced by Eurogate, the country's largest passive housing block, with its green area, artificial lake, gardens and play spaces. These green spaces were completed and opened to the public in 2013 as Leon-Zelman-Park and Ziakpark.[2]

Environmental protection and nature conservation have long been an integral part of the functioning of several City of Vienna departments and agencies.[3] The S.T.E.P. 1994, S.T.E.P. 2005 and S.T.E.P. 2025 urban development plans for the city include greenspace expansion as part of programming for ecological improvements and in terms of the rest, leisure and health needs of the population. The 2025 plan goes further to define greenspace as a right under the framing of *Grünraumgerechtigkeit* (greenspace justice), designed for everyday life, well distributed across all city areas and as instrumental in creating an inclusive and pluri-identity society.[4] The most prominent aspect of greening policy in 1990s Vienna was the development and enlargement of the city's greenbelt through the 1995 Vienna Green Belt Plan, which outlines measures for the protection and enhancement of the 19,260 hectares of open natural and agricultural areas, parks and cemeteries making up Vienna's "green lung."[5]

Small parks and greenspaces have also been constructed, mostly in the unused gaps between buildings and on vacant lots in the city since the 1980s. These *Baulückenparks* are meant to alleviate the density of highly built-up city center areas by providing small natural spaces for respite. In 1990, the small Minna-Lachs-Park for children was created in the dense Mariahilf neighborhood and was later enlarged with the demolition of an adjacent building. Several other such spaces, including the 2011 resident-inspired renovation of Siebensternpark, were constructed or expanded following the demolition of aging residential structures and have often been designated as spaces for children's play. The creation of these green patches of infill has created stronger connections to existing green spaces.

A dozen or so parks have undergone restoration since 1990. In the early 2000s, the 50 Orte Programm and 50 Orte+ established a focus on public space regeneration and enhancement, marking a turning point in the city's policy to focus on inner-city areas and their existing green spaces. Another dozen new city parks were constructed between 1990 and the early 2000s, often in conjunction with large infrastructure projects in transportation or social housing. For example, the underground construction of the U3 subway line was accompanied by the creation of four different parks between 1991 and 2003.[6] Some of these projects involved public participation, some co-designed spaces for a particular demographic such as children, youth or the elderly, while others made previously private parks public.[7] Since the 2000s, the greening of new city districts has been another key emphasis within greenspace construction. The southern part of the city and the area north of the Danube River are receiving several new green and leisure areas as part of development and redevelopment programs. For example, in 2008 the Rudolf-Bednar Park opened in a new district, which will continue adding housing and office space until 2025. The park is considered the "green heart" of the area and is the largest park built in Vienna since 1974.[8]

A complex housing and land policy architecture sustaining an equitable green city

Today, Vienna's policies are ambivalent between limiting and promoting change. A robust set of decommodification measures to protect housing rights coexists with liberalizing policy changes to attract middle- and upper-income families to the city (Kadi 2015; Cucca 2019). There are two key pillars that particularly stand out as sustaining the bulk of the decommodification measures: the regulation of the housing market and land use politics.

The Viennese housing market is characterized by its segmentation into the private and publicly subsidized sectors (Kohlbacher and Reeger 2016) and

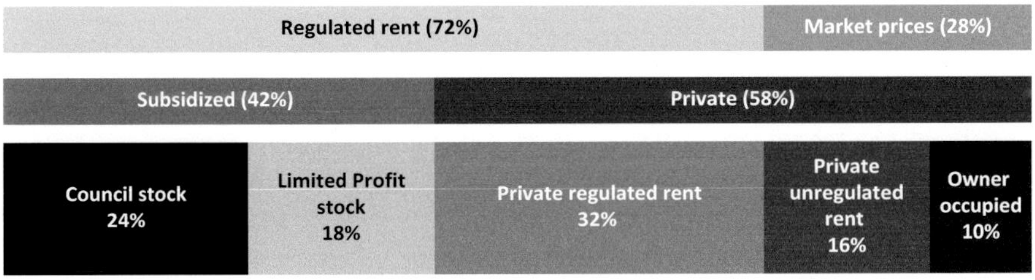

FIGURE 20.2 Organization ofViennese housing market.

further subdivided into private unregulated rental housing, private regulated rental housing, council rental housing, limited-profit rental housing and owner-occupied housing (Kohlbacher and Reeger 2016). Sixty-two percent of the city's population lives in subsidized housing, which represents 42% of the housing stock, distributed between 24% for council housing (220,000 units) and 18% for cooperative or limited-profit housing[9] (200,000 units). Within the privately provided sector, private regulated rental housing accounts for 32% of the housing stock, followed by 16% of the stock being private unregulated rental housing (Kadi 2015) and a relatively low (10% of the housing stock and 17% of the population) but constantly growing owner-occupied housing sector.

Controls over housing construction, restoration, allocation and affordability

The high budget allocated for subsidized housing is a key instrument for the provision of decommodified housing. At a national level, 1% of the national income tax is earmarked for construction (approx. €250 million per year). On the federal level, federal budget subsidies for subsidized housing amount to approximately €350 million. Our interviews reveal that in Vienna, the former is usually assigned to limited for-profit housing developments and the latter is handed out in the form of loans to limited-profit and for-profit developers with a 1% interest rate, covering about 15% of the construction costs and supporting around 75% of new developments. Although the current conservative-led (F.P.Ö.) government coalition in Austria is promoting home-ownership by redirecting the national tax funds to the construction of dwellings for sale at a national level, Vienna keeps favoring subsidized renting regimes through its right to set its own criteria for self-funded housing subsidies guaranteed by the federal constitution. Another decisive organism is the municipal program *Wohnfonds Wien* (Housing Fund Vienna). Founded in 1984, it manages the municipal land and the *Wiener Bodenbereitstellungs und Stadterneuerungsfonds* (Vienna Land Procurement and Urban Renewal Fund), that contains the urban renewal/restoration subsidies. The explosion of land prices in such a small federal state as Vienna, its isolation within a currently politically conservative-neoliberal national environment and the complex political dynamics between Vienna and neighboring conservative political states are amongst the main concerns for the preservation of such comprehensive housing protection tools with regards to land management. Such challenges led to a new building code approved in 2018 that defines a new inclusionary zoning program allocating 60% of new residential land for subsidized housing. Some opponents to this policy criticize that it interferes with the rights of private ownership. Commenting the oppositions to the new measure, a representative from the municipal housing department Wiener Wohnen expressed:

> It is quite tricky because the conservative side argues "you're interfering with our property", and we say "but it's our land, you still owe us a lot of money and we have to make sure people get decent housing". We'll see how it works out in five years. I think it's a very brave step forward because the main driver of cost is land. You can't replicate land ... it's a unique resource and no matter what city it is, Amsterdam or Berlin, land prices are exploding and in the end the tenants or the homeowners pay for that.

Such controls are put in place to ensure that land, rather than treated as a commodity, is treated as a public good that every resident should benefit from to the same extent.

Despite housing decommodification trends, the Vienna council stopped most of its council housing construction in 2004, resuming only in symbolic numbers in 2019. The rhythm of council housing construction began to slow in the 1970s when the Vienna management corporation started promoting public-private partnerships in planning procedures and competencies. This trend was exacerbated by the European Union's fiscal framework which, following the Maastricht criteria for public investment, limited public debt and public investments and made it virtually impossible for cities and municipalities to invest in social housing.

Due to the standstill of council housing construction since 2004, the limited-profit sector is currently responsible for most new subsidized affordable housing. When more than 300 housing units are planned, *Wohnfonds Wien* selects which developer will get the land based on a tender procedure called *Bauträgerwettbewerb* (developers' competition). The land allocated by *Wohnfonds Wien* is either owned by the municipality (via *Wohnfonds Wien*) or the proportion of private land reserved for subsidized housing via the inclusionary zoning program. The developers' competitions are based on a four-pillar system whereby each subsidized housing project is judged along four sets of criteria: social sustainability, planning, ecology and costs. These pillars reflect the municipal mid- and long-term ambition to expand the affordable housing stock while considering the environmental impact of the urban form and the access to green and recreational spaces:

> Today our urban planning works according to clear ecological and social principles. It is a permanent process, especially in times of profound change. The climate crisis is at the center of our actions, always in connection with a social policy that enables a decent and affordable life for everyone.
>
> *(Birgit Hebein, Vice Mayor and City Councilor for Planning[10])*

Both nonprofit and commercial developers are entitled to take part in these competitions. The winning team has the right to buy the construction site at an a priori fixed price and is also granted housing subsidies for up to 35% of the total construction costs.

The rent prices of new developments are limited in two ways. In the case of the limited-profit housing association being the developer, the national Limited Profit Housing Act (*Wohnungsgemeinnützigkeitsgesetz*) strictly regulates the rents of limited-profit housing, capping them just to cover the costs of land, construction, administration, financing and long-term maintenance. In addition, it stipulates that limited profits must be reinvested in the purchase of land, housing refurbishment or new construction. In compensation for these restrictions, housing associations are exempted from corporate tax. Furthermore, when part of the construction costs is covered by federal subsidies, the federal Subsidies Act limits the maximum rent of limited-profit and for-profit housing constructed with federal loans[11] to cover the costs during the duration of the repayment period. It also

FIGURE 20.3 Limited-profit housing 2019. (Photo by: Carmen Pérez del Pulgar.)

FIGURE 20.4 Subsidized housing WohnPark Alterlaa 1983. (Photo by: Carmen Pérez del Pulgar.)

mandates that one-third of the constructed housing units (a stock called smart housing) are allocated by the municipal housing department (*Wiener Wohnen*) according to their own criteria. After the repayment of loans, which typically takes 35 years, for-profit housing is freed from any regulation, while for limited-profit providers the national Limited Profit Housing Act continues to apply.

A socially oriented and green urban renewal

Wohnfonds Wien coordinates a soft urban renewal program that subsidizes the renovation of old building stock (e.g., connection to central heating or district heating systems, renovation of facades) and the upgrading of deteriorated public spaces, if the owners agree to maintain existing rental contracts and rents are frozen for the ten years following renovation.

> The government of the city of Vienna made clear that whenever urban renewal takes place, it is only allowed to increase rent on a very, very low level.
>
> *(Representative from the municipal housing department Wiener Wohnen)*

This "soft" approach to urban renewal is the result of social protests in the 1970s that began in Spitellberg (District 7) against a "hard" urban renewal strategy that tackled urban decay by demolishing entire neighborhoods. A more social- and resident-oriented urban renewal strategy was institutionalized in 1984, in which tenants would collaborate in the renovation processes of their buildings in order to mitigate cultural or social displacement.

As part of this program, many green urban renewal projects have been implemented, including small urban gardening projects, park restorations and the waterfront revitalization of the Danube canal. Since the 2000s, the ecological retrofitting of old buildings (i.e., improved thermal insulation) and entire urban quarters was prioritized in order to increase quality, thermal isolation and energy standards.

Although the soft urban renewal program is mostly praised as a success and is one of the largest in the world, with more than 201,000 dwellings renovated up until 2004 (Franz 2006), it has also raised criticism. A local activist reports that critics see rent caps and regulations as outdated, claiming that "you're just delaying everything by 10–20 years and then gentrification happens. *Stadterneuerung* [urban renewal] is a gentrification factor." Indeed, after the initial ten-year rent freeze, rents can be legally increased. Furthermore, the rent freeze only applies to contracts existing prior to renovation and thus is not applicable to new contracts

FIGURE 20.5 Restoration of former gas tanks. (Photo by: Carmen Pérez del Pulgar.)

FIGURE 20.6 Limited-profit housing, 2019. (Photo by: Carmen Pérez del Pulgar.)

signed after renovation, including those starting before the ten years moratorium, whose rent can be based on market prices alone.

As a result, urban renewal and ecological retrofitting seem to be driving displacement and residential segregation in a layered fashion. In other words, while some households remain protected by the remaining decommodification policy, others are experiencing the erosion of their benefits through recommodification (Kadi 2015). As a result, the overall stock of affordable housing is shrinking in Vienna, and poor tenants not able to afford the new rents are forced to move to often poorly insulated units and less environmentally friendly and renovated neighborhoods, facing combined social and environmental injustices (Cucca 2019).

The federal Subsidies Act that freezes rents of subsidized housing (limited-profit and for-profit) for the duration of the repayment period received similar criticisms. At best, these measures only seem to delay rather than prevent gentrification. Additionally, many developers and private investors prefer to turn down subsidies in order to avoid the regulations of both the soft urban renewal program and the Subsidies Act and retain flexibility in rent adjustments through what is seen as a private-led method of urban renewal and gentrification (Franz 2006, 194).

Another important aspect that determines access to affordable housing is the eligibility criteria for subsidized housing. *Wiener Wohnen* is the municipal company that administers, renovates and manages the most affordable options in Vienna's housing market. It allocates council housing and smart housing—the latter of which comprises one-third of the housing stock built with public subsidies[12]—via the Vienna Living Ticket (*Wiener Wohn Ticket*), which residents register for to enter a waiting list that currently takes about seven to eight years. Once granted, council or smart housing is the most affordable housing option in Vienna's housing market and requires a maximum entry down payment of €10,000 from a new tenant.

There are a set of minimum conditions required to register and some bonus conditions that move citizens up to a higher position on the waiting list. Minimum conditions include being at least 18 years old; being registered at an address in Vienna for a minimum of 2 years; being an Austrian citizen or equivalent; and maximum income, which in comparison to other social housing systems, is relatively high (e.g., for 2019, the joint annual net income of a one-person flat could not exceed the maximum thresholds of €46,450[13]). Some of the bonus conditions include social criteria such as living in overcrowded conditions, being a young, first-time mover or an elderly person with special needs. Furthermore, there is a contentious bonus condition that favors residents who have lived in the city for more than five years, moving them automatically up to a higher position on the waiting list. Though aimed at protecting longtime residents, this seriously penalizes low-income newcomers, especially migrants in urgent need of affordable housing.

Limited-profit associations can allocate two-thirds of their housing stock according to their own criteria (one-third is allocated via the Vienna Living Ticket). Rents in the limited-profit sector are affordable due to the restrictions imposed by the federal Subsidies Act and the national Limited Profit Housing Act, but there is a higher down payment of about €30,000. As a result, limited-profit housing ends up being a very convenient solution for middle-income households but not for low-income households. On the flip side, limited-profit housing associations do not include any "local" bonus criteria—such as the ones imposed by the Vienna Living Ticket—and are thus a more viable option for newcomers.

The last significant legal framework safeguarding the city's housing affordability is the regulation of the private rental market. Private rental housing comprised about 280,000 units in Vienna in 2011 (Kadi 2015), approximately two-thirds of which are regulated by the National Rent Act. The Act regulates two independent dimensions: the duration of the tenancy agreement and the magnitude of the rent. It was liberalized in 1984, in 1986 and again in 1994. First, the norm of unlimited rental contracts was replaced with the possibility of fixed-term contracts. While old contracts maintain an unlimited term and can even be passed across relatives, many new contracts are fixed term. Second, the former category rent setting (*Kategoriemietzins*) was replaced with the benchmark

TABLE 20.1 Summary of anti-gentrification policies and measures

Legal framework mechanism	*1 % national income tax*	*Inclusionary zoning*	*Developers competition*	*Soft urban renewal*	*Limited Profit Housing Act*	*Subsidies Act*	*National Rent Act*	*Vienna Living Ticket*
Goal	Funding of housing promotion	Two-thirds of residential land allocated to subsidized housing	City development according to social sustainability, planning, ecology and costs	Urban renewal without displacement	Limitation of rents of limited-profit housing	Limitation of rents of subsidized housing	Private rent regulation	Allocation of council housing
Risk	Increasingly assigned to right to buy schemes			Does not avoid gentrification, only delays it	High down payment. Not accessible for low-income households	Does not avoid gentrification, only delays it	Lax criteria. In practice enables gentrification and rents increase	Discriminatory against low-income newcomers

rent-setting system (*Richtwertmietzinssystem*). The former category rent-setting regulation was based on the categorization of each housing unit ranging from high quality/amenities A to lowest quality/amenities D and its associated price per square meter. In the new benchmark rent-setting system, a unit is compared to a fictive standard housing unit and rent increases can be added based on improvements vis à vis the fictitious unit or the location of an apartment. However, as an activist explains, apartments in the city center have seen strong price increases due to the competition given that this new system

> offers a lot of room for interpretation of what is standard. As a renter you just basically have no idea why you're paying the rent you're paying. Also, the location factor was introduced by this system, which is a classic gentrification factor.

An indicator of the new rent-setting system's lack of effectiveness is the rapid increase of private investment in the regulated private housing stock. Since the liberalization of the National Rent Act, the previously unattractive, low-quality/low-rent regulated private rental sector has become an attractive investment product and high-priced sector whose rents increased 30% more than in the rest of the city between 2001 and 2010 (Kadi 2015). As of 2020, all buildings recently built in the private sector do not meet any rent control regulations.

How long will historic social victories remain in place in a recommodified housing market?

The repositioning of Vienna as a central globalizing city and regional connector after the end of the Cold War has exposed the city to vast flows of capital and investment that have affected Vienna's ability and willingness to address emerging social equity challenges. From the 1990s onward, the housing sector has shifted from council to subsidized to regulated and now to a loosely restricted private sector. Consequently, even the ratio of rental versus property units is changing, with housing increasingly seen as an investment commodity. Most new leases are now time-limited rather than unlimited, and private rent has increased over 50% in the last decade (Kadi and Verlic 2019). Gentrification can no longer be denied in parts of Vienna, and the cutbacks in council housing, together with the upcoming expiration of moratoriums on rent freezes, do not suggest a future reversal of that trend.

These reforms, rather than radically eradicating available housing benefits altogether, are creating a dual trend: a widening gap between the rights and entitlements of longtime residents and newcomers (Kadi and Verlic 2019). Those who have lived in the city for longer remain fairly well protected, either by the subsidized sector or by the private regulated sector with unlimited contracts and regulated rents predating liberalization in 1994. By contrast, low-income newcomers (e.g., recent migrants or refugees) with no access to council housing and not enough resources for limited-profit housing are forced to seek housing in the private market, facing rocketing prices and shorter-term leases. Furthermore, the fact that low-cost apartments available in the private rental market tend to have poor thermal insulation, are more affected by air pollution and noise and are located in crowded districts with few green facilities raises questions of environmental justice (Cucca 2019).

The state of affairs—in which longtime residents are relatively unaffected while the condition of outsiders deteriorates—is creating a political fragmentation of the working class, by which

> those [locals] who are more afraid of losing their standard of living are moving further to the right of the political spectrum, and those who have a higher incentive to fight for an improved standard of living [low-income outsiders] lack political rights to begin with.[14]
>
> *(Candidate Links Party to the City Council)*

Some measures are indeed designed to protect a somewhat regressive, self-closing and defensive sense of place. As a landing place for refugees and low-income labor migrants, Vienna planners and public officials need to re-assess who the "new" citizens, entitled residents or "locals" of their city are and whose right to stay should be protected. Finding answers to these questions would redirect Vienna toward its original goal of making the city a space for emancipation, organized on the basis of an egalitarian concept of justice.

Is the most livable city becoming exclusive?

Vienna can serve as a model for what an equitable and sustainable city could look like. It is also an example of municipal welfare resistance in a pressingly neoliberal environment. To date, current housing reforms have not managed to affect a considerable part of Vienna's housing and social infrastructure, which reveals the resilience of a complex and polycentric policy architecture. Organized in the form of overlapping policy layers (Kadi 2015), many decommodification layers—though heavily weakened—remain as social bastions. Nevertheless, the well-documented representation of Vienna as an exceptional case of housing rights fails to grasp recent commodification trends and the social, cultural and residential exclusion of some groups in some neighborhoods. As of 2020, a unique land and housing regime, geopolitical repositioning and demographic reconfiguration have created a distinct set of challenges threatening a green, social and inclusive Vienna.

Keywords

a) the urban development pattern of the city and neighborhood: fast recent growing; a rising global city; livable city planning

b) the urban greening of the city and/or neighborhood: parks and gardens; waterfront redevelopment, eco-districts, ecological retrofitting; greenbelt
c) the inequalities at stake: historic affordable and public housing infrastructure; recent dualization of the access to affordable housing, (re)-commodification of housing, unequal climate protection

Notes

1 The Donaupark project transformed a former landfill into a 100-hectare in conjunction with the 1964 International Garden Show. (See: https://www.wien.gv.at/english/environment/parks/donaupark.html) The 86-hectare Kurpark Oberlaa was built in 1974 on a decommissioned clay pit and opened to the public later that year after the 1974 International Garden Show. (See: https://www.wien.gv.at/umwelt/parks/anlagen/kurparkoberlaa.html) For the 1980s creation of the Donauinsel island-park and channel as a flash flood protection solution, see: https://www.wien.gv.at/umwelt/gewaesser/donauinsel/geschichte/
2 Co-financed by the City as a social housing development, Eurogate is just one of a number of low-energy and passive developments in the city. Thousands of such structures are publicly subsidized and built annually. For more information on Eurogate and the low-energy standards for social housing, see: https://www.irishtimes.com/business/commercial-property/vienna-is-active-on-passive-public-housing-1.675473
3 The City has an Executive City Councilor for the Environment who is responsible for nearly a dozen departments charged with protecting and enhancing the environment. Since 1993, the City has also had an Environmental Ombudsman office in place to "assert and protect the interests of the environment on behalf of the Viennese population." See the introductory chapter of the 2004/2005 Vienna Environmental Report.
4 Translated from: "Dabei liegt ein zentrales Augenmerk auf der „Grünraumgerechtigkeit": Alle Bürgerinnen und Bürger der Stadt haben das gleiche Recht auf hochwertige Versorgung mit Grün- und Freiräumen." See the STEP 2025 Fachkonzept: Grün und Freiraum.
5 For information on the greenbelt and other Viennese green spaces https://www.wien.gv.at/english/environment/protection/reports/pdf/green-04.pdf
6 Rochuspark (1991), Kardinal-Nagl-Park (1991), Reithofferpark (1997) and Haugerpark (2003) were built alongside the U3 line, (Connolly, Trebic, Anguelovski,Wood, Thery (Eds) 2018, p. 247).
7 Parks such as Parkanlage Krötzlergasse and the Schloss Neugebäude – Unterer Garten were renovated and opened up to the public in the 2000s. See https://www.wien.gv.at/umwelt/parks/anlagen/kroetzlergasse.html and https://www.wien.gv.at/umwelt/parks/anlagen/neugebaeude.html for more details.
8 The park planning project was carried out through an EU-funded international contest starting in 2005. https://www.wien.gv.at/umwelt/parks/anlagen/rudolf-bednar-park.html
9 Limited-profit housing associations have existed since 1930 in Austria and can be owned by local or regional public bodies, trade unions, religious institutions, banks, cooperatives, companies or private persons.
10 Interview available at www.ots.at, own translation.
11 Federal loans for housing have 1% interest, are granted for up to 35 years and cover approximately 35% of the construction costs.
12 For the duration of the repayment of the public loans.
13 Other thresholds are two persons: €69,220, three persons: €78,330, four persons: €87,430 and each additional person + €5,100.
14 https://www.jacobinmag.com/2020/09/red-vienna-austria-social-democrats-links-city-council

References

Cucca, Roberta. 2019. "Taking Contextual Differences into Account in Green Gentrification Research. The Case of Vienna." *Sociologia Urbana e Rurale*, 119: 46–58. https://doi.org/10.3280/SUR2019-119004.

Franz, Yvone. 2006. "Gentrification Trends in Vienna." In *Urban Sprawl in Europe: The Ignored Challenge*, edited by David Ludlow, 187–208.

Kadi, Justin. 2015. "Recommodifying Housing in Formerly 'Red' Vienna?" *Housing, Theory and Society* 32 (3): 247–65. https://doi.org/10.1080/14036096.2015.1024885.

Kadi, Justin, and Mara Verlic. 2019. *Gentrifizierung in Wien. Perspektiven Aus Wissenschaft, POlitik Und Praxis*. Arbeiterkammer.

Kohlbacher, Josef, and Ursula Reeger. 2016. "HOUSING INTEGRATION OF IMMIGRANTS IN VIENNA – EQUAL OPPORTUNITIES ?" *ISR-Forschungsberichte*, January. https://doi.org/10.1553/ISR.

Maloutas, Thomas. 2012. "Contextual Diversity in Gentrification Research." *Critical Sociology* 38 (1): 33–48. https://doi.org/10.1177/0896920510380950.

Schwarz, Werner Michael, Georg Spitaler, and Elke Wikidal, eds. 2019. *Das Rote Wien (1919–1934)*. Birkhäuser Verlag.

21

CAN COMMUNITY MOBILIZATION BE INCLUSIVE OF THE BLACK COMMUNITY IN ITS FIGHT AGAINST GREEN GENTRIFICATION?

Margarita Triguero-Mas, Mario Fontán-Vela, and Taliah Dommerholt

Portland is a pioneer of sustainability planning. Yet its traumatic history of racial exclusion and discrimination together with the inequitable distribution of environmental amenities and increasing (environmental) gentrification raise concerns about procedural justice linked to environmental interventions in the city. Although community groups are leading sustainability projects that acknowledge the risks of (environmental) gentrification—for example in Cully neighborhood—the formal inclusion of the African American community in these projects is still insufficient. In response to both environmental gentrification and challenges of racial inclusion, various new anti-gentrification policies and regulations have emerged. While some of these policies are the result of community mobilizations that include certain African American groups, this community is still largely neglected in the decision-making process.

Portland's white sustainability based on privilege and historical exclusion

Portland, Oregon, is one of the whitest cities in the U.S., with a 71% white population[1] out of a total of 653,000 residents. Non-white residents are divided between 10% Hispanic/Latinx, 8% Asian and 6% Black/African American residents[2] and a small population of Native American, Alaska Native and Pacific Islanders.

Portland has a history[3] of exclusion and discrimination[4] mainly affecting African American residents, but also Native Americans, immigrants and refugees. One of the widely known examples of Portland's racial injustice and institutional racism[5] towards African American residents is the case of the Albina District, located in inner North-Northeast Portland.

DOI: 10.4324/9781003183273-26

Albina District has historically been subjected to industrial pollution and environmental hazards. Before 1891, the area was mostly populated by Black railway company workers and prior to 1940, the area was home to approximately half of the Portland Black community because of housing choice restrictions imposed by the local government, real estate industry and private landlords. The flooding of Vanport and Guild's Lake—two other Portland Black community enclaves—in the late 1940s and early 1950s, together with redlining policy implementation, pushed the Black community further into the Albina District. Consequently, by the end of 1950s, approximately 40% of Albina was inhabited by Black/African American Portlanders. However, over the next few decades, Albina was subjected to the destruction of key African American landmarks and homes related to the construction of Interstate 5, Highway 99 and Memorial Coliseum (the latter destroyed commercial establishments and nearly 500 homes). Furthermore, the expansion of Emanuel Hospital destroyed what was considered the heart of the Black community in Albina and Portland, including several businesses and more than 1,100 housing units. In the following years, Albina became an abandoned and dilapidated area until the 1990s, when the district revitalization began. The Portland Development Commission offered favorable loans to home buyers and business owners from outside Albina to relocate to the district. Low prices, its convenient location and the Victorian homes that remained attracted mostly white young Portlanders.[6] At the same time, housing affordability issues became a problem in the district and many African American residents were displaced to other neighborhoods, physically dispersing the Black community. However, despite all the economic, social, psychological and spiritual trauma suffered by the African American community, today most of Portland's African American community still claims a strong connection with Albina District, even those who no longer reside in the area.

While Portland is characterized by a history of exclusion and discrimination strongly affecting the Black community, it is also recognized as a pioneer of urban sustainability in the U.S. and holds a strong reputation for its livability and climate change mitigation initiatives (Goodling, Green and McClintock 2015; Krumholz, Hexter and Bates 2018). For example, although Portland's population increased by 38%[7] from 1990 until 2017, carbon emissions decreased by 38% on a per person basis. It branded[8] itself as a sustainable and green city[9] and has kept up a high intensity/long duration of greening rhetoric since 1990 (Garcia-Lamarca et al. 2019). According to its Parks 2020 Vision, it aimed to provide "nature in the city" and be "the walking city of the West."[10]

Portland's sustainability successes, however, have not been equally distributed, and are increasingly critiqued for sustaining white privilege and exclusion.[11] Industrial pollution and environmental hazards have been and are still higher where unprivileged residents (such as Hispanic, African American or Asian) reside and disproportionally impact groups who live below the poverty line (Oregon Health Authority 2016; Smith 2009). For example, the Albina District has historically been an area of high environmental risks, given that it is bordered by the Williamette River (listed as a Superfund site by EPA in 2000), multiple industrial zones (including Swan Island District, the largest Portland industrial zone) and the Columbia Slough (a toxic waterway and one of the most hazardous sites in Portland) (Stroud 1999). These settings have been linked to soil and water pollution, including heavy metals (such as lead), P.C.B.s, pesticides, cyanide, toxins, carbon monoxide, nitrogen oxides, sulfur oxides, particulate matter, benzene, ozone and noise (Stroud 1999) that may be transferred to the food chain and to humans. The district is also crossed by the I-5 freeway and Interstate-84, both of which are important sources of air and noise pollution due to traffic volume. In addition to exposure to environmental pollutants, in 2015 it was estimated that 600,000 Portland residents still lacked secure access to fresh and healthy food, convenient transit, neighborhood safety and affordable housing (Goodling, Green and McClintock 2015).

Environmental amenities and sustainability projects have also been associated with large-scale urban renewal projects that have attracted tech companies and middle-class residents. Urban Renewal Areas (U.R.A.s) throughout the city, including the North Macadam and River District, actively promote economic growth by creating attractive living and working areas. These mixed-use areas include amenities such as walking paths, parks, trails and an aerial tram.[12] Portland's rising popularity[13] has resulted in increasing real estate prices and eviction to such an extent that in October 2015 the city declared a housing emergency.[14] Since then, gentrification[15] and the displacement of long-term minority residents (particularly Hispanic/Latinx and Black/African American) have intensified (see Figure 21.1). Gentrification in Portland is still partly driven by green developments (Goodling, Green and McClintock 2015) that began in the north and northeast parts of the city and have since moved east.

These processes of gentrification and displacement have also exacerbated inequities in the city and particularly affected Portland's Black/African American community, who have experienced the most severe displacement of all minority groups. Recent studies show that one-third of Black/African American long-term residents have been displaced from their homes in Northeast Portland in the last ten years (Krumholz, Hexter and Bates 2018). Moreover, homelessness is on the rise, having begun in the early 1980s with cuts in federal funding for affordable housing. Homelessness data from Portland, Gresham and Multnomah County for 2019 indicate that 4,015 people were unsheltered, of whom almost 40% belonged to communities of color,[16] a staggering number considering that only roughly 30% of the area population is non-white. This situation has led to social tensions over cohabitation between long-term residents and homeless people in neighborhoods such as Lents,[17] a highly diverse neighborhood located in East Portland.

In addition to dealing with social and racial inequalities, Portland also faces issues of procedural (environmental) justice. Given that decisions around environment and development are made mostly by white male experts, low-income residents and people of color are consistently left out of the decision-making

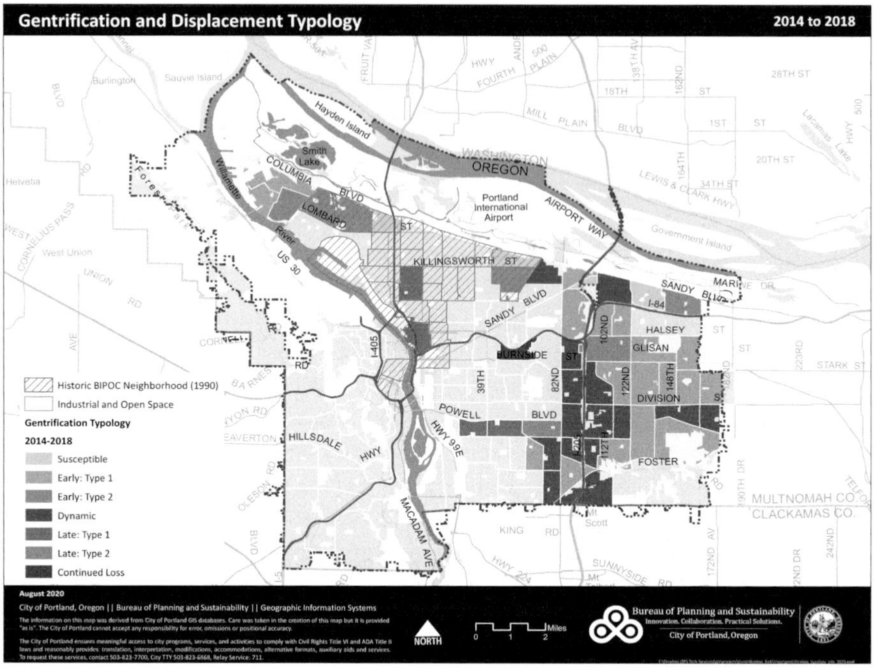

FIGURE 21.1 Classification of gentrification typologies according to changes between 2014 and 2018. The map shows late gentrification in some neighborhoods located in the limit between North-Northeast Portland and in some from Southeast Portland. Signs of early gentrification are present in nearly all neighborhoods in East Portland and parts of other neighborhoods like Cully. (Source: City of Portland Bureau of Planning and Sustainability.)

process. Moreover, some of the information used to take these decisions clearly underrepresents Latinx/nonwhite residents.[18] As a community organization member expressed,

> In the past and to some degree now, environmental decisions have been a very insular practice. A few elected officials, a few agency staff, a few NGOs, environmental non-profits, etc., were driving the agenda. Almost none of those who took the decisions were represented, or were accountable to low-income people and people of color.

Cully neighborhood: building community-led development amidst gentrification

Some neighborhoods have responded to environmental injustices by leading sustainable development initiatives to counteract environmental gentrification threats. An emblematic case of that is Cully, one of the most racially diverse neighborhoods in Portland with the smallest proportion of non-Hispanic white population[19] (52% non-Hispanic/Latinx white against 24% Latinx/Hispanic, 12% Black, 6% Asian and 6% other).

Together with the rest of East Portland, Cully has a disproportionately high concentration of subsidized, low-income and working-class renters (Goodling, Green and McClintock 2015). Previously a farming area, Cully was not fully annexed into the city of Portland until 1985, and as a result of its late annexation, parts of the neighborhood still endure transportation system deficiencies, as illustrated by the widespread absence of pedestrian sidewalks.[20] Some areas also lack standard infrastructure, including storm sewers, street lighting and paved roads. These characteristics, in addition to the high number of large lots and garden plots, give the neighborhood a less intensively built and more rural atmosphere than the rest of Portland. But despite being less urbanized, its residents have historically been exposed to hazardous air and acoustic pollution (e.g., from an asphalt manufacturing plant, street traffic and the nearby airport) while lacking environmental amenities such as green spaces for recreation. In 1985, the tiny Sacajawea Park became the first park in Cully, only followed by the creation of the Whitaker Ponds Nature Park in 1998 and the more recently built K^{h}unamokwst and Cully Parks in 2015 and 2018 respectively.[21]

Community-led development in Cully plays a role in advocacy and decision-making regarding housing and businesses that should remain in the neighborhood, and in the creation of community housing and environmental development alternatives. Examples of Cully's community housing activism include the Normandy Apartments[22] and Oak Leaf.[23] The Normandy Apartments, which local activists unsuccessfully attempted to save from high rent increases, eventually priced out long-term residents. In contrast, Oak Leaf managed a successful campaign to avoid the sale of a trailer park to a developer and that is now managed and renovated by a nonprofit organization instead.

Cully community's involvement in deciding which businesses should remain in the neighborhood is illustrated by the cases of Las Adelitas,[24] 42nd Avenue[25] and Cully Boulevard.[26] In 2015, the community network Living Cully (constituted by Hacienda C.D.C., Habitat for Humanity Portland/Metro Northwest, Verde and N.A.Y.A.) successfully purchased a site previously home to a strip club and adult video store. Now, after its complete demolition, a new plan to build 141 units of affordable housing and a community center with green features (Las Adelitas project) by 2020 awaits funding. Las Adelitas is touted as a model for both anti-displacement strategy-building and green investments for front-line climate-affected communities.[27] The areas around 42nd Avenue and Cully Boulevard are undergoing the most commercial development. Both streets were given Neighborhood Prosperity Initiative (N.P.I.) zone grants that have helped support already existing businesses and those owned by vulnerable local populations, particularly women of color.

As part of co-creating housing alternatives, a group of Cully neighbors is partnering with six community organizations to explore whether a new model of urban revitalization that uses an adaptation of Tax Increment Financing (T.I.F.) would be feasible while being controlled by community groups. Moreover, environmental alternatives and interventions are being reconceptualized using a community-centered perspective[28] that emphasizes community ownership and focuses on the needs and benefits for underprivileged residents. As expressed by one environmental activist,

> We always felt that environmental issues and environmental policies could do more than protect the environment; they could build community, create jobs, educate youth, support entrepreneurship ... and the only way to figure out how to do that was to make investments at the grassroots level so that folks could say what they felt an environmental investment or environmental work in their community could do ... With time and resources they [low-income people and people of color] could define a new environmental movement that was community-based and would create deep co-benefits in community.

FIGURE 21.2 Thomas Cully Park, from left to right and top to bottom: gathering space designed for the Native American community, playground area with water pump, general view of the park, vegetable garden. (Photo by: Margarita Triguero-Mas.)

Many activists are inspired by green Keynesian economics (the basis of the Green New Deal framework) to create a green workforce that will build green infrastructure and clean energy projects while guaranteeing training and jobs for local residents. Cully has been branded as an EcoDistrict and recognized for its efforts to work towards equity, resilience and climate protection. Embracing "sustainability as an antipoverty strategy,"[29] local organizations are pairing sustainability investments with job opportunities and high-quality affordable housing, aiming to prevent green gentrification and help the community gain control of the local housing market. For example, Cully community activists, driven by the need for secure access to housing and sustainable livelihoods, have launched two emblematic initiatives: the "Living Cully weatherization + home repair project 2.0" and the Thomas Cully Park. The weatherization project aims to lower energy bills by upgrading homes to low-cost energy systems and to repair household mold, failing roofs and heat systems.[30] Thomas Cully Park is a new park built thanks to a coalition of neighborhood organizations called Let Us Build Cully[31] (see Figure 21.2) that transformed a landfill into a green space.

Representation and inclusion in community activism

In response to gentrification and displacement, various forms of activism and resistance have emerged within the larger Portland African American community. Partly thanks to this work, neighborhoods like Cully—where gentrification is still in its early stages—have received significant resources. But community-led environmental development and resistance against environmental gentrification in Cully are strongly dependent on the Hispanic/Latinx, Native American and white populations, while the integration of the African American population in community-led movements remains difficult.

The main community organizations in Cully are: (1) Hacienda Community Development Corporation (C.D.C.), which provides affordable housing, homeownership support, economic improvement and educational opportunities particularly for the Hispanic/Latinx population of Cully; (2) Verde, a group that focuses on building environmental wealth for all the different population groups of the neighborhood; (3) the Native American Youth and Family Center (N.A.Y.A.), an organization that aims to provide educational opportunities, leadership development, elderly and residential support and financial security training for the Native community to reduce poverty, hunger, family violence and homelessness rates; and (4) the Cully Association of Neighbors, which focuses on improving livability in Cully neighborhood. The needs and experiences of the Hispanic/Latinx community are the focus of several of these community organizations, allowing this community to build communal cohesion in the neighborhood. In fact, out of all Cully communities, the Latinx/Hispanic community probably has the highest presence and leadership in mobilizing and shaping the demands of the anti-gentrification and environmental movements.

Although community organizations have traditionally underserved the African American community of Cully, they are beginning to focus on bridging the gap and broadening representation of other racial and ethnic groups. An activist in charge of understanding how Cully community organizations can engage with and be more supportive of the African American community explained that a very small number of African Americans are involved in community development and expressed their frustration with this issue. The activist explained that lack of affordable housing is often blamed on the Latinx/Hispanic community, whom Cully residents perceive as the ones being secured affordable housing by community organizations that leave other community groups aside. Accordingly, the activist suggested that this attitude does not "relate so much to African Americans' true feelings as to this pitting, this financial weaponizing of the money and resources causing divide" between the African American community and other communities in Cully, such as Latinx/Hispanic.

Institutional and community action to tackle Portland challenges

The neighborhood green development work in Cully is a clear example of the difficulty in including the African American community in the decision-making process regarding green developments and (environmental) gentrification issues and that stem from Portland's history of environmental injustice, displacement, gentrification and institutionalized racism against the Black/African American population.[32] As some indicate, "the voice of the Black community has been prevented, discounted and generally ignored"[33] in Portland.

These barriers are amplified by struggles over the translation of technical documents and processes linked to housing, equity and sustainability into everyday terms.[34] Moreover, African American residents and their advocates often suffer hurdles at the institutional level such as limited awareness of the persistent and emerging issues that African American residents experience, hindering public processes and policies that serve the African American community (Krumholz, Hexter and Bates 2018). According to some, Portland should require local institutions to include a diversified staff whose broad range of experiences and identities allow them to understand and more effectively address issues like the affordable housing gap between Black/African Americans and white people (Krumholz, Hexter and Bates 2018).[35]

Despite the lack of inclusion, the African American community has not remained silent and has become involved in anti-gentrification policy work and community mobilization. The people-of-color-led organization Community Alliance of Tenants (C.A.T.), for instance, lobbied to pass a ban on no-cause evictions. The new Master Plan of Portland, a 25-year comprehensive plan adopted in 2016, included some of the anti-displacement and affordable housing policies advocated by the Anti-Displacement P.D.X. (A.D.P.D.X.) coalition of 30 community organizations. The strategy, which called on outside advocates to urge planners to include equity,[36] led to the plan's inclusion of several demands made by disadvantaged and underserved populations of Portland, such as their meaningful participation in decision-making processes and their consultation when assessing impacts of any major change in Portland city (Krumholz, Hexter and Bates 2018). The A.D.P.D.X. also establishes a working advisory group composed of staff planners and A.D.P.D.X. members.

The wider community has also influenced new anti-gentrification policies and bans in Portland, including bills that limit new single-family home construction and strengthen renters' rights. In 2019, the State of Oregon passed a bill[37] ordering all municipalities with over 10,000 residents to legalize duplexes, triplexes, cottage clusters and attached townhouses in areas currently zoned as single-detached homes. This bill partly results from Portland citizens' demands for zoning changes. Moreover, a group of housing rights organizations, including Portland Tenants United, worked to pass the Fair Access in Renting (F.A.I.R.) ordinance in June 2018 to force owners to use a first-come-first-serve system for rental applications and limit the use of credit and criminal histories as criteria for denying rental applications. Another notable success was a state ban on no-cause evictions for tenants who have lived in a rental for at least a year, building upon the advocacy of the CAT. Lastly, in early 2020, the Portland Housing Bureau allocated funding for 7 new apartment buildings in Albina, out of which 531 units were designated as "Preference Policy" units, to allow for the return of displaced residents affected by gentrification and reconnect them with neighborhood networks.

However, Portland municipal representatives and technicians are ill-equipped to tackle environmental gentrification. As one member from the City of Portland Bureau of Planning and Sustainability expressed:

> I think what we're struggling with now is to prioritize these [environmental] improvements in communities that need them, without it resulting in gentrification. So now we're trying to figure out how we can put in anti-displacement rules … how we can make sure that the new sidewalks and new parks doesn't mean that lots of money comes in and the people who are living there get pushed out. And I think we're struggling with that right now.

In short, environmental gentrification is acknowledged but yet unresolved by local planners.

In contrast to the lack of a citywide anti-environmental gentrification strategy, there are some specific strategies in place at the neighborhood level. In addition to Cully's anti-gentrification approach—based on pairing sustainable investments with job opportunities and high-quality and affordable housing—initiatives from Jade District,[38] an ethnically and racially diverse neighborhood located in central East Portland, are also notable. Here, the community-led initiative Portland Ballot Measure 26-201[39] (known as the Portland Clean Energy

Initiative[40]) taxes large retailers in order to fund things like green job training programs and energy-efficient home upgrades that prioritize low-income residents and communities of color. All these initiatives, however, still lack the inclusion of the African American community.

The anti-environmental gentrification strategies of Cully and Jade District are generally perceived as strategies adaptable to Portland city and beyond. However, they seem to follow a growth-obsessed model of environmentalism (Dillon 2014; Lang and Rothenberg 2017; Long 2016) where social aspects, care, public health and wellbeing remain absent—clearly demonstrated by the exclusion of the African American community—at the expense of the green growth paradigm.

Towards more inclusionary anti-gentrification policies

As one of the whitest and most sustainably conscious cities in the U.S., Portland faces major challenges in addressing advanced (environmental) gentrification, the inequitable distribution of benefits of urban greening and the lack of African American voices in community development. Although some environmental groups are making progress by framing environmental interventions as anti-gentrification tools and enacting various city and state anti-gentrification policies, the inclusion and participation of the African American community in these processes remains insufficient, reproducing a growth-obsessed model of environmentalism that lacks the inclusion of social aspects.

Keywords

a) the urban development pattern of the city and neighborhood: fast-growing city; tech city, business, historic racial segregation; large-scale urban renewal; sustainability planning
b) the urban greening of the neighborhood/city: community-led greening; green adaptative infrastructure; workforce development-centered greening; livable city planning
c) the inequalities at stake: environmental gentrification; historic environmental injustices; lack of Black/African American inclusion

Notes

1 https://www.census.gov/quickfacts/fact/table/portlandcityoregon/PST045219 and https://www.jstor.org/stable/10.5406/jamerethnhist.36.1.0063?seq=1#page_scan_tab_contents
2 https://www.youtube.com/watch?v=NxHa9l625OM
3 https://www.nytimes.com/2008/05/29/us/29portland.html
4 https://static1.squarespace.com/static/5501f6d4e4b0ee23fb3097ff/t/5935cb971e5b6c0cd0c97458/1496697798853/Rebuilding+Community+-+Low+Res.pdf
5 https://scholarship.claremont.edu/cgi/viewcontent.cgi?referer=https://www.ecosia.org/&httpsredir=1&article=1071&context=pitzer_theses
6 https://www.blackpast.org/african-american-history/albina-portland-1870/

7 https://www.portland.gov/bps/climate-action/2017-carbon-emissions-and-trends
8 http://www.bcnuej.org/projects/green-trajectories/
9 https://scholarship.claremont.edu/cgi/viewcontent.cgi?referer=https://www.ecosia.org/&httpsredir=1&article=1071&context=pitzer_theses
10 https://www.portlandoregon.gov/parks/40182
11 https://www.paalf.org/paalf-peoples-plan/
12 https://prosperportland.us/portfolio-items/river-district/?portfolioCats=30
13 https://www.theatlantic.com/business/archive/2016/05/the-real-roots-of-portlands-housing-crisis/482988/
14 https://www.opb.org/news/article/portland-oregon-housing-affordability-emergency/
15 https://www.portland.gov/bps/adap/gentrification-and-displacement-studies
16 https://pdxscholar.library.pdx.edu/rri_facpubs/63
17 https://www.oregonlive.com/portland/2017/10/livability_in_portlands_neighb.html
18 https://www.mdpi.com/2071-1050/11/16/4336/htm
19 https://statisticalatlas.com/place/Oregon/Portland/Race-and-Ethnicity
20 http://www.cullyneighbors.org/history-of-cully/
21 https://www.portlandoregon.gov/parks
22 https://www.oregonlive.com/portland/2017/01/northeast_portland_families_fa.html
23 https://nextcity.org/features/view/when-green-infrastructure-is-an-anti-poverty-strategy
24 https://141homes.org/
25 http://www.42ave.org/
26 https://cullyblvdalliance.org/
27 https://www.motherjones.com/environment/2019/02/the-green-new-deal-is-already-happening-in-one-portland-neighborhood/
28 https://grist.org/article/the-green-new-deal-is-already-at-work-in-one-portland-neighborhood/
29 http://www.livingcully.org/about-living-cully/
30 https://digitalcommons.humboldt.edu/cgi/viewcontent.cgi?article=1106&context=etd
31 https://letusbuildcullypark.org/
32 http://right2root.com/people-place & https://scholarship.claremont.edu/cgi/viewcontent.cgi?referer=https://www.ecosia.org/&httpsredir=1&article=1071&context=pitzer_theses
33 https://www.paalf.org/paalf-peoples-plan/
34 https://www.paalf.org/paalf-peoples-plan/
35 https://www.paalf.org/paalf-peoples-plan/
36 https://pdxscholar.library.pdx.edu/cgi/viewcontent.cgi?article=1252&context=usp_fac
37 https://www.sightline.org/2019/06/30/oregon-just-voted-to-legalize-duplexes-on-almost-every-city-lot/
38 http://jadedistrict.org/
39 https://crosscut.com/2018/11/portland-passed-climate-justice-initiative-should-seattle-follow-suit
40 https://www.usnews.com/news/cities/articles/2019-01-30/portland-oregon-taxes-big-businesses-to-fund-clean-energy-jobs

References

Dillon, Lindsey. 2014. «Race, Waste, and Space: Brownfield Redevelopment and Environmental Justice at the Hunters Point Shipyard: Waste, Race and Space». *Antipode* 46 (5): 1205–21. https://doi.org/10.1111/anti.12009.

Garcia-Lamarca, Melissa, Isabelle Anguelovski, Helen Cole, James JT Connolly, Lucía Argüelles, Francesc Baró, Stephanie Loveless, Carmen Pérez del Pulgar Frowein, and Galia Shokry. 2019. «Urban Green Boosterism and City Affordability: For Whom Is the 'Branded' Green City?» *Urban Studies*, diciembre, 004209801988533. https://doi.org/10.1177/0042098019885330.

Goodling, Erin, Jamaal Green, and Nathan McClintock. 2015. «Uneven Development of the Sustainable City: Shifting Capital in Portland, Oregon». *Urban Geography* 36 (4): 504–27. https://doi.org/10.1080/02723638.2015.1010791.

Krumholz, Norman, Kathryn Wertheim Hexter, and Lisa K. Bates, eds. 2018. *Advancing Equity Planning Now*. Ithaca: Cornell University Press.

Lang, Steven, and Julia Rothenberg. 2017. «Neoliberal Urbanism, Public Space, and the Greening of the Growth Machine: New York City's High Line Park». *Environment and Planning A: Economy and Space* 49 (8): 1743–61. https://doi.org/10.1177/0308518X16677969.

Long, Joshua. 2016. «Constructing the Narrative of the Sustainability Fix». *Urban Studies* 53(1): 149–172.

Oregon Health Authority. 2016. «Draft Environmental Justice Issue Paper». Portland, OR.

Smith, Chad L. 2009. «Economic Deprivation and Racial Segregation: Comparing Superfund Sites in Portland, Oregon and Detroit, Michigan». *Social Science Research* 38 (3): 681–92. https://doi.org/10.1016/j.ssresearch.2009.03.005.

Stroud, Ellen. 1999. «Troubled Waters in Ecotopia: Environmental Racism in Portland, Oregon». *Radical History Review* 74: 65–95. https://doi.org/10.1215/01636545-1999-74-65.

22

ENACTING JUST URBAN GREEN FUTURES

Promising policy and planning tools and regulations for Europe and North America

Emilia Oscilowicz, Sarka Hajtmarova, and Isabelle Anguelovski

Throughout this book, we have examined green imaginaries and resulting landscapes of urban environmental injustices in American, Canadian and European cities while also demonstrating the inherent connection of urban environmental injustice with housing vulnerability and community displacement. Through our 21 tales, we have argued that achieving a just green city will require guaranteeing housing and land rights before greening projects to ensure the ability of lower-income, minority and immigrant residents to remain in place. It will also require prioritizing inclusivity goals in planning for green development. Our research has revealed that two policy implementation goals must thus be achieved: on the one hand, housing stability and emancipation for working-class and minority residents and, on the other hand, green equitable (re)development through inclusive planning, or (ideally) a combination of both.

In the North American context, socially vulnerable, often racialized communities have been successfully mobilizing for more equitable greening and environmental justice. This mobilization has come as a direct response to an enduring legacy of historical and political racial injustice and a relative dearth of public housing. Yet, it is important for North American cities to strengthen their support to existing community green justice groups while also prioritizing funding for more and improved affordable and public housing. In the European context, urban residents have generally benefited from well-funded social housing programs accompanied by top-down municipal efforts to provide enhanced green spaces to European residents. Lacking, however, is support for civic groups and residents in establishing neighborhood leadership and a robust organizational structure for alternative housing and equitable green planning, particularly as affordable and public housing becomes increasingly privatized. In Figures 22.1–22.3, we summarize key takeaway points for future policy and

DOI: 10.4324/9781003183273-27

Key takeaways for North American policymakers, planners, and community activists

Resist and overturn highly neo-liberal positionalities (focusing on market mechanisms) and opt for public institution-led planning practices, with opportunity for rent control, and inclusive zoning

Hold developers accountable to mandated fees, taxes, and requirements such as minimum amounts of green space, minimum number of affordable units, and inclusion of affordable units built on site.

Require developers to research, consider neighborhood gentrification pressures and outcomes prior to development proposals.

Require developers to submit (and fund) an anti-displacement neighborhood benefits agreement based on community proposals.

Utilize a variety of funding schemes and land use tools to help support the financial hardships of vulnerable tenants/ homeowners as well as non-profit and affordable housing developers.

Implement historical and special zoning protections with intention and significant detail to the vulnerable communities that will be affected by such land use regulations.

Include community participation in every planning decision, particularly in contexts with a deep racial dimension.

Advocate for the preservation, improved maintenance and upgrading, and increased funding of public and affordable housing currently threatened by the increased commodification and neoliberalization of the housing market.

FIGURE 22.1 Key takeaways for North American policy-makers, planners and community activists.

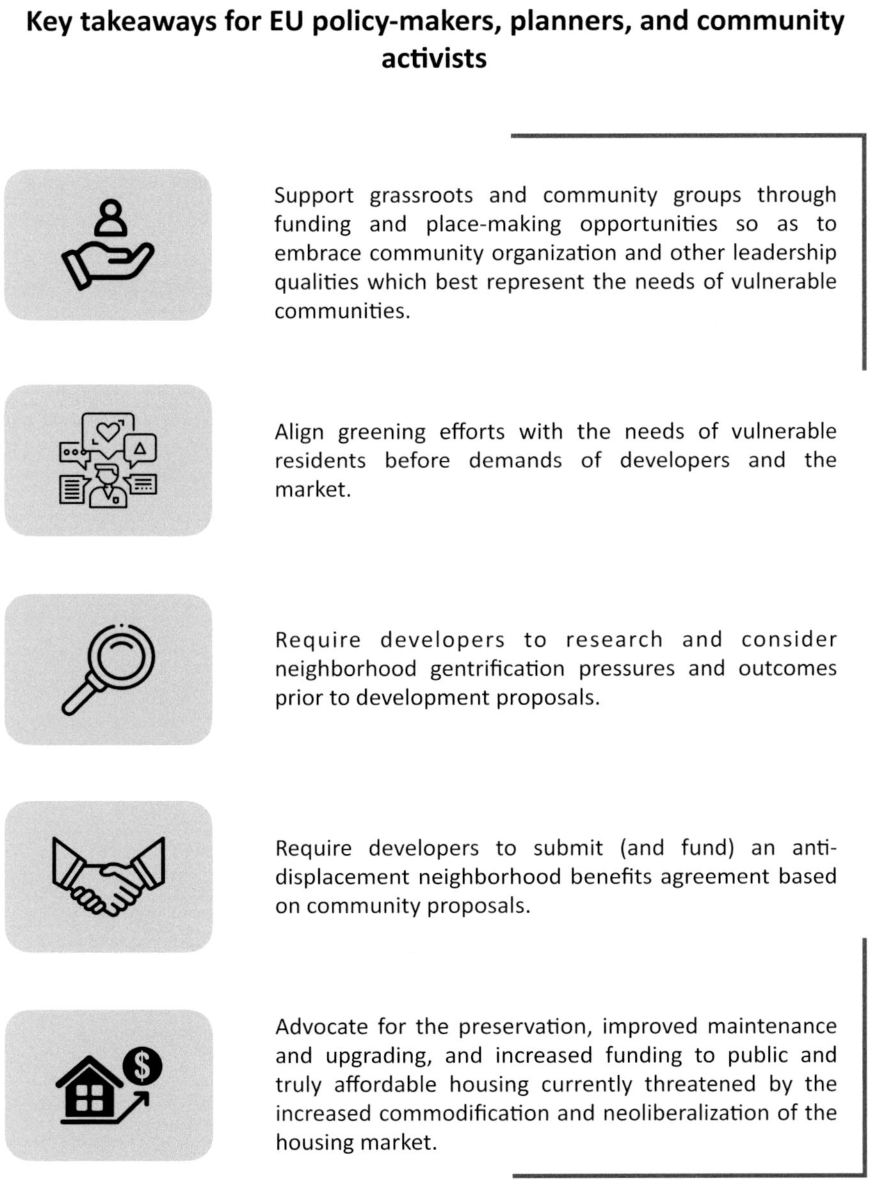

FIGURE 22.2 Key takeaways for E.U. policy-makers, planners and community activists.

FIGURE 22.3 Key takeaways for all global policy-makers, planners and community activists.

planning directions. Some of these points are more context focused while others can be more applied in a more general manner.

In the following sections, we examine in greater detail some specific policy tools that have been and could be used in a number of contexts to address the dimensions of green injustices and inequalities identified in the previous sections of this book. We ask, *what are promising policy and planning directions for the equitable green city? What are alternative routes for more equitable, democratic and emancipatory green cities?* We recognize that not all tools might work in all contexts due to different histories of urban development and local or national policy-making systems. Furthermore, we emphasize that not all gentrification is explicitly connected to greening and not all greening leads to gentrification. The tools we present in this chapter are selected from a report by the Barcelona Lab for Urban Environmental Justice and Sustainability—"Policy and Planning Tools for Urban Green Justice"—which itself draws on the extensive field work and data analysis presented through this book (see Oscilowicz et al. 2021). The report includes 30 anti-displacement and anti-gentrification tools and 20 equitable green development tools. This selection is not exhaustive by any means; however it can act as a library of possible bold policy actions that respond to the needs of marginalized groups and to the civic mobilization of residents, community activists and nonprofits working for urban green justice.

Policy and planning tools towards green justice

We have identified four types of policy and planning tools that can contribute to greater urban green justice. *Land use tools* define the purpose of a designated urban space. These can be managed by local or higher levels of government and determine in what ways, for what reasons or in what manner a space may or may not be utilized. *Developer-centered tools* demand that private, public or nonprofit real estate developers meet certain regulatory or financial obligations, and can apply to any type of land use development, both greening and housing. *Financial schemes* refer to different local or higher government mechanisms to optimize the provision of affordable housing or green space, and can be broken down into three separate focuses based on the stakeholder concerned: homeowner, tenant and community. Next we present some examples of those tools, separating anti-displacement and anti-gentrification from equitable green development.

ANTI-DISPLACEMENT AND ANTI-GENTRIFICATION

1. **Land use tools**

Inclusionary zoning

Inclusionary zoning (I.Z.) is a planning ordinance in which developers set aside a share of new housing to be affordable for people with low to moderate incomes.

City-mandated I.Z. typically requires developers to include a percentage of new residential development for affordable units as determined by the affordable housing guidelines set out by individual municipalities. In some cases, developers can choose to provide affordable units off-site rather than on-site or pay fees in-lieu that would benefit other affordable housing projects. In the U.S., non-mandated developer-voluntary I.Z. also exist, and can depend on incentives, such as increased density allowances or zoning variances. Financially, inclusionary zoning removes the financial burden of providing affordable housing from government or non-profit providers as it utilizes private developers' funds for the development of units. I.Z. policies also can indirectly benefit the socio-urban fabric of a community. On the other hand, difficult enforcement, a consequential lack of accountability and the limited suitability of the I.Z. policies to many urban redevelopment sites, especially so in the U.S. context, can lead to a limited real provision of affordable units.

A successful and bold practice of I.Z. is illustrated by Nantes, which has committed to incorporating a share of 33% social housing and 24% affordable housing in all new real estate projects. This municipal promise followed the French-national housing act of 2000, called Solidarité et Renouvellement Urbain (S.R.U.). S.R.U. is a countrywide social policy that aims to support mixed-income housing by requiring that all French towns above 3,500 residents have a minimum of 20% social housing in their housing stock, later changed to 25% in 2004. Adherence to the policies is monitored at a local level by the Nantes Development department with a second tier of accountability supervised by the Department of Housing, with steep fines for developers or local government if infraction is observed.

Community land trusts

Community land trusts (C.L.T.) offer the opportunity for a nonprofit trust to purchase and own unused or under-maintained land with the goal of permanently providing or preserving affordable housing within a neighborhood. C.L.T.s are explicitly community-focused and prioritize helping residents become homeowners of affordable housing in areas threatened by increased rent costs and gentrification. Finally, C.L.T.s can also revitalize public space and improve a sense of belonging to the community through functional landscaping and the provision of community gardens. In a community land trust, residents build financial equity by owning their home, while the land remains in the property of the trust. Community leaders in the Boston neighborhood of Dudley Square have successfully incorporated a C.L.T. as a means of providing affordable housing to low-income residents as well as reducing negative externalities associated with formerly contaminated or vacant parcels in the neighborhood. This land trust emerged in the 1980s after members of the local grassroots campaign "Don't dump on us," which fought for a sustainable, just and safe urban community through the revitalization of 1,500 vacant and contaminated lots. Much of this vacant land was purchased and redeveloped by the Dudley Land Trust, which has since developed 227 affordable housing units managed by nonprofit housing providers as well as a green play space, a community garden and greenhouse and commercial and office space. The Dudley Land Trust model has been regarded nationally as a strong anti-displacement model and emulated by dozens of community groups around the country. Although the development of a C.L.T. often encounters legal and financial barriers and the rate of displacement is often faster than the preservation or construction of affordable units, it is praised for its ability to permanently preserve affordable housing, offer access to new amenities and build community political power.

2. **Developer-centered tools**

Density bonuses to encourage affordable housing

Density bonuses allow developers to build more floor space in a specific area by exchanging and dedicating it to community amenities such as libraries, parks and childcare centers. Similar to inclusionary zoning, a density bonus may be exchanged for the developer's commitment to include a certain percent or number of affordable housing units within a parcel. Other times, bonuses can be offered under predetermined non-negotiable rules. In order to protect against over-development, municipalities often place a limit on the extra floor space that can be added to a development, usually a maximum of 125 to 150% of the building's base zoning height limit or floor space area. By reducing the financial impact, developers may be more willing in proposing developments in areas that density bonusing or inclusionary zoning applies. It is also a tool that can ensure that a certain percent of affordable housing units is included in a new housing complex, including a mix of housing forms such as medium-density townhouses and duplexes. However, developers may choose to only develop in certain zones of the city where profits are highest. In practice, the city of Austin, Texas, has leveraged density bonuses as a tool to circumvent out-of-date, low-density-focused zoning codes which force many residents to live in car-dependent housing far from the city center. In an attempt to densify Austin while providing more housing in the city center, the city has developed ten density bonus programs each with its own set of criteria regarding developer qualification based on existing zoning regulations. These density bonuses are meant to allow developers to provide expedient affordable housing units to residents on smaller parcels without having to make an exception to the zoning code, a lengthy and costly legal process.

3. **Financial schemes**

For homeowners

Tax on foreign ownership/vacant units

Tax on foreign ownership of a property is a financial scheme implemented by policy-makers at the municipal, provincial, state or federal level to disincentivize property value speculation and consequential distortion of the residential housing market. In order to protect existing housing stock and affordability for local residents, some cities apply a one-time purchase tax or annual property tax on homes owned by foreigners and/or non-residents. This tax may also be extended to manage vacancy when the property is used as a secondary residence. Through this preventative taxing system, more affordable and family-size homes may be made available to local residents. Additionally, taxes on foreign homeownership may indirectly increase local development and construction of housing as lower property values are more accessible to developers seeking profits on their margins. Vancouver is consistently ranked as one of the most expensive real estate markets in the world and the most expensive market in Canada where considerable real estate capital in the city is sourced from foreign owners, with 19.2% of all condos owned by a non-resident of Canada and 6.5% of housing empty or underused. To improve access to affordable housing, the Province of British Columbia implemented the *Additional Property Transfer Tax for Foreign Entities and Taxable Trustees* in 2018 which applies a tax of 20% towards the fair market value of the property. The City of Vancouver also passed the 2017 *Empty Homes Tax (E.H.T.)* as a 1.5% tax (increased to 3% in 2021) on property value of vacant units which goes directly to support affordable housing initiatives. The application of both taxes has helped to slow property speculation and mildly corrected housing prices, as demonstrated in the 17% decrease of housing prices in the affluent municipality of West Vancouver.

For homeowners

Housing tax credit programs

Housing tax credit programs are utilized as a financial incentive for private developers and investors to provide more lower-income housing by subsidizing the financing, acquisition, construction and rehabilitation of affordable rental housing for low- and moderate-income tenants. Such housing tax credit programs are most homogenous among states in the United States, exemplified by the Low-Income Housing Tax Credit (L.I.H.T.C.) created by the 1986 Tax Reform Act. In the U.S. context, the federal government issues tax credits to state and territorial governments for which state housing agencies award the credits to private developers of affordable rental housing projects through a competitive process. However, tax exemptions, like other tax relief programs, can negatively impact some lower-income neighborhoods as they often lead to higher property tax rates, especially under locally funded programs where jurisdictions raise taxes to offset the drop in the tax base from the exemptions. For example, the state of Colorado has implemented a low-income housing tax credit providing over $2 billion in equity for affordable rental housing in Colorado since program conception in 1986. The program allocates tax credits through the Colorado Housing Finance Authority (C.H.F.A.) awarding $16.1 million to 14 developments in 2020 that will support the construction of 940 affordable rental housing units.

For renters

Rent control

Rent control is a legal restriction that consists of limiting the amount that a landlord can demand for leasing a home or renewing a lease. In general, rent control laws have two related goals: to maintain existing affordable housing and to limit disruptions caused by rapid rent increases. Rent control on apartments can create an incentive for residents to remain, instead of moving into a volatile non-rent-controlled unit. Moreover, such laws might also have secondary goals, such as protecting tenants from unjust eviction, decreasing tenant turnover or enhancing mixed-income neighborhoods. These policies can be tailored to avoid wicked effects by ensuring that landlords receive enough compensation to maintain their properties and earn a reasonable profit while also contributing to tenant stabilization, reduced prices and positive physical and mental health outcomes. In Oregon, state-wide rent control acts as one of the responses in place to address climbing rents, especially in its largest city, Portland. Before the 2019 Senate Bill 608, landlords were only obliged to deliver a 30-day notice prior to eviction and there was no limit to how much or how often rents could increase. In response, the Senate Bill 608 established a maximum amount of rent increase, determined by the state at the beginning of each year for each leasing-period. However, this measure has its limits, as the rent cap does not apply to buildings less than 15 years old, nor to government-subsidized rents, and landlords can still evict tenants if they violate the terms of their lease, or if they plan to upgrade or demolish a building.

For communities

Support to housing co-ops

Co-op housing is a form of tenure that falls between rental housing and private homeownership where members are not landlords nor tenants, but instead work together to govern their housing association. There are three different types of housing co-ops: market-rate (membership and shares can be bought or sold at the market price), limited-equity (includes restrictions on profit of co-op shares) and leasing cooperatives (whereby no equity can be claimed from the sale or exchange of a unit). Co-ops offer their members many advantages that are often unavailable in other forms of housing including security of tenure, active participation in decision-making processes within the community and a life within a diversified community of people as well as economic advantages including affordability, tax deductions, consumer action and limited liability. In Amsterdam, social housing constitutes almost 40% of the city's housing stock, and the city now looks to co-op housing to provide further support and living opportunity for residents in need of affordable housing. Through the Housing Co-operatives Action Plan, a loan-fund of 50 million euros, with 10–15-year repayment terms, supports the development of housing co-operatives. By 2025, the council estimates that 7,000 co-operative housing dwellings will be available to residents in need.

4. **Legal tools and policies to grant rights to place**

Formal recognition of the right to stay or right to return

Formal recognition of the right to stay or right to return policies are designed to ensure that the original residents and those with long-term connections to gentrifying neighborhoods are able to receive affordable housing within their communities. These strategies include assistance to renters, homebuyers and existing homeowners. The city may provide its own payment assistance for home buyers that have been historically prevented from purchasing homes or have faced financial pressures that are associated with gentrifying neighborhoods. Cities using this tool have found that despite their positive intention to provide some form of reparation for past harm and losses, these policies can be difficult to implement. Because these policies rely on an adequate supply of housing options, they are limited in the number of people they can support (unless the programs directly involve the construction of new housing). Cities are challenged in determining requirements and evaluating needs for those who are eligible to receive assistance. Some cities, such as Portland, have application processes that involve point systems and proof of ancestral connection to the neighborhood. In 2020, the Portland Housing Bureau also allocated funding towards seven new apartment buildings designed to allow for the return of some displaced residents affected by gentrification. Finally, many of these policies assume that residents want to stay in or return to their gentrifying communities without addressing the challenge residents will face as they continue to live in their changing neighborhoods.

EQUITABLE GREEN DEVELOPMENT

1. **Land use tools**

Eco-district zoning

Eco-districts, also known as écoquartiers, eco-cities, zero- or low-carbon and carbon-positive cities, are neighborhood-scale developments that address climate mitigation and adaptation through sustainable planning strategies. The U.N.-Habitat (2014) defines five essential principles of sustainable neighborhood and eco-district planning: adequate space for streets and efficient street network, high density, mixed land use, social mix and limited land use specialization. To create low-carbon and resilient districts, these principles can include the employment of technologies in green building, smart infrastructure, renewable energy and nature-centered solutions. A great example of an eco-city is Copenhagen which was selected as the European Green Capital in 2014 and strives to be the first carbon-neutral city in Europe through adoption of the Urban Nature

Strategy. Given that the neighborhood of Nørrebro currently has the fewest square meters of green space per inhabitant in the city, this strategic plan prioritizes community-led greening projects through the development of 26 urban gardens.

2. **Developer-centered requirements**

Minimum green space in new development projects

This measure sets a minimum amount or percentage of green infrastructure in new development projects. This policy tool aims to ensure that cities improve access to green spaces, especially so on sites targeted for new real estate development projects. By targeting developers directly, it ensures that developers themselves compensate for limiting the availability of urban open spaces and relieves the municipality of the cost. However, when implemented on a voluntary basis, increasing green space as part of private development can intensify gentrification pressure or create exclusive or privatized access to greening. In 2019, the City of Lyon, in order to ensure greater accessibility to urban green spaces (as currently only 10% of it is accessible to the public), introduced the P.L.U.-H. 2019 Local Urbanism and Housing Plan. This policy seeks to encourage urban revitalization, the protection of natural areas, the preservation of historic heritage and housing affordability, along with environmental awareness. P.L.U.-H. regulations determine the minimum area of open space on each piece of land and the type of greening required, such as tree planting or community gardens.

3. **Financial schemes**

Funding green resilient infrastructure in socially and environmentally vulnerable neighborhoods

Green resilient infrastructure funding from municipalities or private actors can be directed towards socially and environmentally vulnerable neighborhoods. This tool is crucial to bridge the gap in green infrastructure between affluent and impoverished or vulnerable neighborhoods and help to build the adaptation capacity of those neighborhoods using nature-centered projects. However, this policy could conversely be a driver of green climate gentrification by attracting wealthier residents or new private developments to these green-protected neighborhoods. In Philadelphia, the Office of Sustainability launched the Beat the Heat initiative in 2018, wherein the city partners with lower-income and minoritized communities in the hottest neighborhoods to expand tree-planting and green stormwater infrastructure. In the historically Latinx and Black neighborhood of Hunting Park, local stakeholders have been partnering with nonprofit and municipal programs to gain access to trees at low or no cost and motivating residents to join the initiative. Esperanza's NeighborCare program, a mini grant that provides up to $1,000 in materials for residents to green, clean and beautify their blocks, has benefited 112 blocks so far and aims to achieve the goal of 30% canopy cover by 2025 established by the Greenworks sustainability plan. This project also works to help residents overcome trauma from previous sidewalk and house damages provoked by historic trees built in their neighborhood.

4. **Bringing together environmental and food justice**

Municipal ordinances for urban agriculture

In many U.S. neighborhoods, food deserts often coincide with food mirages, especially in gentrifying areas. A promising planning policy tool that introduces new green spaces and simultaneously increases a city's food security is new planning ordinances for urban agriculture, that is, urban gardens, farms and rooftop gardens. In 2013, the City of Boston adopted Article 89 to the Boston Zoning Code, allowing Boston residents to grow and access healthy foods, provide new opportunities for local business growth and develop knowledge and education about healthy eating. Article 89 allows farming outright, meaning no variances—and thus public hearings—from the existing zoning are required. One such urban farm was created by several neighbors in 2015 in East Boston. Now, Eastie Farm produces healthy, locally grown and culturally relevant foods, and offers educational programs on composting, sustainable growing techniques and environmental stewardship. Through these practices the farm improves food access, builds new community ties and supports community resilience (Table 22.1).

For further analysis

Despite significant challenges presented by urban green inequality and their drivers, different sets and combinations of policy tools, planning regulations, financial incentives and participatory schemes can promise more just green urban futures. Yet, cities are still confronted with tradeoffs and wicked challenges when deciding about policy priorities and corresponding sets of tools. In the U.S. and Canada, many planners seek to address housing affordability and shortage crises through the development of new social housing, but must also navigate a historical and cultural preference for homeownership over rental tenancy, with simultaneous stigmatization of and lack of funding for affordable and social housing. In Western European cities, despite early strides toward strong social housing culture, more recent harsh financial austerity measures are slashing available public budgets for both affordable housing and equitable greening projects. For cities where an intensive tourism economy pervades in both geographical contexts, attempts by the local government to provide new green amenities can often be appropriated to benefit the needs and uses of visitors and tourists, particularly when private development or financing become involved. These patterned scenarios and their wicked challenges present significant challenges in

TABLE 22.1 Complete list of 50 policy and planning tools for urban green environmental justice divided into (a) anti-gentrification/displacement policy responses (30) and (b) more green amenities/open public amenities (20). Available help_outlineat: http://www.bcnuej.org/wp-content/uploads/2021/04/Toolkit-Urban-Green-Justice.pdf

(a) ANTI GENTRIFICATION / DISPLACEMENT POLICY RESPONSES (30)							
Land use		*Developer requirements/ support*	*Housing focused financial schemes aimed at homeowners*		*Housing focused financial schemes aimed at renters*	*Community focused financial schemes*	*Other anti-gentrification regulations/ ordinances*
Inclusionary zoning	Regulations on touristic/short-term rental apartments (e.g. AirBnB)	Obligation for developers to include affordable units in development	Property tax payment support for homeowners	Loans for home ownership for middle/ low-income residents	Rent control	Investment incentives in specific areas - Opportunity Zones, Strategic Development Zone, Special Interest Zone, Business Improvement District)	Formal recognition of the right to stay/place OR right to return
Up-zoning	Moratorium on new businesses, hotels, and other hospitality industry permits	Density bonuses to encourage affordable housing	Homestead tax credit/tax exemption	Development tax paid by developers towards an affordable housing trust fund	Rent subsidies or vouchers	Regular support to housing coops and other non-profit housing organizations	Municipality or tenant opportunity to purchase act

(Continued)

TABLE 22.1 Continued

(a) ANTI GENTRIFICATION / DISPLACEMENT POLICY RESPONSES (30)							
Land use		*Developer requirements/ support*	*Housing focused financial schemes aimed at homeowners*		*Housing focused financial schemes aimed at renters*	*Community focused financial schemes*	*Other anti-gentrification regulations/ ordinances*
Rezoning of green space to residential space	Defence of single family homes or minimum lot size	Support for developers to develop empty lots or buildings into affordable housing	Limitations or freezes to property tax increases	Transfer tax on luxury property with funds directed towards affordable housing			Public workforce housing
Preservation Districts or Historic Districts	Community Land Trusts		Housing tax credit programmes	Tax on foreign ownership of units			Equitable economic development plans with opportunity for equitable access to jobs
Land bank			Tax on vacant housing units				Equity scoring/ evaluation/ index of urban projects

(b) MORE GREEN AMENITIES / OPEN PUBLIC AMENITIES (20)				
Land use		*Developer requirements/support*	*Financial schemes*	*Other greening regulations/ordinances*
Conservation areas to preserve green spaces	Eco-district zoning and climate change mitigation	Fee paid by developer directed to green funding	Specific national financial schemes to fund green infrastructure or parks	Improved maintenance and security of public space
Interim green spaces on vacant land	Rezoning of urban land to green space	Minimum amount of green space in new development projects	Green bonds	Clarification over administrative roles for green space management-security
Repurposing of streets of green transit areas	Opening of private green space to the public		Regulations to support the development of urban agriculture	Improved food security or sovereignty
Measuring/ mapping access to green space	Ambitious green space development plan		New/improved urban green amenities	
Waterfront redevelopment	Green amenity planning in large-scale urban developments		Green resilient infrastructure funding in socially and environmentally vulnerable neighbourhoods	

achieving socio-environmental justice. However, the successful application of a mix of anti-gentrification and equitable green development tools, dependent on the specific urban, cultural and historical context, can be a means for planners to advance toward a just and green urban future.

Reference

Oscilowicz, Emilia, Emilia Lewartowska, Ariella Levitch, Jonathan Luger, Sarka Hajtmarova, Ella O'Neill, Aina Planas Carbonell, Helen Cole, Carla Rivera Blanco, and Erin Monroe. 2021. "Policy and Planning Tools for Urban Green Justice." Barcelona: BCNUEJ. Available at http://www.bcnuej.org/wp-content/uploads/2021/04/Toolkit-Urban-Green-Justice.pdf

CONCLUSION

A new tale for the green city?

James J.T. Connolly, Isabelle Anguelovski, Melissa García-Lamarca, and Emilia Oscilowicz

Urban greening is often thought of as a tool for aligning developmental and environmental goals, but it is also a tool for magnifying the city. It exposes and expands almost invisible dimensions of our hyperlocal environment. It draws our attention toward the ways that our everyday environment is conditioned and patterned by cultural, social and political preferences. Greening has become one of the strongest mechanisms for transforming these preferences from a figurative guide for action into the literal cities we inhabit. Given its rise to prominence and its ability to expose challenges in creating socially just urban environments, reflective practice requires synthesizing the tales being told through urban greening around the world. In doing so, we can begin to tell a new tale, a tale that learns from reflection.

The chapters of *The Green City and Social Injustice* expose several parallel tracks on which the race is run to comply with the idealized green city across the Global North. Along the way, some cities such as Vienna or Nantes are trying desperately—despite the pressure to attract capital that pushes them in a different direction—to hold onto a pre-globalized and more social (or socialist, in European municipalities) tradition, wherein local governments seek a healthy balance between neighborhood-based quality of life and preservation of the natural ecosystem. But also, these chapters show that some cities such as Atlanta, Dallas or Milan do not have that tradition in mind. As they undertake massive state-led postindustrial transformation, "urban problems" come to be seen as anything (and anyone) that does not fit the mold of global markets or so-called avant-garde urbanism, and greening becomes a way of masking or removing those problems. Cities like Dublin or Portland are conflicted over the extent to which they should undo a rich historic inventory of green spaces in order to accommodate development, while cities like Barcelona are turning every stone to find developed areas that can be cleared and converted into green spaces. In

DOI: 10.4324/9781003183273-102

places like Boston, Montréal or San Francisco there are deep social and environmental contradictions by which greening and luxury development become either juxtaposed or superposed onto one another under a logic of green, resilient urbanism and redevelopment.

Within these parallel tracks for city greening, there are several pathways through which green environmental injustices unfold. These pathways combine historic links between land use, institutionalized racism and class relations. They also express an ever-deeper expanse between the haves and have-nots, which has left urban geographies economically and politically polarized. Greening is pushing this polarization to even greater depths. It is one of the key tools for rebranding racialized neighborhoods or ghettoized concentrations of minority and low-income groups into areas of privilege for upper-class and white residents—such as occurred in neighborhoods surrounding Atlanta's Beltline, areas of Seattle's South Park or Amsterdam Noord. These pathways are also defined by impediments around mobilizing at a grassroots level for control over local green agendas. Such mobilizations are often focused on wresting control away from those that would stand idly by, watching as the benefits of greening initiatives flow away from residents who often pioneered local environmental initiatives, fueling their startups with blood, sweat and tears. In the tension between top-down branding and bottom-up decommodification, particularly well-illustrated by the tales of Milan, Barcelona, Copenhagen, Boston and Montréal, the branding often wins and the inequities of the city swallow up the non-monetary benefits of urban greening, leaving many to wonder what the purpose of greening was in the first place.

While cities are moving toward the green moniker on several parallel tracks and injustices are intensified along multiple pathways in the process, there are lessons to be learned from what we observe across many cities that can guide us toward a vision for equitable urban green praxis. The five sections of the book demonstrate how green injustices manifest and how they are contested in five different contexts. Each of these contexts demonstrates a different degree to which greening begins with social justice as at least a primary goal, if not *the* primary goal.

Greening injustices across five contexts

Part 1, "The social costs of glitzy green urbanism," considered two extravagant new greening projects and two less spectacular but more extensive new green urban developments that heighten exclusivity and existing injustices. On the one hand, Milan's so-called Gate of Green, including vertical garden luxury skyscrapers, and Valencia's Parc Central, marketed as the city's new green lung, both employed starchitects, global real estate developers and international competitive logics to build the cachet and glamor of new green space design and high-end housing development. On the other hand, high-profile green urbanism can also come in the form of citywide initiatives that refocus the identity of the city on green goals. In Bristol, this manifested in the efforts to become a European Green Capital and in Amsterdam as a tech-, arts- and design-driven green transformation of Amsterdam Noord. In all four cases, speculative greening and eco-urbanism agendas brought about the privatization of green space, rising housing prices, the destruction of heritage, cultural and physical displacement and participatory processes characterized by activists as "social-washing." Overall, these agendas have laid the groundwork for an exclusive present and future urban transformation. While this greening was driven by the city or private sector, we also saw how community organizers and activists in Milan and environmental groups in Bristol created new green spaces that embody values of responsibility, cooperation and the co-construction of urban imaginaries and fought for policy change and action to address historic racial inequality, affordability in new green developments and lack of quality green spaces for working-class communities.

In Part 2, "Compounded risks and impacts of urban greening in post-industrial environments," we highlighted cases of areas in four American cities and one Scottish city characterized by historically toxic landscapes and working-class, often racialized, residents with ongoing and disproportionately high health risks. More recently, the extensive land decontamination underway has rapidly transformed decades of neglect and disinvestment. The newly cleaned-up areas—either in extremely close proximity to their respective city centers (West Dallas, Glasgow's East End, Cleveland's Detroit Shoreway) or vastly more accessible due to transit stops or new infrastructure (San Francisco's Bayview-Hunters Point and Seattle's South Park)—are now sites with enormous development pressure. While green and blue infrastructure development visions abound, the type of creative, smart and/or green city model that is often being pursued leaves behind, excludes or even abolishes long-term working-class and racialized residents while encouraging an influx of more affluent and whiter residents through high-end housing developments. Resistance and alternatives by mobilized vulnerable residents, however, are present especially in Dallas, Seattle, San Francisco and Cleveland, pressing for more equitable green development approaches and interventions.

Part 3, "(Re)creating unjust racialized landscapes in the green city?," uncovers the racialized dimensions of greening in Atlanta, Washington D.C., Boston and Austin, all cities with a long history of racial and class segregation. Residents in East Boston, East Austin, Washington D.C.'s Historic Anacostia and the predominantly Black neighborhoods surrounding Atlanta's Beltline endured neglect, disinvestment and abandonment for decades until new resilient, green, climate interventions arrived in or next to the neighborhood (East Boston waterfront, East Austin's smart growth plan and Atlanta's Beltline) or right across the river (Navy Yards, Washington, D.C.). In all four cases, the dominant development processes are reinforcing and retrenching racial relations while generating the political, cultural and economic displacement of Black and Latinx folk, as it is largely wealthy and white newcomers who can afford to live amidst new green amenities. This type of urban greening effect is perhaps most acute in the cities outlined in Part 3 but is also a common strand across many cities in this book. The re-creation of racialized environments is the impetus behind a recent scholarly push to address the deeply embedded white privilege in greening through mechanisms of abolitionist values rooted in an ethics of care and healing as well as land recognition, redistribution and control by historic residents of color (Connolly and Anguelovski 2021; Ranganathan and Bratman 2019).

Through looking at neighborhoods in Montréal, Dublin, Barcelona and Philadelphia, Part 4 of this book, "The complex entanglement of greening and multiple other gentrification pressures," illustrates how greening is intertwined and made more complex in its relationship to other urban dynamics related to food, tourism, creative industries, Opportunity Zones and the studentification of cities. Montréal's Saint-Henri is not only being transformed by large-scale, top-down and small-scale, community-led greening interventions, but also, in its foodscape, what was once affordable, culturally loved and supportive of community needs has been replaced by expensive and gourmet options. Similarly, despite being a non-binding document, Dublin's *Liberties Greening Strategy* has been used as a tool for the legitimization and marketing of new high-end student housing, luxury hotel development and small apartments for wealthy international IT workers, which has added another trigger to gentrification while excluding working-class residents from new (semi)-privatized greening. Meanwhile, comprehensive greening efforts throughout Barcelona are increasingly at risk of being appropriated by global tourism and corporate investments as residents are socially and physically displaced. As well, Philadelphia's Hunting Park, a model neighborhood for equitable climate resilience planning, faces shifting and combined social and environmental riskscapes as efforts of Latinx and Black residents to improve neighborhood environmental quality are concentrated on an Opportunity Zone, whereby new tax incentives meant to boost private development may not necessarily generate benefits for vulnerable residents. We see in this section how greening is thus one piece of a larger gentrification and displacement puzzle.

The book's fifth and final section, "(Fragile) green justice victories and grey zones in the just green city," demonstrates both the efforts and pitfalls of historic and present attempts to create greener and more equitable urban fabrics in American and European cities. Copenhagen, Nantes and Vienna are all known for their deep social and environmental welfare roots that decades ago centered on progressive tax and land use policies as well as the provision of large amounts of public housing and green space. While a significant stock of decommodified housing partially remains today, in recent decades it has been eroded through the neoliberal restructuring of housing systems that has come to dominate global housing development trends and through national or international competitive urbanism logics. Trade-offs between social justice, rapidly rising land prices, urban renewal and environmental innovations are exemplified by Copenhagen's Nørrebro neighborhood. Similarly, Nantes has made huge strides in becoming a green and equitable city over the past 30 years due to an enforced commitment to providing significantly affordable and social housing in all new eco-developments, yet tensions exist about what type of greening is being created and who may participate and decide. Finally, despite Portland being a sustainability planning pioneer in North America, in practice the city sustains white privilege as industrial pollution, food deserts and the lack of green amenities remain a reality for most Black, Latinx and Native American populations. Yet the grassroots-led development in Portland's Cully neighborhood, the focus of the chapter, illustrates how sustainability can be reframed as an anti-poverty and climate justice strategy, with community mobilization sparking important simultaneous wins around housing, ethical and local economic opportunity and inclusive green development.

These five sections comprise a list of the basic elements that become visible when we look carefully through the magnifying glass of urban greening at city-making processes, but the tale they tell on the whole is better expressed within cross-cutting themes. In the introductory chapter of the book, we described the analytical lenses that we used as we examined the case studies. Below, we revisit these lenses in order to discern the extent to which they surfaced as thematic trends across all cases.

Green locally unwanted land uses and the formation of green privilege

One common dynamic seen in many cities demonstrates the counter-intuitive trend wherein the motivations for and ultimate effect of urban greening initiatives become suspect, rendering them green locally unwanted land uses (GreenL.U.L.U.s). In nearly all cases, there is a strand of suspicion around the alliances between municipal planners, elected officials and developers promoting a local urban "green turn" (Anguelovski et al. 2019) in order to enhance their competitive position amongst global, cosmopolitan cities. The suspicion reflects an internalized sense of a trend observed elsewhere (e.g., Heynen, Perkins and Roy 2006) that the green rents accumulated by upper-class residents and the real estate-financial complex around them come at a social cost for minorities and low-income groups. Thus, the formation of GreenL.U.L.U.s posits that green space planning, design and management intersect with cultural politics of race, class and nature, and in turn create new socio-natural identities in and around cities benefiting newcomers and investors in the form of rent capture.

Our analysis reveals that parallel tracks of urban greening and the multiple pathways through which urban green injustice unfolds generate numerous forms of GreenL.U.L.U.s. This mode of green privilege is being enacted through a *diversity of materialities* (i.e., from the conversion of vacant lots to gardens; transportation and industrial infrastructure into linear parks and other green spaces; green resilience and adaptation along climate-threatened coastlines and waterfronts). It also has several *associated environmental amenities* that become part of the green package (self-branded healthy, organic and local food stores in the process of food gentrification; or sustainable and active transit infrastructure in transit-oriented development projects and others). As well, it occurs across a variety of *urban spatialities* (historic urban centers, post-industrial and military areas, peripheral zones, and coastlines and waterfronts). Finally, GreenL.U.L.U.s are championed by an array of institutional, physical and individual *green actors* (elected officials branding a new green city; planners and designers locating and projecting greening; investors, developers and real estate agents profiting from and financing greening; residents moving to green areas while others are pushed out; and the green infrastructure, materials, plants and broader green ecology and economy allowing greening to be spatially enacted).

Last, many of the tales—San Francisco, Dallas, Boston, Seattle and Dublin in particular—illustrate the real problem expressed by GreenL.U.L.U.s: when privileged green areas result in the production of new green inequalities, enduring environmental risks and exposure are not abated. Rather, urban greening in this circumstance leaves long-term environmental and health inequities unaddressed as new parks, greenways and other renaturing programs rise up along the polluted waterfronts or next to waste sites of American and European global cities. While it is true that gentrified residents moving to greener yet environmentally contaminated areas face environmental risks as well, their social position allows them to organize, protest and if necessary, move out more easily than marginalized residents with limited political power, income or financial assets. Early gentrifiers also tend to pave the way for and legitimize large-scale developments for upper-class groups and for a second, more acute, gentrification wave. In that process, lower-income groups are pushed away to browner, more distant landscapes (Gould and Lewis 2017).

Dispossession, displacement and the racialization of nature

Another common trend throughout the book picks up on how new aesthetics, designs and formalizations of nature codify and restrict who has access to the benefits of the green city. Beautification projects can produce what Goodling and others call houselessness (Goodling 2021) by creating visual landscapes and green spaces appealing to middle-class and white consumption and leisure practices precisely because they send a clear signal that the area is off-limits to others. In contrast, homeless encampments and environmental practices become portrayed as undesirable and criminal objects to be removed to make way for new businesses and market opportunities (Goodling 2021; Speer 2019; Dooling 2009). Here, the reuse, survival and collective appropriation of space by vulnerable groups become subjugated by a green makeover (Speer 2019; Anguelovski, Irazábal and Connolly 2019). Green spaces are thus the locus of socio-spatial tensions (Dooling 2009) wherein park planners and new users shape a more manicured and normalized form of greening while other uses of the spaces—which are often the only ones vulnerable residents have access to—become ruled out and marginalized. In other cases, permitted yet temporary uses of urban spaces for greening become washed away by new formal greening and other land uses in processes of "clashing temporalities of support and care" (Kotsila et al. 2020).

As many of the cases presented in this volume show, this sometimes quiet and creeping form of dispossession is increasingly linked to racialized uneven development and the racialization of urban nature within seemingly depoliticized and ahistorical urban green planning. Collectively, this type of greening is illustrative of settler colonialism, that is of practices of land grabbing, frontier-driven value capture, landscape manipulation and invisibilization of minority land practices and uses (Safransky 2014; Dillon 2014). Under this regime of accumulation by green dispossession, immigrant and minority practices become overlooked and pushed aside when the preferences of white culture conflict to reflect the deep history of colonial settlement and a colonizing approach to the organization of society. As greening plows through neighborhoods such as Anacostia in Washington D.C., East Boston in Boston, South Park in Seattle, Bayview-Hunters Point in San Francisco and in different racialized contexts such as Ciutat Vella, Barcelona, or Nørrebro, Copenhagen, local histories of the racial settlement are obscured by racial amnesia that strips from the public consciousness an awareness of the deeper legacies and impacts of environmental and social inequities (Tuck 2009) in the communities where green projects are taking place. Put simply, greening is not race-neutral and can evict alternative narratives, practices and identities for the benefit of new, whiter residents and cultural uses.

The cases presented in *The Green City and Social Injustice* broadly point toward race and racism as the primary and ultimate predictor and indicator of unequal green inequities (Mohai and Saha 2015; Pulido 2017). In many of the neighborhoods examined, racism—and here race-based environmental inequalities—is produced and reproduced over time, and calls for the exposure of pervasive and malleable racialized processes and racialized inequities in regard to urban greening (Anguelovski et al. 2019; Connolly and Anguelovski 2021). In Europe, this process is embodied in the isolation, ghettoization, territorial stigmatization, advanced marginality and increased legal targeting of immigrants along a race-class axis (Wacquant 2014). In the U.S., many minorities and immigrants have a conflicted relationship with nature because natural spaces have not always been integrated, welcoming and safe amenities for them, especially cognizant of racist discourses and practices that have excluded them (Park and Pellow 2011; Rigolon and Németh 2018). In short, we take from our observations a sense of the entanglement of urban greening, segregation and exclusion and white privilege and supremacy as nature becomes increasingly an expression of white culture in gentrifying neighborhoods of North America and Europe.

Banking on nature

Finally, a thread that is woven throughout most of our cases involves the ways in which urban nature is packaged and presented to enable its possible monetization and mobilization for finance capital. Most overtly, there are many examples wherein city governments justify greening within policy processes primarily based upon an expectation of financial return on investment. This approach sets aside the usual impetus of environmental and social benefits and forces greening down a path wherein monetized returns can be assured, while concern for equity is neglected. This process—the financialization of urban greening—undergirds much of what drives the injustices observed above.

In our 21 tales of urban greening, we show how planning for and investment in various forms of greening work to cordon off space so it can be more easily packaged and monetized within a sustainable, smart, resilient city narrative (Connolly 2019). Eventually, the accumulated effect of this cordoning is the production of injustices for residents who are themselves ignored within the hegemony of urban greening and the financialization of nature around it. Such a process is particularly visible in the cases of Dublin, Milan, Valencia, Austin, Seattle and Portland where a network of investors and real estate stakeholders come together to fund large-scale yet increasingly exclusive greening, often to justify and legitimize their real estate operations. In this process, urban nature is a product and a resource tamed, transformed, re-created, financed and regulated within a dominant market-centered ideology (Castree 2008; Knuth 2016) of economic development that (only) highlights all the gains and values harnessed via greening for a few, excluding the effects on the many. As Marcuse (2009) reminds us, this reinforces a system wherein, "The law locks up the man or woman who steals a goose from off the common, but leaves the greater villain loose who steals the common from the goose."

This is a thread of increasing importance within urban greening because new financial instruments and tools—the recent emergence of green bonds as a particular example—are being widely utilized by planners and developers alike. In many cases the result is to lock green urban development into a logic and practice of future monetary value creation under the guise of "doing good" (Knuth, 2016; Garcia-Lamarca and Ullstrom, 2020). Thus, the chapters here help us unpack how emerging strategies of urban "green grabbing" (Fairhead, Leach and Scoones 2012) operate within the broader context of green capitalism (Prudham 2009; Wallis 2010). This process is characterized by a systemic attempt to "solve" economic and ecological crises through integrating ecological conditions—in this case greenspaces—into the circuits of capital accumulation, towards new and supposedly sustainable growth trajectories (Anguelovski et al. 2019).

A new tale for the green city

The Green City and Social Injustice endeavors toward a deep engagement with the ways in which urban greening amplifies injustices without ignoring the many social, health and environmental benefits that also come with greening. Rather, it directs us toward the need to generate a new storyline for green cities. This storyline must develop outside of the growth of green privilege built on the back of racialized and financialized models of urban greening. It must value all people's right to a healthy environment before some people's desire for monetary gain. In our view, this new tale of the green city comes about through an antisubordination, intersectional and relational approach.

An *antisubordination* mode of planning forces a reflection on effect rather than intent. If the effect is to reinforce or exacerbate existing social inequalities, the result is undesirable regardless of intent. Green planners can no longer only operate on good intent; instead, antisubordination green planning requires ongoing reflection, adjustment and collaboration. In Washington D.C., for example, organizers of the proposed 11th Street Bridge greening initiative try to account for historic racism, property market effects and gentrification. Importantly, the initiative is evolving over time and continues to change and grow as perceived and observable effects of initial interventions become clear. Following principles of reparative and preventative justice, antisubordination greening calls for three deep changes: (1) the questioning of experiences of domination, racial stratification and oppression and of the institutions that sustain them; (2) the construction of emancipatory spaces and geographical formations that can secure land, resources and safe access to nature for marginalized groups; (3) and the creation of new institutional arrangements, practices and policies that such groups can control.

An *intersectional* mode of greening embraces the complexity of ways in which injustices are reproduced across race, ethnicity, gender, class and identity. This new pathway for urban greening embraces green space creation that embodies values of responsibility, duty-sharing and the co-construction of urban imaginaries. Recognizing the multi-faceted and intersectional network of outcomes in urban social and environmental justice that result from greening, this new pathway calls for planners and policymakers to prioritize anti-displacement tools related to appropriate and affordable housing provision and maintenance in conjunction with green (re)development tools in order to achieve a green and just city. Here, green planners have a particularly important role to play in placing housing rights and justice at the center of—and not as an adjunct to—new greening projects. In the previous chapter, we explored how different municipalities have engaged with combinations of both anti-displacement and equitable urban greening tools, in both the North American and the European contexts, in order to address green inequalities that have surfaced as a result of post-industrial redevelopment, unregulated real estate speculation, unjust racialized landscapes and neighborhood gentrification pressures. In that chapter and the 50 policy tools highlighted in an accompanying report, we emphasized participatory community involvement, regulation of neoliberalized housing markets, further financial support for affordable and social housing programming, among others.

A *relational* mode of urban greening turns the focus away from high-profile mega-projects designed to grab the attention of global competitors, and toward the everyday lives of every resident in the city. It seeks out greening agendas that meet the simple needs of people with basic demands in order to allow for socio-cultural growth amongst residents. It helps ensure that working parents have places of communal support for childcare as we see in cities like Barcelona, while also building grassroots efforts to steward the local environment from a desire to make an area more livable, as we see in cities like Philadelphia. Here, we highlight the role of planners in valuing multiple formal and informal uses for green space and the role of marginalized or invisibilized groups in occupying green spaces. Without such spatial occupation and claim, risks of socio-cultural displacement are rampant.

Taken together, antisubordination, intersectional and relational modes of greening add up to a radical new praxis capable of rewriting the tale of the green city. This book has demonstrated that in many cases, greening is highly associated with the displaceability of residents or susceptibility to exclusion from neighborhood and housing resources, but also points to a more radical potential that is always present and championed, even if not by those with the greatest resources. The leap is not so great to realize this potential.

References

Anguelovski, Isabelle, Clara Irazábal-Zurita, and James J. T. Connolly. 2019. "Grabbed urban landscapes: Socio-spatial tensions in green infrastructure planning in Medellín." *International Journal of Urban and Regional Research* 43 (1): 133–156.

Anguelovski, Isabelle, James Connolly, Melissa Garcia Lamarca, Helen Cole, and Hamil Pearsall. 2019. "New scholarly pathways on green gentrification: What does the urban "green turn" mean and where is it going?" *Progress in Human Geography* 43 (6): 1064–108.

Castree, Noel. 2008. "Neoliberalising nature: The logics of deregulation and reregulation." *Environment and Planning A* 40 (1): 131–152.

Connolly, James J. T. 2019. "From Jacobs to the Just City: A foundation for challenging the green planning orthodoxy." *CITIES*, 91: 64–70.

Connolly, James J. T., and Isabelle Anguelovski. 2021. "Three histories of greening and whiteness in American cities." *Frontiers in Ecology and Evolution* 9 (2021): 101.

Dillon, Lindsey. 2014. "Race, waste, and space: Brownfield redevelopment and environmental justice at the hunters point shipyard." *Antipode* 46 (5): 1205–1221. doi: 10.1111/anti.12009.

Dooling, Sarah. 2009. "Ecological gentrification: A research agenda exploring justice in the city." *International Journal of Urban and Regional Research* 33 (3): 621–639.

Fairhead, James, Melissa Leach, and Ian Scoones. 2012. "Green grabbing: A new appropriation of nature?" *Journal of Peasant Studies* 39 (2): 237–261.

García-Lamarca, Melissa, and Sara Ullström. 2020. ""Everyone wants this market to grow": The affective post-politics of municipal green bonds." *Environment and Planning E: Nature and Space*: 2514848620973708.

Goodling, Erin. 2021. "Urban political ecology from below: Producing a "peoples' history" of the Portland Harbor." *Antipode* 53 (3): 745–769.

Gould, Kenneth A., and Tammy L. Lewis. 2017. *Green Gentrification: Urban Sustainability and the Struggle for Environmental Justice*. New York: Routledge.

Heynen, Nik, Harold Perkins, and Parama Roy. 2006. "The political ecology of uneven urban green space." *Urban Affairs Review* 42 (1): 3–25. doi: 10.1177/1078087406290729.

Knuth, Sarah. 2016. "Seeing green in San Francisco: City as resource frontier." *Antipode* 48 (3): 626–644.

Kotsila, Panagiota, Kathrin Hörschelmann, Isabelle Anguelovski, Filka Sekulova, and Yuliana Lazova. 2020. "Clashing temporalities of care and support as key determinants of transformatory and justice potentials in urban gardens." *Cities* 106: 102865.

Marcuse, Peter. 2009. "From justice planning to commons planning." In Marcuse, Peter, James Connolly, Johannes Novy, Ingrid Olivo, Cuz Potter, and Justin Steil, eds. *Searching for the Just City: Debates in Urban Theory and Practice.* New York: Routledge.

Mohai, Paul, and Robin Saha. 2015. "Which came first, people or pollution? Assessing the disparate siting and post-siting demographic change hypotheses of environmental injustice." *Environmental Research Letters* 10 (11): 115008.

Park, Lisa Sun-Hee, and David Pellow. 2011. *The Slums of Aspen: Immigrants vs. The Environment in America's Eden.* New York: New York University Press.

Prudham, Scott. 2009. "Commodification." In *A Companion to Environmental Geography,* 123–142. Malden, MA.

Pulido, Laura. 2017. "Geographies of race and ethnicity II: Environmental racism, racial capitalism and state-sanctioned violence." *Progress in Human Geography* 41 (4): 524–533.

Ranganathan, Malini, and Eve Bratman. 2019. "From urban resilience to abolitionist climate justice in Washington, DC." *Antipode* 53 (1): 115–137.

Rigolon, Alessandro, and Jeremy Németh. 2018. ""We're not in the business of housing:" Environmental gentrification and the nonprofitization of green infrastructure projects." *Cities* 81: 71–80.

Safransky, Sara. 2014. "Greening the urban frontier: Race, property, and resettlement in Detroit." *Geoforum* 56: 237–248.

Speer, J. 2019. "Urban makeovers, homeless encampments, and the aesthetics of displacement." *Social & Cultural Geography* 20 (4): 575–595.

Tuck, Eve. 2009. "Suspending damage: A letter to communities." *Harvard Educational Review* 79 (3): 409–428.

Wacquant, Loïc. 2014. "Class, race and hyperincarceration in revanchist America." *Socialism and Democracy* 28 (3): 35–56.

Wallis, Victor. 2010. "Beyond" green capitalism." *Monthly Review* 61 (9): 32–48.

INDEX

Note: Page numbers in *italics* indicate figures and page numbers in bold indicate tables in the text

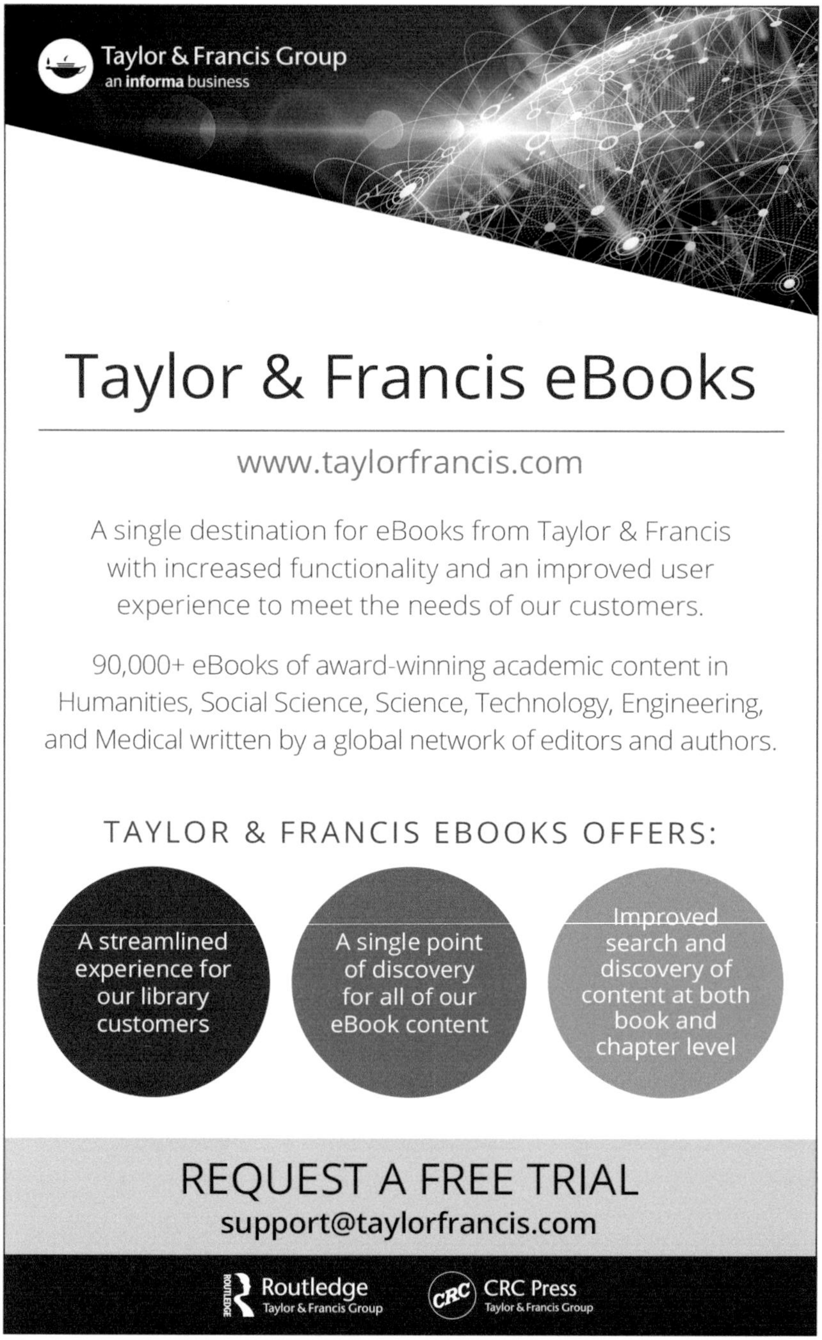
Taylor & Francis Group
an informa business
Taylor & Francis eBooks
www.taylorfrancis.com
A single destination for eBooks from Taylor & Francis with increased functionality and an improved user experience to meet the needs of our customers.
90,000+ eBooks of award-winning academic content in Humanities, Social Science, Science, Technology, Engineering, and Medical written by a global network of editors and authors.
TAYLOR & FRANCIS EBOOKS OFFERS:
A streamlined experience for our library customers
A single point of discovery for all of our eBook content
Improved search and discovery of content at both book and chapter level
REQUEST A FREE TRIAL
support@taylorfrancis.com
Routledge
Taylor & Francis Group
CRC Press
Taylor & Francis Group